# Molecular Mechanisms
# of Membrane Traffic

# NATO ASI Series

## Advanced Science Institutes Series

*A series presenting the results of activities sponsored by the NATO Science Committee, which aims at the dissemination of advanced scientific and technological knowledge, with a view to strengthening links between scientific communities.*

The Series is published by an international board of publishers in conjunction with the NATO Scientific Affairs Division

| | | |
|---|---|---|
| A | Life Sciences | Plenum Publishing Corporation |
| B | Physics | London and New York |
| C | Mathematical and Physical Sciences | Kluwer Academic Publishers |
| D | Behavioural and Social Sciences | Dordrecht, Boston and London |
| E | Applied Sciences | |
| F | Computer and Systems Sciences | Springer-Verlag |
| G | Ecological Sciences | Berlin Heidelberg New York |
| H | Cell Biology | London Paris Tokyo Hong Kong |
| I | Global Environmental Change | Barcelona Budapest |

## NATO-PCO DATABASE

The electronic index to the NATO ASI Series provides full bibliographical references (with keywords and/or abstracts) to more than 30000 contributions from international scientists published in all sections of the NATO ASI Series. Access to the NATO-PCO DATABASE compiled by the NATO Publication Coordination Office is possible in two ways:

- via online FILE 128 (NATO-PCO DATABASE) hosted by ESRIN,
  Via Galileo Galilei, I-00044 Frascati, Italy.

- via CD-ROM "NATO Science & Technology Disk" with user-friendly retrieval software in English, French and German (© WTV GmbH and DATAWARE Technologies Inc. 1992).

The CD-ROM can be ordered through any member of the Board of Publishers or through NATO-PCO, Overijse, Belgium.

Series H: Cell Biology, Vol. 74

# Molecular Mechanisms of Membrane Traffic

Edited by

## D. James Morré

Department of Medicinal Chemistry
HANS Life Science Research Building
Purdue University
West Lafayette, IN, USA

## Kathryn E. Howell

C & S Biology
University of Colorado
School of Medicine
Denver, CO, USA

## John J. M. Bergeron

Department of Anatomy and Cell Biology
McGill University
Montreal PQ H3A 2B2, Canada

Springer-Verlag
Berlin Heidelberg New York London Paris Tokyo
Hong Kong Barcelona Budapest
Published in cooperation with NATO Scientific Affairs Division

Proceedings of the NATO Advanced Research Workshop on Cell-Free
Analysis of the Functional Organization of the Cytoplasm: Molecular
Mechanisms of Membrane Traffic, held at Airlie, Virginia, USA, May 9–13, 1992

ISBN 3-540-53096-7 Springer-Verlag Berlin Heidelberg New York
ISBN 0-387-53096-7 Springer-Verlag New York Berlin Heidelberg

Library of Congress Cataloging-in-Publication Data. Molecular mechanisms of membrane traffic /
edited by D. James Morré, Kathryn E. Howell, John J. M. Bergeron. p. cm. – (NATO ASI series. H, Cell
biology; vol. 74) "Proceedings of the NATO Advanced Research Workshop on Cell-Free Analysis of
the Functional Organization of the Cytoplasm: Molecular Mechanisms of Membrane Traffic, held at
Airlie, Virginia, USA, on May 9-13, 1992"–T.p. verso. Includes bibliographical references and indexes.
ISBN 3-540-53096-7 (Berlin: alk. paper). – ISBN 0-387-53096-7 (New York: alk. paper) 1. Cell
membranes–Congresses. 2. Biological transport–Congresses. I. Morré, D. James, 1935- . II. Howell,
Kathryn E., 1939- . III. Bergeron, John J. M., 1946- . IV. NATO Advanced Research Workshop on Cell-
Free Analysis of the Functional Organization of the Cytoplasm: Molecular Mechanisms of Membrane
Traffic (1992: Airlie, Va.) V. Series: NATO ASI series. Series H. Cell biology; vol. 74.
QH601.M65   1993   574.87'5–dc20   93-1200

© Springer-Verlag Berlin Heidelberg 1993
Printed in Germany

Typesetting: Camera ready by authors
31/3145 - 5 4 3 2 1 0 - Printed on acid-free paper

# Preface

The study of membrane traffic in reconstituted cell-free systems has generated an unprecedented amount of new information on the biochemistry, molecular biology and genetics of membrane-based molecular events that underly normal and abnormal cellular function. Many of the individual steps have now been isolated and dissected in simple systems that permit detailed molecular analyses of transport mechanisms and their regulation. Reconstituted events of intercompartment transport include inter-membrane recognition, and controlled membrane fusion-fission reactions.

Among the many advances is the growing awareness of a remarkable evolutionary conservation of many of the components involved in the many steps of membrane traffic; this realization has accelerated greatly the pace of progress in the field.

This book provides a collection of participant contributions from the 1992 Summer Research Conference, "Molecular Mechanisms of Membrane Traffic," jointly sponsored with NATO by the American Society of Cell Biology. The conference was held May 9-13, at the Airlie Conference Center in the Virginia countryside, near Warrenton.

The conference was attended by 158 scientists. A unique feature was the high proportion of young scientists among the participants. Approximately 65% were students, postdoctoral fellows and young investigators. Each attendee contributed to the conference with either a platform or poster presentation.

The major focus of the conference was on new data that begins to provide, for the first time, a detailed picture of the molecular events of vesicular membrane traffic. Both the small molecular mass GTP-binding and trimeric G proteins were implicated in practically all membrane traffic steps and provided a major theme of the conference.

We thank the plenary speakers who contributed so much to the success and continuity of the program, and Dottie Doyle and the staff of the American Society of Cell Biology National Office for exquisite program organization and local arrangements. Appreciation is extended as well to Prof. Placido Navas, Chairperson of the NATO Advisory Committee, and Sarah Craw for capable assistance in the organization of this volume.

Financial support for student scholarships provided by the National Science Foundation, National Institutes of Health and the Keith Porter Endowment for Cell Biology is gratefully acknowledged. Additional support was provided by the Upjohn Company, Abbott Laboratories, Bristol-Myers Squibb Company, Corning Incorporated, Estee Lauder, Inc., JEOL U.S.A., Inc., Merck Research Laboratories, Monsanto Company and Rainin Instrument Company, Inc.

D. J. Morré

K. E. Howell

J. J. M. Bergeron

# CONTENTS

Preface

## ENDOPLASMIC RETICULUM, STRUCTURE, FUNCTION AND TRANSPORT TO GOLGI COMPLEX

## LYSOSOMES AND DEGRADATION

## MOLECULES IDENTIFIED IMPORTANT IN MEMBRANE TRAFFIC

# CELL-FREE ANALYSIS OF ER TO GOLGI APPARATUS VESICULAR TRAFFIC

D. James Morré and D. M. Morré
Department of Medicinal Chemistry & Department of Foods and Nutrition
Purdue University
West Lafayette, Indiana 47907

## SUMMARY

Membrane traffic between the endoplasmic reticulum (ER) and the Golgi apparatus has been analyzed using a cell-free system in which transfer is via 50 to 70 nm transition vesicle intermediates. The latter have been isolated and characterized. Vesicle formation is temperature- and ATP-dependent and inhibited by N-ethylmaleimide. Transfer to Golgi apparatus is specific for cis Golgi apparatus membranes as acceptor. Heterologous transfer has been demonstrated among animal and plant systems. Nuclei with intact or semi-intact membranes will replace ER as the donor. The ability to form transition vesicles is restricted to a subpopulation of the ER, called the transitional or cis Golgi apparatus-associated ER. The cell-free systems employ a fraction enriched in this ER subpopulation. Both lipids and proteins are transferred. Transferred proteins and lipids are processed at the Golgi apparatus suggestive of membrane fusion of the transition vesicle donor with membranes of the Golgi apparatus acceptors. An ATPase with unique properties has been purified from transitional ER of liver. Other proteins under study and potentially involved in the transfer process include a 55 kDa GTP-binding protein of the transitional ER and a putative docking protein of the cis Golgi.

## INTRODUCTION

The development of cell-free systems has progressed in a relatively orderly fashion over several decades. The beginnings trace to the first isolation of microsomes by Albert Claude in the mid 1940s for which achievement he shared the Nobel prize with George Palade and Christian de Duve in 1975. The subfractionation of microsomes was subsequently supplanted by preparation of specific cell components, Golgi apparatus, endoplasmic reticulum, plasma membrane, etc., relatively intact and functional, without

NATO ASI Series, Vol. H 74
Molecular Mechanisms of Membrane Traffic
Edited by D. J. Morré, K. E. Howell, and J. J. M. Bergeron
© Springer-Verlag Berlin Heidelberg 1993

resorting to an intermediate preparation of mixed microsomes.

The next plateau was initiated by the applications of cell-free systems to inter Golgi apparatus membrane trafficking by Rothman and collaborators (Balch et al. 1983, Balch et al. 1984, Balch and Rothman 1985, Dunphy and Rothman 1985, Rothman 1987a, 1987b, Rothman and Orci 1990, Wattenberg 1991, 1992). Here, crude Golgi apparatus isolates were utilized to first investigate vesicular stomatitus G (VSV G) protein processing and later to identify cytosolic factors required for inter-compartment transfer.

The bulk flow of lipids and proteins between the ER and the Golgi apparatus is mediated by 50-70 nm transition vesicles (Ziegel and Dalton 1962, Morré et al. 1971, Mollenhauer et al. 1976) that form from transitional elements of the ER (Morré et al. 1971) and fuse with membranes of the cis face of the Golgi apparatus. The formation and fusion of ER-derived transition vesicles have been reconstituted in cell-free systems from rat liver (Morré et al. 1986, Nowack et al. 1987, Paulik et al. 1988), cultured cells (Balch et al. 1986) and yeast (Haselbeck and Shekman 1986). Semi-intact cells permeable to macromolecules also have been employed (Beckers et al. 1984, Balch and Keller 1986). As with cell-free systems for analysis of Golgi transport, the cell-free ER to Golgi apparatus systems permit an approach to molecular characterization of membrane traffic not attainable with intact or even permeabilized cells.

## Reconstitution of ER to Golgi Membrane Traffic in a Cell-Free System

Our approach to reconstitution of ER to Golgi membrane traffic has been to utilize highly purified cell fractions relatively free from contamination by membrane compartments other than a pre-determined donor or acceptor. This requirement has been obviated in other cell-free systems by the use of processing events considered to be restricted to specific compartments to monitor transfer. While ensuring specificity and eliminating the requirement for relatively pure fractions, the processing criterion at the same time limits study of membrane traffic to single membrane constitutents such as VSV G protein.

The advantage of reconstitution with highly purified cell fractions is that the system is general and can be applied to the study of trafficking of membrane proteins and membrane lipids with equal facility. The disadvantage is that since one measures transfer only and not processing, other criteria must be developed to establish physiological relevance and fidelity to the situation *in vivo.*

## Isolation of transitional ER donor

In liver, the vesicles which transport membranes to the Golgi apparatus originate from specialized part-rough, part-smooth ER domains designated as transitional ER (Palade 1975). Because they are depleted in ribosomes, they may be separated from the bulk of the rough ER on sucrose gradients.

The procedure we have developed is to start with a microsomal supernatant from which the Golgi apparatus, plasma membrane, mitochondria and nuclei have been removed in previous fractionation steps (Morré et al. 1986). This material is then applied to a discontinuous sucrose gradient and the regions of part-rough, part-smooth ER are removed from the 1.3 M/sucrose homogenate interface.

The composition of the transitional ER-enriched fraction based on analysis of enzyme markers (Table 1) reveals less that 5% total contamination

Table 1

Specific activities of marker enzymes for transitional ER[a]

| Marker enzyme activity | Total homogenate | Transitional ER |
|---|---|---|
| 5'-Nucleotidase (Plasma membrane) | 1.5 ± 0.4 | 1.9 ± 0.7 |
| Succinate INT-reductase (Mitochondria) | 2.6 ± 1.1 | 0.2 ± 0.1 |
| Galactosyl transferase (Golgi apparatus) | 2.4 ± 1.4 | 3.2 ± 1.3 |

[a]Units of specific activity are $\mu$moles/h/mg protein for succinic dehydrogenase measured as succinate INT-reductase, $\mu$moles inorganic phosphorous released/h/mg protein from 5'-AMP or glucose-6-phosphate for 5'-nucloetidase and glucose-6-phosphatase, respectively and nmoles galactose transferred to ovomucoid/h/mg protein for glactosyltranferase. The fraction in parentheses refers to the fraction in which the enzyme was assumed to be specifically located for purposes of estimating purity.

by Golgi apparatus, plasma membrane and mitochondria. Its ER nature is verified by specific activities of NADPH cytochrome c reductase and glucose-6-phosphatase comparable to those of the purified fractions (Table 2). Only the RNA content is different, being one-third that of the conventional rough ER (Table 2).

Table 2

Activities and constituents characteristic of ER comparing the transitional ER fraction with conventional rough ER fractions of rat liver[a].

| ER fraction | Glucose 6-phosphatase | NADPH-cytochrome c reductase | RNA/Protein |
|---|---|---|---|
| Transitional ER | 8.2 ± 1 | 30  ± 12 | 0.5 |
| Rough ER | 9.0 ± 3.5 | 28  ± 8 | 0.9 |
| Total homogenate | 1.8 ± 0.6 | 4.2 ± 1.9 | - |

[a]Units of specific activity are $\mu$moles inorganic phosporous released/h/mg protein for glucose-6-phosphatase and $\mu$moles NADPH reduced/h/mg protein for NADPH-cytochrome c reductase.

To establish that the transitional ER was, in fact, transitional ER, and that its function in transition vesicle formation could be reconstituted, the fractions enriched in transitional ER were incubated at 37°C with and without cytosol and an ATP-regenerating system and the preparations were checked for morphological evidence of vesicle formation (Morré et al. 1986). Buds resembling transition vesicles were formed (Fig. 1). Quantitation of 50-70 nm vesicle profiles (Fig. 2) revealed that their formation was both temperature- and ATP-dependent. Similar buds were not formed by conventional ER from the same gradients.

Fig. 1. Electron micrograph of a portion of an isolated membrane preparation consisting of part-rough-part-smooth ER. The fraction was treated for 15 min at 37°C with ATP plus an ATP-regenerating system and a concentrated cytoplasmic protein extract as described (Nowack et al. 1990). Numerous 50 - 70 nm vesicles and blebs (arrows) with solid interiors and nap-like surface coats (not clathrin-coated) resembling transition vesicles seen *in situ* were present. Scale bar = 0.5 μm

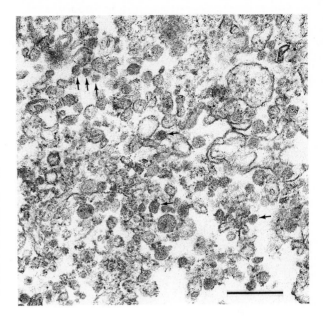

Fig. 2. Quantitation of 50 - 70 nm vesicles formed by isolated transition vesicles in the presence and absence of ATP at 4° and 37°C. Diameters of all vesicles in the preparation were measured using an image analyzer. Only the vesicles in the 50 - 70 nm size range were increased specifically by ATP addition.

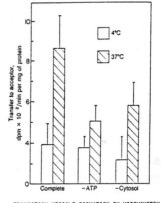

TRANSITION VESICLE FORMATION BY MORPHOMETRY

| Incubation conditions (15 min) | 45 to 60 nm vesicles % of total | 37° - 4° |
|---|---|---|
| Complete 4° | 0.4 ± 0.1 | - |
| Complete 37° | 1.1 ± 0.6 | 0.7% |
| -ATP, - Cytosol 37° | 0.6 ± 0.4 | 0.2% |

## Isolation of cis Golgi apparatus acceptor

Intact Golgi apparatus from rat of greater than 85% purity (Table 3) were the starting material for the acceptor fraction (Morré 1971). These were then unstacked enzymatically (Morré et al. 1983) and resolved into cis,

Table 3

Composition of Golgi apparatus and reference fractions of rat liver determined by electron microscope morphometry[a]

| Cell components present in fractions | Fraction composition, profiles/100 profiles | | |
|---|---|---|---|
| | ER | Golgi apparatus | Plasma membranes |
| ER | 87 ± 4 | 6 ± 2 | 2 ± 3 |
| Golgi apparatus | 1 ± 1 | 86 ± 4 | n.d. |
| Plasma membrane | 1 ± 1 | 1 ± 2 | 89 ± 5 |
| Mitochondria | 6 ± 3 | 1 ± 1 | 7 ± 2 |
| Lysosomes | 1 ± 1 | 0.5 | n.d. |
| Peroxisomes | 0.5 | trace | n.d. |
| Other + Unidentified | 3 ± 2 | 5 ± 3 | 2 ± 2 |

[a]Values are from electron micrographs photographed at random and averaged from 3 to 5 different membrane preparations (9 to 15 electron micrographs total) ± standard deviations among the different preparations. Each value represents an analysis of 4,000 to 10,000 intercepts.

medial and trans subfractions by preparative free-flow electrophoresis (Morré et al. 1984). Trans Golgi apparatus markers (thiamine pyrophosphatase, $H^+$-ATPase, luminal acidification, sialyl-and galactosyltransferase and an NADH-ferricyanide oxidoreductase resistant to fixation) all were located in the fractions of greatest electrophometric mobility (Morré et al. 1983, 1984, Hartel-Schenk et al. 1991, Brightman et al. 1992). Cis markers (osmium tetroxide reduction, nucleoside phosphate diphosphatase, NADH-cytochrome $b_5$ reductase) all were concentrated at the opposite part of the separation, in fractions of lesser electrophoretic mobility. NADPase activity, which marks medial cisternae in liver, was located in the mid portions of the separations (Navas et al. 1986).

Cis Golgi apparatus have been employed to advantage as acceptor fractions for cell-free analyses of ER to Golgi apparatus transport of both lipids (Moreau et al. 1991, Moreau and Morré 1991) and proteins (Paulik et al. In Preparation). With liver, efficient specific or ATP-dependent transfer and processing was seen only with cis-derived fractions. Little or no specific or ATP-dependent transfer or processing was obtained with trans fractions as acceptor and transitional ER-derived vesicles as donor.

## Isolation of the transition vesicle intermediate

The transition vesicle intermediates formed by the isolated transitional ER, once formed, are released spontaneously and do not require harsh extraction conditions to prepare. The released vesicles have been isolated and concentrated by preparative free-flow electrophoresis (Paulik et al. 1988). The vesicles exhibit a slower electrophoretic mobility than the bulk of the ER and collect as a shoulder region trailing the unvesiculated transitional ER in the electrophoretic separation.

## Reconstitution of cell-free ER to Golgi apparatus transfer

In the reconstituted system we have developed, the donor fraction is in solution and normally is radioactively labeled. The manner and location of labeling may vary depending on the constituents under study. To label membrane proteins, liver slices were incubated with [$^{35}$S]methionine or [$^{3}$]H- or [$^{14}$C]leucine. To label lipids, [$^{3}$H]$^{-}$ or [$^{14}$C]acetate was used. Phosphatidylcholine was labeled specifically by direct labeling of the isolated transitional ER with cytidine-5′-diphospho-[$^{14}$C]choline. Radiolabeled sterols and ceramide were incorporated directly into the isolated transitional vesicle membranes to measure transfer.

To facilitate the assay, the acceptor fraction was immobilized on nitrocellulose strips by equilibration with stacked Golgi apparatus or with cis Golgi apparatus cisternae to form a monolayer of attached membranes (Nowack et al. 1987). Unoccupied sites after equilibration with acceptor membranes were occupied by incubation of the loaded strips with albumin, carbonic anhydrase or other proteins that lack acceptor activity.

In addition to the transitional ER the donor suspension contained an ATP-ATP regenerating system patterned after that of Rothman and colleagues (e.g. Balch et al. 1983, Balch and Rothman 1985, Dunphy and Rothman 1985) and a >10 kDa cytosol fraction prepared from a high-speed rat liver supernatant by centricon microconcentration. Incubations were at 37°C and 4°C in parallel. The loaded acceptor strips after blocking with albumin were washed in buffer and introduced into the donor suspension at various times after the addition of ATP. Transition vesicle formation appeared to limit the rate of transfer whereas total transfer was limited by the amount of cis Golgi acceptor.

Some characteristics of the transitional ER to Golgi apparatus cell-free transfer are summarized in Table 4. The process is inhibited by N-ethylmaleimide and is sensitive to GTP-$\gamma$-S as is inter-Golgi apparatus transfer (Glick and Rothman 1987, Malhotra et al. 1988, Melançon et al. 1987).

Table 4

Characteristics of cell-free ER to Golgi apparatus transport system with isolated liver fractions.

---

Transitional ER (TER) donor
- enriched in ER markers
- free of PM and GA contamination
- ATP-dependent budding
- TER-specific
- ~6% transfer to cis GA

Transition vesicles (TV) intermediate
- ~60% TV, TER only major contaminant
- rapid transfer to cis GA compartment
- transfer is ATP-dependent
- TV formed below 18°C are fusion incompetent
- >60% transfer to cis GA

Golgi apparatus acceptor
- cis, medial and trans subfractions by free-flow electrophoresis
- cis specific
- finite number of TV attachment sites (saturation kinetics)
- GTP-$\gamma$-S inhibited step

---

Cell-free ER to Golgi apparatus transfer exhibits both specific and unspecific components (Fig. 3). Transfer is in proportion to the number of 50 - 70 nm vesicles formed (0.5 to 2% of the total membrane present) and is expressed as percent transfer per 1 $cm^2$ nitrocellulose acceptor strip. There is an unspecific, acceptor-independent transfer of about 0.5% per strip which occurs both at 4°C and in the absence of acceptor. This transfer normally is determined from a blank strip loaded only with albumin.

Fig. 3. Composite figure illustrating the kinetics of ER to Golgi apparatus transfer in rat liver. Illustrated are the temperature- and ATP- independent and temperature- and ATP-dependent components of transfer. The latter are specifically inhibited by N-ethylmaleimide (NEM) and correlate with cis-vesicle-mediated mechanism.

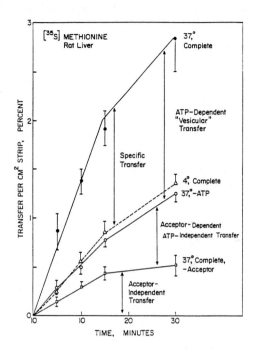

The bulk of the transfer is acceptor-dependent. Transfer mediated by transition vesicles correlates best with that part of the transfer at 37°C that is dependent upon ATP (Fig. 2). This temperature- and ATP-dependent transfer is termed specific transfer. Specific transfer is observed in the reconstituted system with ER as donor and total Golgi apparatus or cis Golgi apparatus-enriched fractions as acceptor and with no other combination of membranes (Nowack et al. 1987, Morré et al. 1991).

A third component of transfer observed primarily with lipids is that which is acceptor-dependent but acceptor unspecific. This transfer is observed at 37°C in the absence of ATP but not at 4°C either in the absence of ATP or in its presence. It is temperature-dependent but ATP-independent. Cis and trans Golgi apparatus function equally well and Golgi apparatus fractions can be replaced by rough ER or plasma membranes. This transfer appears to represent a non-vesicular transfer of lipids from donor membranes to the acceptor membranes immobilized on nitrocellulose. Therefore, the parameter that appears to most accurately measure transition vesicle formation from the donor fraction and the specific attachment (not necessarily fusion) of the vesicles to the acceptor immobilized on nitrocellulose is the ATP-dependent component of specific (temperature-dependent) transfer.

That the vesicles originating from the donor fuse with the acceptor membranes is suggested by specific processing of the $man_9$ form for VSV G transferred from ER of infected CHO cells to the processed $man_5$ form at the cis Golgi apparatus (Paulik et al. In Preparation). Lipid processing also has been demonstrated (Moreau and Morré 1991). Cis cisternae of rat liver Golgi apparatus contain an A-type phospholipase associated with the luminal interiors of the cisternae. Phospholipids of membrane vesicles are inaccessible to this phospholipase unless the membranes fuse. Processing of phosphatidycholine derived from transitional ER to lysophosphatidylcholine through the action of this phospholipase occurs in the cell-free system (Fig. 4). Specific ATP-dependent transfer and processing of phosphatidylcholine in the cell-free system is seen only with cis Golgi apparatus as acceptor and not with trans Golgi apparatus as acceptor.

Fig. 4. ATP-dependent phosphatidylcholine (PC) transfer and processing in the ER to Golgi apparatus cell-free system. Formation of lysophosphatidyl-choline (LPC) from the transfused PC provides a measure of processing. From Moreau and Morré (1991).

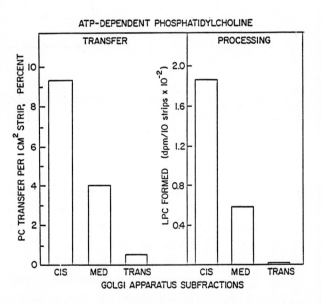

The functional role of a cis Golgi apparatus phospholipase, if any, is presently unknown although the local destabilizing influences of the lysophospholipid products could be important to transition vesicle fusion. That the lysophospholipids normally do not reach high levels in the Golgi apparatus with devastating consequences may be the result of an accompanying acyltransferase activity that occurs throughout the Golgi apparatus and whose function is to reconvert the lysoPC back into PC by acyl transfer from acylcoenzyme A (Lawrence et al., this volume).

One major advantage of cell-free analysis in general is that it allows for the dissection of the molecular events of membrane trafficking in a manner amenable to identification and characterization of specific catalytic and regulatory molecules involved. In our work, the cell-free system consisting of purified donor and acceptor fractions allows not only isolation of the transition vesicle intermediate (Paulik et al. 1988) but a clear separation of membrane budding and fusion events. Some of the characteristics of the two separable steps are summarized in Table 5.

Table 5

Partial reactions of ER to Golgi apparatus membrane transfer ammenable to separation and individual study using purified fractions in a cell-free system from rat liver. TV = Transitional ER-derived transition or tranfer vesicles.

| | ER | TV | TV | CisGA |
|---|---|---|---|---|
| ATP-dependent | | + | - | |
| Temperature-dependent | | + | | - |
| NEM-dependent | | + | | + |
| GTP-γ-S-inhibited | | - | | + |

Vesicle budding is the major ATP- and temperature-dependent step. Once the vesicles are formed, transfer is no longer temperature- or ATP-dependent (Paulik et al. 1988). Because of the central role of ATP in the budding of transition vesicles, we have begun an investigation of the major ATPase of transitional ER. The ATPase activity of the transitional ER exhibits responses to known ATPase inhibitors that distinguish it as unique. It is insensitive to vanadate, ouabain and nitrate and is insensitive to oligomycin at concentrations that inhibit completely mitochondrial ATPase. These same inhibitors also have no effect on transition vesicle budding from transitional ER. The ATPase is inhibited by cobalt chloride which is also a potent inhibitor of transition vesicle budding. The ATPase has been isolated and purified to homogeneity (Zhang and Morré, this volume). A 25 amino acid sequence obtained by digestion of the 100 kDa $M_r$ peptide with cyanogen bromide exhibits 80% identity to p97-ATPase (Peters et al. 1990). This 25 amino acid sequence shares 55% identity with cell division control protein (CDC 48) (Fröhlich et al. 1991). Both p97-ATPase and CDC 48 display homology to the mammalian N-ethylmaleimide sensitive fusion protein (NSF) and yeast Sec 18p which are essential for fusion in secretory process (Wilson et al. 1989). The ATPase is an intrinsic membrane protein of ca 190 kDa $M_r$.

Identification of membrane proteins involved in the mechanism of membrane attachment and fusion of transition vesicles with cis Golgi apparatus has proven more difficult. the attachment and fusion step apparently requires GTP hydrolysis and is blocked by GTP-$\gamma$-S and N-ethylmaleimide. A number of candidate proteins are under investigation and one with interesting properties has emerged.

The protein under investigation migrates as a 38 kDa $M_r$ peptide on SDS-PAGE and has been purified from cis Golgi cisternae from rat liver. The protein is concentrated in cis Golgi apparatus fractions enriched by preparative free-flow electrophoresis and is low or undetectable in trans Golgi apparatus fractions prepared in parallel. Rabbit polyclonal antisera raised to the p38$_{cis}$ protein when preincubated with transitional ER fractions have no effect on transfer but transfer is greatly reduced or blocked when the cis Golgi apparatus membranes on the acceptor strips are preincubated with the antisera.

One possibility under consideration is that the p38$_{cis}$ is part of a transition vesicle docking or receptor complex that normally limits vesicle attachment/fusion at the cis Golgi. Kinetic evidence is provided from transfer studies for a limited number of docking or attachment sites at the cis Golgi. The attachment and subsequent fusion of transition vesicles to cis Golgi is saturatable. Active transitional ER preparations will support several sequential rounds of acceptor strip additions each of which saturate after 15 to 30 min of transfer under conditions where acceptor is rate limiting. The p38$_{cis}$ peptide crossreacts on Western blots with antibodies to a consensus GTP-binding sequence (Cys-<u>Gly-Ala-Gly-Glu-Ser</u>-Gly-Lys-Ser-Thr-Ile-Val-Lys-Gln-Met) (GTP-binding consensus domain underlined) of G$\alpha$ common (Brightman et al., this volume). This is the first evidence that p38$_{cis}$ may represent as well a guanine nucleotide binding protein of the cis Golgi apparatus with some role in explaining the essential requirement for guanine nucleotides in transition vesicle transfer to cis Golgi apparatus, and effects of GTP-γ-S (Wattenberg 1991, 1992, Davey et al. 1985, Gruenberg and Howell 1989).

Thus the isolation of well defined and highly purified cell fractions not only has aided our understanding of membrane trafficking by providing important enzymatic markers and compositional data but now finds applications to the cell-free analysis of membrane traffic. Their application becomes especially important to the investigation of intrinsic membrane proteins involved in membrane trafficking as applied to cytosolic factors. The latter have been approached quite successfully using isolated but mixed cell fractions where transfer is assessed with processing assays (Balch et al. 1983, 1984) or using permeabilized cells (Beckers et al. 1984, Balch and Keller 1986, Simons and Virta 1987). However, mixed fractions or permeabilized cells are less effectively employed than homogeneous cell fractions in the study of the intrinsic proteins involved in bud formation and release from ER or in the subsequent fusion of the released buds with cis Golgi apparatus membranes. It is to the elucidation of the latter steps where the defined cell-free system described in this report seems to be most advantageously directed.

## Acknowledgements

Work supported in part by grants from the NIH GM and the Lilly Research Laboratories, Indianapolis and Phi Beta Psi Sorority. The many contributions of Mark Paulik to the progress of this work are gratefully acknowledged.

## References

Balch WE, Fries E, Dunphy W, Urbani LJ, Rothman JE (1983) Transport-coupled oligosaccharide processing in a cell-free system. Meth Enzymol 98:37-47
Balch WE, Dunphy WG, Braell WA, Rothman JE (1984) Reconstitution of the transport of protein between successive compartments of the Golgi measured by the coupled incorporation of N-acetylglucosamine. Cell 39:405-416
Balch WE, Rothman JE (1985) Characterization of protein transport between successive compartments of the Golgi apparatus: Asymmetric properties of donor and acceptor activities in a cell-free system. Biochem Biophys 240:413-425
Balch ER, Keller DS (1986) ATP-coupled transport of vesicular stomatitis virus protein. J Biol Chem 261:14690-14696
Balch WE, Elliott MM, Keller DS (1986) ATP-coupled transport of vesicular stomatitis virus G protein between the endoplasmic reticulum and the Golgi. J Biol Chem 261:14681-14689
Beckers CJ, Keller DS, Balch WE (1987) Semi-intact cells permeable to macromolecules: Use in reconstitution of protein transport from the endoplasmic reticulum to the Golgi complex. Cell 50:523-534

Brightman AO, Paulik M, Lawrence JB, Reust T, Geilen CC, Spicher K, Schultz
    G, Reutter W, Morré DM, Morré DJ (1992) A 38 kDa protein resident to cis
    Golgi apparatus cisternae of rat liver is recognized by an antibody
    directed against α subunits of trimeric G-proteins (this volume)
Brightman AO, Navas P, Minnifield NM, Morré DJ (1992) Pyrophosphate-induced
    acidification of trans cisternal elements of rat liver Golgi apparatus.
    Biochim et Biophys Acta 1104: 188-194
Davey J, Hurtley SM, Warren G (1985) Reconstitution of an endocytic fusion
    event in a cell-free system. Cell 43: 643-652
Dunphy WG, Rothman JE (1985) Compartmental organization of the Golgi stack.
    Cell 42:13-21
Fröhliich K-U, Fries H-W, Rüdiger M, Erdmann R, Botatein D (1991) Yeast cell
    cycle protein CDC 48p shows full-length homology to the mammalian
    protein VCP and is a membrane of a protein family involved in secretion,
    perosisome formation, and gene expression. J Cell Biol 114:443-453
Glick BS, Rothman JE (1987) Possible role for fatty acyl coenzyme A in
    intracellular protein transport. Nature 326:309-312
Gruenberg J, Howell, KE (1989) Membrane traffic in endocytosis: Insights from
    cell-free assays. Annu Rev Cell Biol 5: 453-481
Hartel-Schenk S, Minnifield N, Reutter W, Hanski C, Bauer C, Morré DJ (1991)
    Distribution of glycosyltransferases among Golgi apparaatus subfractions
    from liver and hepatomas of the rat. Biochim et Biophys Acta 1115: 108-
    122
Haselbeck A, Shekman R (1986) Interorganelle transfer and glycosylation of
    yeast invertase *in vitro*. Proc Natl Acad Sci USA 83:2017-2021
Lawrence JB, Keenan TW, Morré DJ (1992) Acyl transfer reactions associated
    with cis Golgi apparatus. (this volume)
Malhotra V, Orci L, Glick BS, Block MR, Rothman JE (1988) Role of an N-
    ethylmadeimide sensitive transport component in promoting fusion of
    transport vesicles with cisternae of the Golgi stack. Cell 54:221-227
Melançon P, Glick BS, Malhotra V, Weidman PJ, Serafini T, Gleason ML, Orci
    L, Rothman JE (1987) Involvement of the GTP-binding "G" proteins in
    transport through the Golgi stack. Cell 51:1053-1062
Mollenhauer HH, Hass BS, Morré DJ (1976) Membrane transformations in Golgi
    apparatus of rat spermatids. A role for thick cisternae and two classes
    of coated vesicles in acrosome formation. J Microsc Biol Cell (Paris)
    27: 33-36
Moreau P, Morré DJ (1991) Cell-free transfer of membrane lipids. Evidence for
    lipid processing. J Biol Chem 266:4329-4333
Moreau P, Rodriguez M, Cassagne C, Morré DM, Morré DJ (1991) Trafficking and
    sorting of lipids from endoplasmic reticulum to the Golgi apparatus in
    a cell-free system from rat liver. J Biol Chem 266:4322-4328
Morré DJ (1971) Isolation of Golgi apparatus. Meth in Enzymol 22:130-148
Morré DJ, Mollenhauer HH, Bracker CE (1971) In: Reinert J, Ursprung H (eds)
    Results and problems in cell differentiation: Origin and continuity of
    cell organelles, Springer Verlag, Berlin/New York/Heidelberg, Vol 2, pp
    82-126
Morré DJ, Morré DM, Heidrich H-G (1983) Subfractionation of rat liver Golgi
    apparatus by free-flow electrophoresis. Eur J Cell Bio 31: 263-274
Morré DJ, Creek KE, Matyas GR, Minnifield N, Sun I, Baudoin P, Morré DM,
    Crane FL (1984) Free-flow electrophoresis for subfractionation of rat
    liver Golgi apparatus. BioTechniques 2:224-233
Morré DJ, Paulik M, Nowack D (1986) Transition vesicle formation *in vitro*.
    Protoplasma 132:110-113
Morré DJ, Penel C, Morré DM, Sandelius AS, Moreau P, Andersson B (1991) Cell-
    free transfer and sorting of membrane lipids in spinach. Donor and
    acceptor specificity. Protoplasma 160:49-64

Navas P, Minnifield N, Sun I, Morré DJ (1986) NADP phosphatase: A marker in free-flow electrophoretic separations for cisternae of the Golgi apparatus midregion. Biochim Biophys Acta 881:1-9

Nowack DD, Morré DM, Paulik M, Keenan TW, Morré DJ (1987) Intracellular membrane flow: Reconstitution of transition vesicle formation and function in a cell-free system. Proc Natl Acad Sci USA 84:6098-6102

Nowack DD, Paulik M, Morré DJ, Morré DM (1990) Retinoid modulation of cell-free membrane transfer between endoplasmic reticulum and Golgi apparatus. Biochim Biophys Acta 1051:250-258

Palade GE (1975) Intracellular aspects of the process of protein synthesis. Science 109:247-358

Paulik M, Nowack DD, Morré DJ (1988) Isolation of a vesicular intermediate in the cell-free transfer of membrane from transitional elements of the endoplasmic reticulum to Golgi apparatus cisternae of rat liver. J Biol Chem 263:17738-17748

Paulik MA, Widnell CC, Whitaker-Dowling PA, Minnifield N, Morré DM, Morré DJ (In preparation) Cell-free transfer of VSV G protein from an endoplasmic reticulum compartment of BHK cells to a rat liver Golgi apparatus compartment for $Man_{8-9}$ to $Man_5$ processing.

Peters J-M, Walsh MJ, Franke WW (1990) An abundant and ubiquitous homo-oligomeric ring-shaped ATPase particle related to the putative vesicle fusion protein Sec18p and NSF. EMBO J 9:1757-1767

Rothman JE (1987a) Protein sorting by selective retention in the endoplasmic reticulum and Golgi stack. Cell 50:521-522

Rothman JE (1987b) Transport of the vesicular stomatitis glycoprotein to trans Golgi membranes in a cell-free system. J Biol Chem 262:12602-12610

Rothman JE, Orci L (1990) Movement of proteins through the Golgi stack: A molecular dissection of vesicular transport. FASEB J 4:1460-1468

Simons K, Virta H (1987) Perforated MDCK cells support intracellular transport. EMBO J 6:2241-2247

Warren G, Woodman P, Pypaert M, Smythe E (1988) Cell-free assays and the mechanism of receptor-mediated endocytosis. TIBS 13: 462-465

Wattenberg B (1991) Analysis of protein transport through the Golgi in a reconstituted cell-free system. J Electron Micros Tech 17:150-164

Wattenberg B (1992) Vesicular traffic in eukaryotic cells. In: Yeagle (ed) The structure of Biological Membranes, CRC Press, pp 997-1046

Wilson DW, Wilcox CA, Glynn CC, Chan E, Kuang W-J, Henzel WJ, Block MR, Ullrich A, Rothman JE (1989) A fusion portein needed for transport from the endoplasmic reticulum and within the Golgi stack in both animal cells and yeast. Nature 339:355-360

Zhang L, Morré DJ (1992) Isolation and characterization of the principal ATPase activity of transitional endoplasmic reticulum from rat liver. (this volume)

Ziegel RF, Dalton AJ (1962) Speculations based on the morphology of the Golgi system in several types of protein secreting cells. J Cell Biol 15:45-54

# IDENTIFICATION OF A NOVEL POST-ER, PRE-GOLGI COMPARTMENT WHERE UNASSEMBLED MONOMERS OF OLIGOMERIC PROTEINS ACCUMULATE

T. C. Hobman, L. Woodward, M. Komuro and M. G. Farquhar
Division of Cellular and Molecular Medicine and
Center for Molecular Genetics
9500 Gilman Drive
University of California, San Diego
La Jolla, CA 92093-0651

## Introduction

It has been clear for some time that newly synthesized proteins exit the ER via vesicular carriers that are derived from transitional or part rough/part smooth elements of the ER and pass through pre-Golgi elements before reaching the cis side of the Golgi stack (Farquhar, 1985; Farquhar, 1991; Pfeffer and Rothman, 1987). What is not so clear is the number and nature of the pre-Golgi compartments that are variously referred to as pre-Golgi, intermediate or salvage compartments. The existence of an intermediate compartment was originally suggested by Saraste and Kuismanen (1984) based on their observation that when cells infected with Semiliki Forest virus were incubated at 15°C, the envelope glycoproteins accumulated in smooth vacuoles located near the cis side of the Golgi. It was also shown that some viruses bud from smooth ER, pre-Golgi compartments (Tooze et al., 1984; Ulmer and Palade, 1991). Pelham (1989) proposed that one of the functions of this intermediate compartment is to act as a salvage or recycling compartment where resident ER proteins that escape from the ER are specifically and continuously retrieved and returned to the ER by a receptor that recognizes KDEL or closely related sequences. Pelham and co-workers also obtained experimental evidence indicating that this intermediate compartment may contain N-acetyl glucosamine-1 phosphotransferase which is responsible for addition of the phosphorylated glcNAc to lysosomal enzymes (Pelham, 1989). Moreover, three proteins--i.e., p53 (Schweizer et al., 1988), p58 (Hendricks et al., 1991; Saraste et al., 1987), and p63 (Schweizer et al., 1991), have been identified that represent putative markers for the intermediate compartment with the

NATO ASI Series, Vol. H 74
Molecular Mechanisms of Membrane Traffic
Edited by D. J. Morré, K. E. Howell, and J. J. M. Bergeron
© Springer-Verlag Berlin Heidelberg 1993

complication that they appear not to be limited to this compartment, but to cycle between the intermediate compartment and the cis Golgi (Saraste and Svensson, 1991). Using the 53 kD protein as a marker, Schweizer and co-workers (Schweizer et al., 1991) have prepared subfractions enriched in this protein. These fractions were largely separated from rough ER proteins (ribophorins, BIP, and PDI) as well as from N-acetyl glucosamine-1 phosphotransferase and phosphodiesterase. For the purposes of our discussion it is important to realize that a certain amount of confusion exists concerning the number and nature of the putative intermediate compartments and their relationship to the Golgi.

Recently we have made observations that have a bearing on this problem while studying the expression of Rubella virus (RV) envelope glycoproteins.

## Use of Rubella Virus Glycoproteins to Study Golgi Traffic and Targeting

Viral proteins have proved to be very valuable tools to study intracellular membrane traffic due to their relative ease of manipulation and the high levels of expression that can be achieved. Many viruses bud from the cell surface, but others bud from intracellular compartments. For example, the IBV coronovirus and RV bud from Golgi subcompartments. We and others have been interested in determining what is the nature of the targeting signal(s) for Golgi proteins, and have been using RV glycoproteins to study this problem. RV virions contain two envelope glycoproteins, E2 and E1, both of which are type I membrane proteins. These two proteins are thought to form a heterodimeric complex within the ER before proceeding on to their final residence in the Golgi.

To study the assembly and targeting of these proteins, we have developed stable transfectants in CHO cells expressing either E1 or E2 alone or E1 and E2 together. We developed stable transfectants because it has been our experience that transient transfection facilitated by reagents such as DEAE dextran results in an increase in autophagic vacuoles and lysosomes.

Stable cell lines were obtained by cotransfecting dhfr-CHO cells with plasmids containing genes for the RV proteins and the plasmid pFR400 (Simonsen and Levinson, 1983) which contains the selectable marker dhfr. Selection was based on the ability of the transfectants to grow in nucleoside-free medium. The cell lines were screened for expression of RV proteins by indirect immunofluorescence and radio-immunoprecipitation.

E I                    Man II

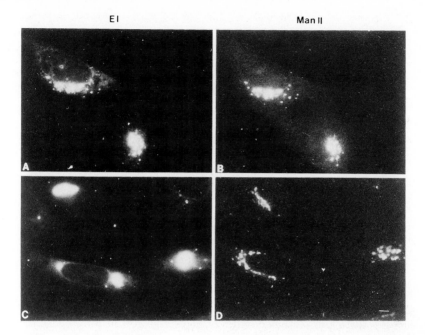

Fig. 1. CHO cells expressing both E2 and E1 (A and B) or E1 alone (C and D). Transfected cells were seeded onto fibronectin-coated chamberslides. Forty h later, cells were fixed and permeabilized with 100% methanol at -20°C, and double labeled for E1 and α-mannosidase II (Man II). In cells expressing both proteins staining for E1 colocalizes with the Golgi marker Man II (B). In cells expressing E1 alone staining for E1 (C) is found in large structures that are not stained for anti-Man II. Bar = 5 µm.


**RV E2 and E1 are Targeted to the Golgi in Cells Expressing both Proteins**


We first examined CHOE2E1 cells expressing both E2 and E1 (Fig. 1A-B). We found that when these two glycoproteins were expressed together, they were targeted to the Golgi. This conclusion was based on the fact that both proteins co-localized with the Golgi marker, α-mannosidase II (manII). Also, when the cells were treated with BFA or nocodazole, agents which specifically disrupt the Golgi, these proteins dispersed along with α-manII (not shown).

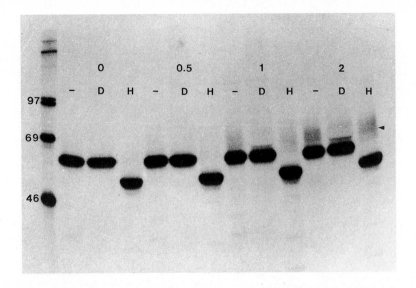

Fig. 2. E1 remains endo H-sensitive and endo D-insensitive. Subconfluent monolayers were pulse-labeled with [$^{35}$S]cysteine for 30 min, chased for the indicated time periods (h), and E1 immunoprecipitates carried out with human anti-RV serum. Samples were divided into three aliquots and treated with or without endo H (H lanes) or endo D (D lanes) or incubated without enzyme (- lanes). The proteins were separated by SDS-PAGE on 10% polyacrylamide gels and fluorographed. Golgi-specific processing of E1 glycans is indicated by an arrowhead. $^{14}$C-labeled protein standards (kD) are included for reference.

## E1 Does not Reach the Golgi in CHOE1 Cells

In CHOE1 cells that express E1 alone (Fig. 1C-D), the situation was quite different: E1 was found in large compact structures located adjacent to the nucleus. The site of localization of E1 did not overlap with that of the Golgi marker manII nor did it overlap with markers for early endosomes (transferrin receptor), late endosomes (mannose-6-phosphate receptor), or lysosomes (lgp120, cathespsin D). Incubating the cells up to 6 h in cycloheximide did not affect the pattern. These findings indicated that when E2 and E1 are expressed together they are targeted to Golgi elements, whereas when E1 is expressed alone it accumulates in a compartment with unusual features that does not appear to correspond to either the Golgi, endosomes or lysosomes.

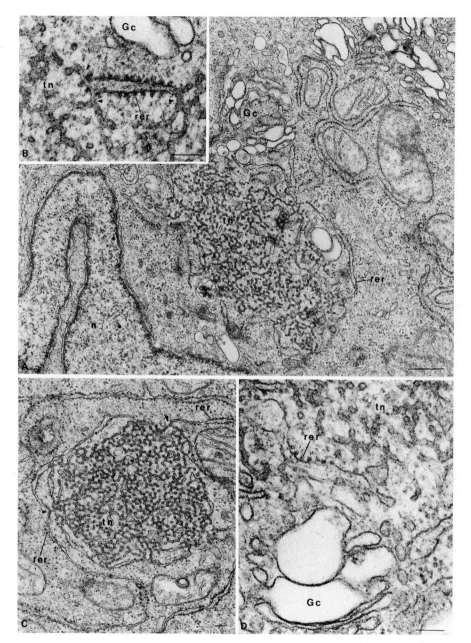

Fig. 3. Electron micrographs showing the tubular network of smooth membranes (tn) that represents the site of E1 arrest in CHOE1 (A, C). Points of continuity (arrowheads) between RER cisternae (rer) and the tubular network are seen in favorable sections (B, D). Bar, A-C 0.5 μm; D = 0.1 μm.

## E1 Remains Endo H-Sensitive and Endo D-Insensitive in CHOE1 cells

To determine whether E1 had left the ER and entered the Golgi, we radiolabled CHOE1 cells with [$^{35}$S]cysteine, carried out immunoprecipitations after different chase periods, and digested the immunoprecipitates with endoglycosidase H (endo H) and D (endo D). Most (80%) of the E1 remained endo H-sensitive after 2 h (Fig. 2). Such a result is usually taken to indicate that a protein has not left the ER, but in reality what this result allows us to say is that it does not reach the middle Golgi. (Glycoproteins become endo H insensitive when they have been acted on by both man II and GlcNAc transferase I, which are middle Golgi enzymes.)

To investigate the possibility that E1 accumulates in the cis region of the Golgi, immunoprecipitates were digested with endo D. At no time could any endo D-sensitive forms of E1 be detected (Fig. 2) (Glycoproteins become endo D sensitive when they are processed by Golgi α-mannosidase I which is a cis Golgi enzyme). Thus both the biochemical and immunofluorescence findings suggest that E1 is arrested in a pre-Golgi compartment and does not reach the Golgi. By contrast, when E1 and E2 are expressed together they become endo H insensitive with a $t_{1/2}$ of 90-120 min (not shown).

## The Pre-Golgi Compartment in which E1 Accummulates Consists of Tubular Networks of Smooth Membranes

To obtain further information on the nature of the site of E1 arrest, we examined cells expressing E1 by electron microscopy. In routine epon sections the CHOE1 cells contained conspicuous networks composed of tubular, smooth membranes lacking ribosomes (Fig. 3 A and C). Typically they were located in close proximity to the Golgi and were surrounded by rough ER. Frequently, in favorable sections, continuity with the rough ER could be detected (Fig. 3B).

After immunogold labeling of ultrathin cryosections (Fig. 4), E1 was detected at high concentration in the tubular networks where it was predominantly associated with the smooth membranes as expected since E1 is a membrane protein. Thus the EM findings indicated that the site of E1 arrest consists of a smooth membrane compartment with distinctive morphologic properties in direct continuity with the rough ER.

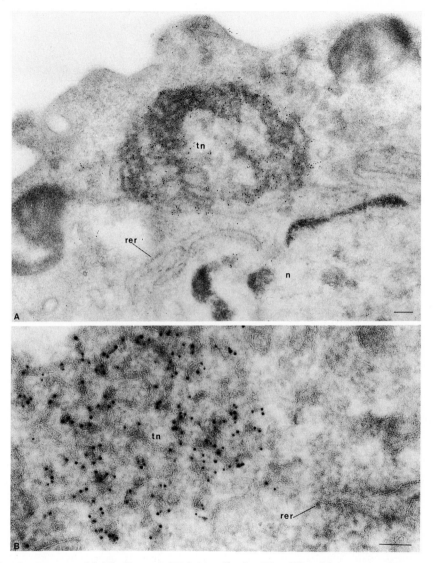

Fig. 4. Imunogold labeling of CHOE1 cells for E1. Ultrathin cryosections were incubated sequentially with mouse monoclonal anti-E1, rabbit anti-mouse IgG, and goat anti-rabbit IgG coupled to 5 nm gold. (A) E1 is present in high concentration in the tubular network, but little labeling of the rough ER (rer) is seen. (B) Higher magnifcation showing the high density of gold particles associated with the tubular, smooth membranes (tn) that represent the site of E1 arrest. Note that a nearby rough ER cisterna (rer) is not labeled for E1. This suggests that E1 rapidly leaves the rough ER and moves to the smooth, tubular membrane compartment where it accumulates. n=nucleus. Bar, 0.1 μm.

**The Site of E1 Arrest Does not Contain ER Membrane Proteins but is Accessible to ER Luminal Proteins**

We next set out to determine if ER proteins are present in the E1-containing tubular network using antibodies raised against various ER marker proteins. We found that an antiserum that recognizes four rough ER membrane proteins (Louvard et al., 1982) did not stain the E1-containing structures (Fig. 5). We obtained a reticular pattern of staining typical for the rough ER; however, no staining of the E1-containing compartment was detected. E1 actually appears to be excluded from these structures (Fig. 5 A and B). Similar results were obtained with monospecific antibodies to ER proteins provided by David Meyer (UCLA) and Bill Dunn (University of Florida). The fact that very little reticular ER staining for E1 is evident suggests that E1 rapidly leaves the rough ER and moves into the tubular network of smooth membranes after its biosynthesis.

The situation was quite different with two KDEL-bearing content proteins, protein disulfide isomerase (PDI) (Tooze et al., 1989) and BIP (Bole et al., 1986). The distribution of these proteins partially overlapped with that of E1; however, neither PDI nor BIP were concentrated in this compartment (Fig. 5 C-F). These results suggest that ER membrane proteins are excluded from the site of E1 arrest, but ER content proteins have access to this compartment.

**The Tubular Network is not Affected by Agents that Disrupt the ER**

Since the tubular network of smooth membranes that represents the site of E1 arrest is in continuity with the rough ER, we carried out experiments to assess its response to agents which are known to disrupt the rough ER. The rough ER is thought to be stabilized in part by a $Ca^{2+}$-dependent protein matrix since depletion of $Ca^{2+}$ stores by thapsigargin (an inhibitor of ER $Ca^{2+}$ATPase) or ionomycin (a calcium-specific ionophore) results in vesiculation of ER membranes and secretion of luminal ER proteins (Koch et al., 1988). In contrast, the Golgi is unaffected by $Ca^{2+}$ depletion. When CHO cells were incubated with 250 nM thapsigargin or ionomycin (5 nM) for up to 5 h, massive vesiculation of the rough ER occurred; however, the E1-containing elements were largely unaffected (Fig. 6). These results indicate that the site of E1 arrest is resistant to agents that disrupt the integrity of RER, and that this compartment is not stabilized by $Ca^{2+}$. Thus the

EI                                    RER

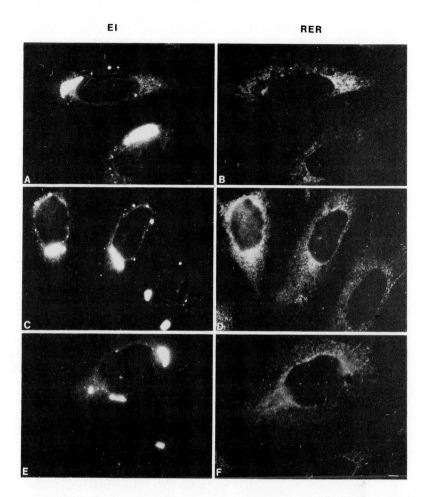

Fig. 5.  Double indirect immunofluorescence on CHOE1 cells with antibodies to resident proteins of the RER.  Cells were incubated with mouse anti-E1 and rabbit or rat antibodies to RER marker proteins.  ER luminal proteins appear to have access to the E1-containing structures, but ER membrane proteins are excluded. A) mouse anti-E1, B) same cells with anti-RER serum (Louvard et al., 1982) which stains four ER membrane proteins.  C) mouse anti-E1; D) same cells with rabbit anti-PDI; E) mouse anti-E1; F) same cells with rat anti-BiP. Bar = 5 μm.

EI　　　　　　　　　　　　　　　RER

control

Thap.
2h

Thap.
4h

Ionomycin
4h

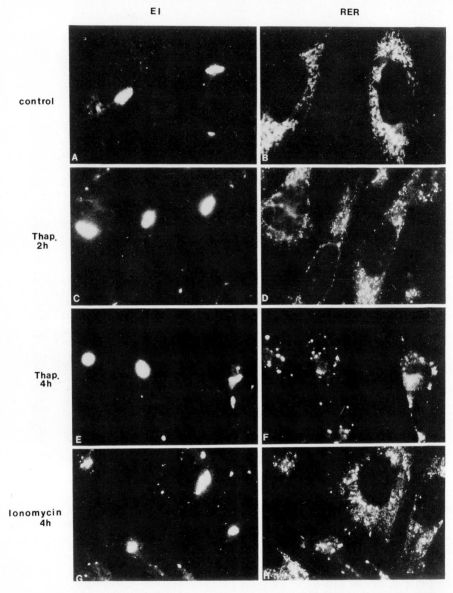

Fig. 6. Ca$^{2+}$ depleting agents do not affect the E1 containing structures. CHOE1 cells were incubated with thapsigargin [Thap] (250 nM) for 2 (C, D) and 4 h (E, F) or with 5 nM ionomycin (G, H). Panels A and B show untreated cells (control) for reference. Panels A, C, E and G show the distribution of E1 using mouse anti-E1 which does not change while cells in panels B, D, F, H were stained using rabbit anti-RER serum (RER) showing disruption of the rough ER. Bar = 5 μm.

properties of the tubular smooth membranes differ from rough ER membranes in both their membrane composition and resistance to $Ca^{2+}$-depletion.

**VSV G Protein Can Enter and Exit the Site of E1 Arrest**

We next wondered if this smooth ER compartment is on the mainline of exocytic traffic or is some sort of an abnormal byway. It could just be that an abnormal compartment is formed to accomodate insoluble aggregates that are presumably too large to enter transport vesicles (Hurtley and Helenius, 1989).

To determine if proteins can enter and exit the tubular networks we infected CHOE1 cells with the temperature sensitive, mutant ts045 strain of vesicular stomatitis virus (VSV) which is extensively used to study ER to Golgi transport. At the restrictive temperature, 39.5°C, the G protein does not leave the ER, but when shifted to 32°C, the protein rapidly exits the ER and is transported to the cell surface. In CHOE1 cells infected with ts045 and held at the restrictive temperature for 2.5 h, VSV G protein was found throughout the rough ER, and by double immunofluorescence labeling, G protein staining overlapped with the staining for E1 (Fig. 7 A and B). By immunogold labeling VSV G protein could also be detected throughout the E1-containing tubular networks (Fig. 8A). Thus despite being unable to enter the Golgi, the G protein had access to the site of E1 arrest at the restrictive temperature.

When the cells incubated at 39°C were shifted to 32°C in the presence of cycloheximide, G protein could be seen to enter vesicular structures surrounding the tubular network and throughout the cytoplasm within 5 min after the temperature shift (Fig. 7 C and D). G protein was seen within similar vesicular structures when infected cells were held at 15°C following G protein buildup at 39.5°C (Fig. 7 K and L). By immunogold labeling VSV G was detected in tubularvesicular structures corresponding to the intermediate or 15° compartment (Fig. 8B) as well as in tubular networks. With continued incubation at 32°C, G protein moved out of the vesicles into the Golgi region (within 20 min) and finally (by 40 min) to the plasma membrane (Fig 7 E-J).

These experiments indicate that: 1) VSV G protein passes through the tubular network and is transported to the Golgi complex en route to the cell surface, and 2) G protein enters structures corresponding in morphology to the 15°C compartment after passing through the site of E1 arrest. We conclude that

E1 G

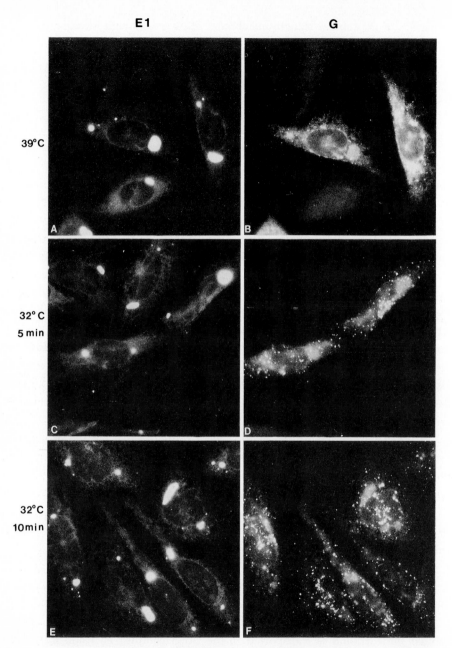

39°C

32°C
5 min

32°C
10min

Fig. 7. Transport of VSV G protein through the tubular network. CHOE1 cells were infected with VSV ts045 at 32°C for 60 min, transferred to 39.5°C for 2.5 h, and processed for immunofluorescence (A, B) or transferred to 32°C for 5 (C, D), 10 (E, F), 20 (G, H) or 40 (I, J) min before assay.

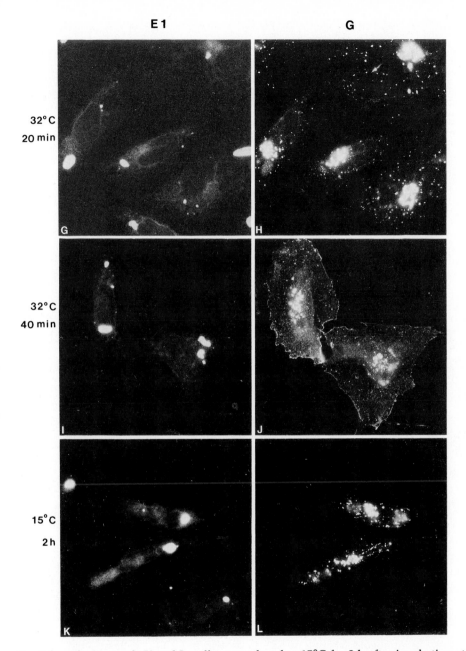

Fig. 7 (cont.): In panels K and L, cells were placed at 15°C for 2 h after incubation at 39°C. Panels (A, C, E, G, I, K) show the distribution of E1 glycoprotein as revealed by human anti-RV serum. The distribution of VSV G protein in the corresponding cells is shown in panels (B, D, F, H, J, L). Bar = 5 μm.

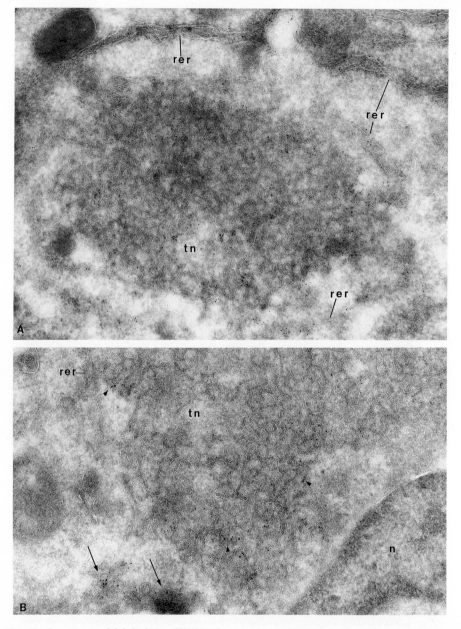

Fig. 8. Immunogold labeling of VSV G protein in CHOE1 cells infected with the mutant ts045 strain of VSV. (A) When cells are held at 39°C for 2.5 h VSV G is found in the rough ER (rer) and in the E1-containing tubular membranes (tn). (B) When cells are held at 39°C and subsequently shifted to 15°C for 5 min, VSV G begins to move out of the E1-containing networks and is seen in vesicular structures (arrows) that correspond to the intermediate or 15° compartment.

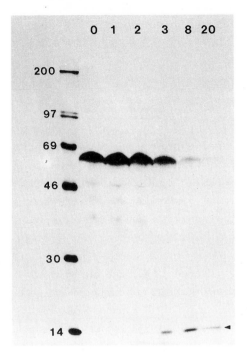

0  1  2  3  8  20

200

97

69

46

30

14

Fig. 9. Degradation of E1 glycoprotein. Subconfluent CHOE1 cells were pulse-labeled for 60 min with 50 µCi of [$^{35}$S]cysteine and chased for the indicated time periods (h) with medium containing excess unlabeled cysteine before lysis. E1 proteins were recovered by immunoprecipitation as above and separated by SDS-PAGE on 10% polyacrylamide gels followed by fluorography. $^{14}$C-labeled protein standards (kD) were included for reference. The 14 kD band (arrowhead) presumably represents a degradation product of E1.

the smooth membrane compartment containing E1 lies along the mainline of exocytic traffic and is proximal to the 15°C compartment.

## Degradation of E1

Recently it has been shown that unassembled subunits of multimeric protein complexes expressed in excess may be rapidly degraded in non-lysosomal, pre-Golgi compartments with the $t_{1/2}$ of the protein varying from minutes to hours. To determine the stability of E1 in the pre-Golgi compartment the level of radiolabeled E1 was examined after prolonged chase periods. Bands on autoradiograms (Fig. 9) were quantitated by scanning densitometery and plotted as a function of time. We found that E1 has a relatively short half life of 6 h. A putative 14 kD degradation product was observed to accumulate during increased chase periods. Chloroquine had little if any effect on the rate of E1 proteolysis (not shown), rendering it unlikely that E1 is degraded in an acidic compartment such as endosomes or lysosomes. Thus the findings suggest that E1 may undergo degradation at its site of accumulation.

## Summary and Conclusions

Our findings indicate that E1 when expressed alone is transport-incompetent and cannot enter the Golgi unless it forms a heterodimer with E2. In this situation it travels to the distalmost regions of the ER where it accumulates in a novel compartment consisting of a tubular network of smooth membranes located at or near the ER exit site. This compartment has distinct properties, but it is in continuity with the rough ER. Kinetically the compartment is located proximal to the 15° block which corresponds to the so-called intermediate compartment (Pelham, 1989; Saraste and Kuismanen, 1984). The site of E1 arrest does not contain ER or Golgi membrane proteins and is not disrupted by agents that disrupt the ER (thapsigargin, ionomycin) or Golgi (BFA, nacodozole), and its morphology is quite different from the ER or Golgi. Moreover, it does not correspond to the intermediate compartment as defined by others, as VSV G protein has access to it at the restrictive temperature and moves out of it at 15°C.

Several other ER-associated specialized pre-Golgi compartments have been described previously (Chin et al., 1982; Rizzolo et al., 1985; Tooze et al., 1984; Ulmer and Palade, 1991). None of these resembles morphologically the structures described here. We consider that the site of E1 arrest may represent a new compartment or a differentiated proximal moiety of the intermediate compartment. Thus our observations raise the possibility that smooth membrane pre-Golgi compartments may consist of two or more subcompartments with different functions.

Furthermore, these experiments demonstrate the existence of another level of control between ER to Golgi transport - i.e., E1 is not confined to the rough ER as are most unassembled subunits. In addition to the implications of our results on our thinking about the nature and number of pre-Golgi compartments our findings raise several new questions, among which are: What features of RV E1 render it competent to leave the rough ER? How is E1 retained and concentrated at the exit site? We know that it is not by tight association with Bip or by formation of covalent aggregates (Hobman et al., 1992). Does this compartment have a counterpart in non-transfected cells? In this regard it is of interest to note that in some cells such as hyperstimulated pancreatic exocrine cells, similar but less highly developed smooth tubular elements are seen in continuity with transitional ER elements. Thus the networks of tubular membranes may have their counterparts among the transitional elements of the ER in secretory cells. These membranes apparently become highly amplified or hypertrophied in the CHOE1 cells due to the high levels of expression of E1. The latter should greatly facilitate the isolation (by cell fractionation and immunoadsorption) of the E1-containing tubular membranes and make it possible to characterize the resident proteins of this novel compartment.

**Acknowledgement:** This research was supported by NIH grant DK17780. Tom Hobman is supported by a fellowship from the Medical Research Council of Canada. Figures 1-6 and 9 are from Hobman et al. (J. Cell Biol. 118:795-812).

# References

Bole D G, Hendershott L M and Kearney J F (1986) Posttranslational association of immunoglobulin heavy chain-binding protein with nascent heavy chains in non-secreting and secreting hybridomas. J Cell Biol 102:629-639.

Chin D J, Luskey K L, Anderson R G W, Faust J R, Goldstein J L and Brown M J (1982) Appearance of crystalloid endoplasmic reticulum in compactin resistant Chinese hamster cells with a 500-fold increase in 3-hydroxy-3-methylglutaryl-coenzyme A reductase. Proc Natl Acad Sci, USA 70:1185-1189.

Farquhar M G (1985) Progress in unraveling pathways of Golgi traffic. Ann Rev Cell Biol 1:447-488.

Farquhar M G (1991). Protein traffic through the Golgi. In: C. Steer and J. Hanover (eds) Intracellular Trafficking of Proteins. Cambridge University Press, Cambridge, 431-471.

Hendricks L C, Gabel C A, Suh K and Farquhar M G (1991) A 58 kDa resident protein of the cis Golgi cisterna is not terminally glycosylated. J Biol Chem 266:17559-17565.

Hobman T C, Woodward L and Farquhar M G (1992) The rubella virus E1 glycoprotein is arrested in a novel, post-ER pre-Golgi compartment. J Cell Biol 118:795-812.

Hurtley S and Helenius A (1989) Protein oligomerization of newly synthesized proteins in the endoplasmic reticulum. Ann Rev Cell Biol 5:277-307.

Koch G L E, Booth C and Wooding F B P (1988) Dissociation and re-assembly of the endoplasmic reticulum in live cells. J Cell Sci 91:511-522.

Louvard D, Reggio H and Graham W (1982) Antibodies to the Golgi complex and the rough endoplasmic reticulum. J Cell Biol 92:92-107.

Pelham H R B (1989) Control of protein exit from the endoplasmic reticulum. Ann Rev Cell Biol 5:1-23.

Pfeffer S R and Rothman J E (1987) Biosynthetic protein transport and sorting by the endoplasmic reticulum and Golgi. Ann Rev Biochem 56:829-852.

Rizzolo L J, Finidori J, Gonzalez A, Arpin M, Ivanov I E, Adesnik M and Sabatini D D (1985) Biosynthesis and intracellular sorting of growth hormone-viral envelope glycoprotein hybrids. J Cell Biol 101:1351-1362.

Saraste J and Kuismanen E (1984) Pre and post Golgi vacuoles operate in the transport of Semliki Forest Virus membrane glycoproteins to the cell surface. cell 38:535-549.

Saraste J, Palade G E and Farquhar M G (1987) Antibodies to rat pancreas Golgi subfractions: Identification of a 58 kD cis-Golgi protein. J Cell Biol 105:2021-2029.

Saraste J and Svensson K (1991) Distribution of the intermediate elements operating in ER to Golgi transport. J Cell Sci 100:415-430.

Schweizer A, Fransen J A M, Bächi T, Ginsel L and Hauri H-P (1988) Identification, by a monoclonal antibody, of a 53-kD protein associated with a tubulo-vesicular compartment at the cis-side of the Golgi apparatus. J Cell Biol 107:1643-1653.

Schweizer A, Matter K, Ketcham C and Hauri H P (1991) The isolated ER-Golgi intermediate compartment exhibits properties that are different from ER and cis-Golgi. J Cell Biol 113:45-54.

Simonsen C C and Levinson A D (1983) Isolation and expression of an altered mouse dihydrofolate reductase cDNA. Proc Natl Acad Sci USA 80:2495-2499.

Tooze J, Fuller S D and Howell K E (1989) Condensation-sorting events in the rough endoplasmic reticulum of exocrine pancreatic cells. J Cell Biol 109:35-50.

Tooze J, Tooze S and Warren G (1984) Replication of coronavirus MHV-A59 in sac cells:determination of the first site of budding of progeny virions. Eur J Cell Biol 33:281-293.

Ulmer J B and Palade G E (1991) Effects of Brefeldin A on the Golgi complex, endoplasmic reticulum and viral envelope glycoproteins in murine erythroleukemia cells. Eur J Cell Biol 54:38-54.

# G PROTEIN REGULATION OF VESICULAR TRANSPORT THROUGH THE EXOCYTIC PATHWAY

W.E. Balch
H. Plutner
R. Schwaninger
E.J. Tisdale
H.W. Davidson
J. Bourne
S. Pind
F. Peter


Departments of Cell and Molecular Biology
The Scripps Research Institute
10666 N. Torrey Pines Road
La Jolla, CA   92037

## INTRODUCTION

A central problem in cell biology is to understand the molecular basis  for the transport of   itinerant soluble and membrane-associated macromolecules through multiple compartments of the exocytic and endocytic pathways of eukaryotic cells.  Transport involves a multiplicity of  components which facilitate the formation of coat protein complexes leading to vesicle budding, and protein complexes involved in vesicle targeting and fusion to downstream compartments. Many gene products involved have been identified through genetic studies in yeast  (reviewed in  Hicke and Schekman, 1990) and through biochemical approaches (reviewed in Balch, 1990b; Mellman and Simon, 1992; Rothman, 1992), although an understanding of the enyzmological mechanisms involved are largely unknown.

NATO ASI Series, Vol. H 74
Molecular Mechanisms of Membrane Traffic
Edited by D. J. Morré, K. E. Howell, and J. J. M. Bergeron
© Springer-Verlag Berlin Heidelberg 1993

Recent work has led to the discovery that multiple GTP-binding proteins are critical for vesicle fission and fusion in the exocytic pathway. One group of GTP-binding proteins, the small GTP-binding proteins belonging to the *ras*-superfamily, now includes SAR1, YPT1, SEC4 and ARF (ADP-ribosylation factor) in yeast, and the ARF and *rab* gene families in mammalian cells (reviewed in Balch, 1990a; Goud and McCaffrey, 1991). We have demonstrated that ARF (Balch et al., 1992) and the rab1 and rab2 proteins are essential for ER to Golgi transport using a combination of molecular, biochemical and morphological approaches (Plutner et al., 1990; Plutner et al., 1991; Tisdale et al., 1992; Plutner et al., 1992a,b).

A second class of GTP-binding proteins, the signal transducing heterotrimeric G proteins composed of $\alpha$, $\beta$, and $\gamma$ subunits (Bourne et al., 1990; Bourne et al., 1991; Ross, 1989; Simon et al., 1991) have recently emerged as key components of the exocytic (reviewed in (Balch, 1992; Barr, 1992)) and possibly endocytic pathways (Columbo, 1992). Their involvement in vesicular transport was presaged by earlier observations that the release of neurotransmitters and fusion of secretory granules in endocrine and exocrine cells may require activated G proteins (Burgoyne, 1987). More recently, both pertussis toxin and $\beta\gamma$ subunits were found to perturb the formation of regulated and constitutive transport vesicles from the trans Golgi compartment (Barr et al., 1991). In support of this observation, $G_{\alpha i3}$ has been found to be localized to the Golgi complex- overexpression of $G_{\alpha i3}$ retards the flow of glycosaminoglycans through the Golgi stack (Stow et al., 1991). At the molecular level, G proteins modulate the association of ARF and $\beta$-Cop with the Golgi complex (Donaldson et al., 1991; Ktistakis et al., 1992). These proteins are putative components of non-clathrin coated vesicles involved in vesicular trafficking between compartments of the Golgi complex (Serafini et al., 1991; Waters et al., 1991; Orci et al., 1991). G proteins may also be required for the association of clathrin and adaptors with the trans Golgi compartment (Robinson and Kreis, 1992).

Using permeabilized cells which reconstitute the transport of vesicular stomatitis virus glycoprotein (VSV-G) from the ER to Golgi in vitro (Beckers and Balch, 1989; Beckers et al., 1987; Beckers et al., 1990; Plutner et al., 1991; Plutner et al., 1992a,b; Schwaninger et al., 1991, 1992), we have provided three lines of evidence using G protein specific reagents (AlF$_{(3-5)}$, mastoparan and $\beta\gamma$ subunits) that export from the ER may be initiated by a G protein (Schwaninger et al., 1992). AlF$_{(3-5)}$, a reagent which selectively inhibits the function G proteins, but not the function of rab and ARF proteins (Kahn et al., 1991), inhibits ER to Golgi transport. Mastoparan uncouples ligand-activated receptors from their cognate $G_\alpha$ subunits (Higashijima, 1990, 1991). In the presence of mastoparan, VSV-G protein is retained by the ER reticulum. Finally, in the presence of excess βγ subunits which can serve as a sink for activated $G_\alpha$ subunits, VSV-G protein cannot exit the ER. While the identity of the relevant G protein(s) regulating transport from the ER and other compartments is unknown, the cumulative evidence strongly suggests a key role G proteins throughout the exocytic pathway.

**Gated flow: a model for G protein function in vesicular transport**

In classical signal transduction, trimeric G proteins in association with their cognate receptors couple a diverse range of extracellular events to a restricted range of intracellular effectors (Ross, 1989). Currently, 16 different $G_\alpha$ subunits interact with a more limited set of βγ isotypes to form the trimeric complex (Simon et al., 1991). G proteins can be activated through a wide range (>100) integral membrane receptors which, upon interaction with extracellular agonists and antagonists, trigger $G_\alpha$ dissociation from βγ and GTP exchange to form the activated $G_\alpha$ subunit. The receptor activated G proteins can function in both inhibitory and stimulatory modes. The general activating amphiphilic cationic peptides such as mastoparan trigger G protein activation by mimicking receptor binding while simultaneously uncoupling the $G_\alpha$ subunit from their receptors (Higashijima et al., 1990; Higashijima and Ross, 1991; Ross, 1989; Weingarten et al., 1990; Okamoto et al., 1991). Activated G proteins in turn enhance the function of a more limited set of intracellular "effector" proteins including phospholipases A2 and C, phosphodiesterases, adenylyl cyclase and ion channels (Bourne et al., 1990; Bourne et al., 1991; Ross, 1989; Simon et al., 1991).

Does the signal transduction model explicitly apply to vesicular transport? Fig. 1 illustrates the standard model for G protein function (Bourne et al., 1990; Bourne et al., 1991; Ross, 1989; Simon et al., 1991) adapted to their potential role in vesicular transport. In this case, **R** (the receptor in Fig. 1), instead of only involving proteins found on the cell surface (the exclusive domain of signal transducing G proteins) would now include proteins found in different organelles of the exocytic and endocytic pathways. The **effectors** (Fig. 1) are the components of the transport machinery involved in the vesicular trafficking and organelle structure/function. The agonist(s) (or antagonist(s)) (**L** in Fig. 1) in the exocytic pathway are transported proteins. These provide signals to activate the G-protein through transmembrane-coupled receptor(s). The "signal transduction" model suggests that transmembrane receptors, their cognate trimeric G proteins and signalling play a critical role in "gating" the flow of protein through the endomembrane system.

Exocytic compartment       Cytoplasm

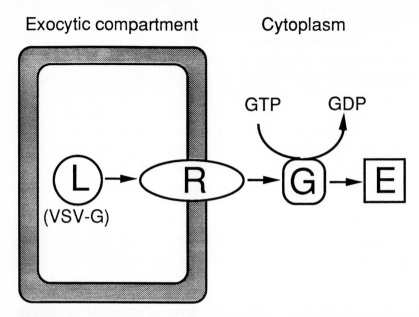

**Figure 1. G protein "gated" flow of protein between exocytic compartments. L** in the diagram is a ligand (in this case a transported protein) which is the agonist or antagonist for a G-protein receptor (**R**). This receptor serves as a link between the transported protein and the cytosolic machinery involved in vesicle fission and fusion. In the endocytic pathway, **L** could also be an extracellular ligand found on the cell surface or an endocytic compartment. **G** in the diagram is a heterotrimeric G protein composed of $\alpha$, $\beta$, and $\gamma$ subunits. **Effectors** are the possible targets of the activated **G** protein. These could include proteins involved in assembly of fission and fusion scaffolds (see text), or potentially, even the classical effectors found in the signal transduction pathway.

### Role of G proteins in vesicular transport

What role do signal transducing G proteins play in regulating vesicular trafficking through the exocytic pathway? How is their function coupled to the function of the ubiquitous small GTP-binding proteins also required for transport? Organelles of both the exocytic and endocytic pathways have defined composition, therefore sorting is crucial to maintain the stable protein "scaffolds" defining organelle structure and to generate the more transient protein "scaffolds" involved in the formation and consumption (fusion) of carrier vesicles.

Do G protein-coupled receptors participate in sorting? The answer is likely to be yes. In contrast to classical G protein coupled-receptors which contain seven transmembrane spanning domains, many proteins with recognized sorting function contain only a single transmembrane spanning segment. These include, for example, the cation independent mannose 6-phosphate/IGF II receptor (CIM 6-

P/IGF II), the epidermal growth factor (EGF) receptor and the insulin receptor regulating glucose transporter distribution. All of these proteins have an intracellular itinerary which confines their circulation to a restricted set of exocytic or endocytic compartments. However, their intracellular itinerary is, in a sense, "ligand"-gated. For example, the CIM-6-P/IGFII receptor sorts soluble lysosomal enzymes from the Golgi to the lysosome (Dahms et al., 1989), whereas the the EGF and insulin receptors are transported in a ligand responsive manner between the cell surface and compartments of the endocytic pathway. Interestingly, all three receptors have been demonstrated to associate with G proteins (Murayama et al., 1990; Nishimoto et al., 1989; Yang et al., 1991). There is also a large group of soluble proteins which are residents of a restricted set of organelles. The ERD2 gene product is believed to be the KDEL receptor involved in recycling of soluble "resident" ER proteins containing the carboxyl-terminal KDEL motif between the ER and early Golgi compartments (Pelham, 1990). The circulation of this protein has been recently shown to be ligand-activated (Lewis and Pelham, 1992). Moreover, ERD2 is a seven transmembrane spanning integral membrane protein, similar in structure to classical G protein receptors.

Since sorting occurs through vesicular trafficking, G proteins could potentially regulate a number of different steps in transport. They could serve to initiate recruitment of coat proteins (vesicle formation), trigger vesicle fission (separation) from the parent membrane, control coat disassembly, and/or regulate vesicle targeting or fusion with the downstream acceptor compartment. Given the signal transduction paradigm, G protein-coupled transmembrane receptors are also likely to be essential components of any step regulated by a cytoplasmic G protein. For example, our data would suggest that a G protein(s) initiates the formation of coat complexes involved in the export of protein from the ER. We have proposed that soluble or membrane-bound itinerant proteins such as the vesicular stomatitis virus glycoprotein (VSV-G) protein serve as agonists (L in Fig. 1), activating a key G protein(s) through a transmembrane G-protein coupled receptor. This ternary complex would initiate an "effector" cascade leading to coat assembly. While this mechanism is similar to the role of ligand-receptor interactions at the cell surface which transduce extracellular signals to ion channels, phospholipases, etc., the effectors in ER export may involve the coat components ARF, β-Cop, and possibly members of the rab family of small-GTP binding proteins. Sustained activation of a $G_{\alpha}$ through its cognate receptor-ligand interaction may favor coat assembly over disassembly, since these reactions may be in dynamic equilibrium prior to completion of the fission event.

In contrast to vesicle fission a topologically different set of events must occur leading to vesicle fusion. Given the evidence that G proteins may also be involved in regulated exocytosis (Burgoyne, 1987) and endosome fusion (Columbo et al., 1992), one might speculate that docking (recognition between the incoming vesicle and the target organelle) could activate the assembly/disassembly of a G-protein regulated fusion complex. Regulation of fusion would be expected to require a high signal to noise ratio to insure fidelity between the endomembrane trafficking pathways, a task ideally suited for networks initiated or terminated by G proteins. The dramatic effects of BFA in collapsing both the exocytic and endocytic pathways may be an example of unregulated fusion between endomembrane compartments when G protein control is disrupted (Klausner et al., 1992; Pelham, 1991).

## Predictions

A signal transduction model has a number of useful predictions. First, transmembrane receptors, similar or different from those normally recognized for classical signal transducing trimeric G proteins will be essential components of the transport machinery sorting protein between compartments. The KDEL receptor and the CIM 6-P/IGF II are two currently recognized examples. Second, the model suggests that G proteins may initiate the assembly and/or disassembly of coat protein complexes required at multiple stages of exocytic and endocytic pathways. While the evidence for their role in coat assembly in vesicular transport is still tentative, the sensitivity of ARF and $\beta$-Cop association with membranes to the G protein specific reagents AlF$_{(3-5)}$ and $\beta\gamma$ subunits provides tantalizing evidence for this role (Donaldson et al., 1991).

A final (and more provocative) prediction is that itinerant (transported) proteins serve as agonists, regulating the cytosolic transport machinery through coupling to the appropriate G protein receptors. Our studies (Schwaninger et al., 1992) were based on the analysis of the temperature-sensitive ts045 VSV-G protein, a transmembrane protein which is transported from the ER to the cell surface via "bulk" flow (Pfeffer and Rothman, 1987). A fundamental tenet of the bulk flow model is that transport of proteins like VSV-G occur via a default pathway and do not require sorting signals (Pfeffer and Rothman, 1987). However, the involvement of G protein-coupled receptors now suggests that sorting signals are indeed required. The requirement for G protein-coupled receptors could readily provide a biochemical foundation for the "quality control" processes which insure that only mature proteins are exported from the ER to the Golgi (Hurtley and Helenius, 1989; Gething and Sambrook, 1992). For example, at the restrictive temperature tsO45 VSV-G is unable to initiate export from the ER due to improper oligomerization (Doms et al., 1989; de Silva et al., 1991). Upon shift to the permissive temperature, VSV-G protein rapidly oligomerizes to its native trimeric structure. Only the mature protein may be able to recognize a G protein-coupled receptor gating export. A similar result may apply to proteins such as the T-cell receptor which must form heterooligomers (Klausner, 1989). Prior to oligomerization, individual subunits are defective in transport because they cannot serve as agonists. If **L-R-G** (Fig. 1) coupling is essential for export from the ER (and/or other compartments), then the concept of bulk flow may no longer serve as a useful model for vesicular trafficking.

## REFERENCES

Balch WE (1990a) Small GTP-binding proteins in vesicular transport. Trends Biochem Sci 15: 473-477

Balch WE (1990b) Molecular dissection of early stages of the secretory pathway. Curr Opin Cell Biol 2: 634-641

Balch, WE (1992) From G minor to G major. Curr Biol 2: 157-160

Balch W E, Kahn RE, Schwaninger R (1992) ADP-ribosylation factor (ARF) is required for vesicular trafficking between the endoplasmic reticulum (ER) and the cis Golgi compartment. J Biol Chem (In press)

Barr FA, Leyte A, Huttner WB (1992) Trimeric G proteins and vesicle formation. Trends in Cell Biol 2: 91-94

Barr FA, Leyte A, Moliner S, Pfeuffer T, Tooze SA, Huttner WB (1991) Trimeric G-proteins of the trans-Golgi network are involved in the formation of constitutive secretory vesicles and immature secretory granules. FEBS 293: 1-5

Beckers CJM, Balch WE (1989) Calcium and GTP: essential components in vesicular trafficking between the endoplasmic reticulum and Golgi apparatus. J Cell Biol 108: 1245-1256

Beckers CJM, Keller DS, Balch WE (1987) Semi-intact cells permeable to macromolecules: use in reconstitution of protein transport from the endoplasmic reticulum to the Golgi complex. Cell 50: 523-534

Beckers CJM, Plutner H, Davidson HW, Balch WE (1990) Sequential intermediates in the transport of protein between the endoplasmic reticulum and the Golgi. J Biol Chem 265: 18298-18310

Bourne HR, Sanders DA, McCormick F (1990) The GTPase superfamily: a conserved switch for diverse cell functions. Nature 348: 125-132

Bourne HR, Sanders DA, McCormick F (1991) The GTPase superfamily: conserved structure and molecular mechanism. Nature 349: 117-127

Burgoyne RD (1987) Control of exocytosis. Nature 328: 112-113

Columbo MI, Mayorga LS, Casey PJ, Stahl PD (In press) Evidence of a role for heterotrimeric GTP-binding proteins in endosome fusion. Science

Dahms NM, Lobel P, Kornfeld S (1989) Mannose 6-phosphate receptors and lysosomal enzyme targeting. J Biol Chem 264: 12115-12118

de Silva AM, Balch WE, Helenius A (1990) Quality control in the endoplasmic reticulum: folding and misfolding of vesicular stomatitis virus G protein in cells and in vitro. J Cell Biol 111: 857-866

Doms RW, Keller DS, Helenius A, Balch WE (1987) Role for adenosine triphosphate in regulating the assembly and transport of vesicular stomatitis virus G protein trimers. J Cell Biol 105: 1957-1969

Donaldson JG, Kahn RA, Lippincott-Schwartz J, Klausner RD (1991) Binding of ARF and β-COP to Golgi membranes: possible regulation by a trimeric G protein. Science 254: 1197-1199

Gething M-J, Sambrook J (1992) Protein folding in the cell. Nature 355: 33-45

Goud B, McCaffrey M (1991) Small GTP-binding proteins and their role in transport. Curr Opin Cell Biol 3: 626-633

Hicke L, Schekman R (1990) Molecular machinery required for protein transport from the endoplasmic reticulum to the Golgi complex. Bioessays 12: 253-258

Higashijima T, Burnier J, Ross EM (1990) Regulation of $G_i$ and $G_o$ by mastoparan, related amphiphilic peptides, and hydrophobic amines. J Biol Chem 265: 14176-14186

Higashijima T, Ross EM (1991) Mapping of the mastoparan-binding site on G proteins. J Biol Chem 266: 12655-12661

Hurtley SM, Helenius A (1989) Protein oligomerization in the endoplasmic reticulum. Annu Rev Cell Biol 5: 277-307

Kahn RA (1991) Fluoride is not an activator of the smaller(20-25 kDa) GTP-binding proteins. J Biol Chem 266: 15595-15597

Klausner RD (1989) Architectural editing: determining the fate of newly synthesized membrane proteins. The New Biologist 1: 308-324

Klausner RD, Donaldson JG, Lippincott-Schwartz J (1992) Brefeldin A: insights into the control of membrane traffic and organelle structure. J Cell Biol 116: 1071-1080

Ktistakis NT, Linder ME, Roth MG (1992) Action of brefeldin A blocked by activation of a pertussis-toxin-sensitive G protein. Nature 356: 344-346

Lewis MJ, Pelham HRB (1992) Ligand-induced redistribution of a human KDEL receptor from the Golgi complex to the endoplasmic reticulum. Cell 68: 353-364

Murayama YT, Okamoto E, Ogata E, Asano T, Liri T, Katada M, Ui JH, Grubb, WS, E, Nishimoto, I (1990) Distinctive regulation of the functional linkage between the human cation-independent mannose 6-phosphate receptor and GTP-binding proteins by insulin-like growth factor II and mannose 6-phosphate. J Biol Chem 265: 17456-17462.

Nishimoto I, Murayama Y, Katada T, Ui M, Ogata E (1989) Possible direct linkage of insulin-like growth factor II receptor with guanine nucleotide-binding proteins. J Biol Chem 264: 14029-14038

Okabe K,Yatani A, Evans T, Ho Y-K, Codina J, Birnbaumer L, Brown AM (1990) $\beta\gamma$ dimers of G proteins inhibit atrial muscarinic K+channels. J Biol Chem 265: 12854-12858

Okamoto T, Murayama Y, Hayashi Y, Inagaki M, Ogata E, Nishimoto, I. 1991. Identification of a $G_s$ activator region of the $\beta2$-adrenergic receptor that is autoregulated via protein kinase A-dependent phosphorylation. Cell 67: 723-730.

Orci, L, Tagaya M, Amherdt M, Perrelet K, Donaldson JG, Lippincott-Schwartz J, Klausner RD, Rothman JE (1991) Brefeldin A, a drug that blocks secretion, prevents the assembly of non-clathrin-coated buds on Golgi cisternae. Cell 64: 1183-1195

Pelham, HRB (1991) Multiple targets for Brefeldin A. Cell 67: 449-451

Pfeffer SR, Rothman JE (1987) Biosynthetic protein transport and sorting by the endoplasmic reticulum and Golgi. Ann Rev Biochem 56: 829-852

Plutner H, Schwaninger R, Pind S, Balch WE (1990) Synthetic peptides of the Rab effector domain inhibit vesicular transport through the secretory pathway. EMBO J. 9: 2375-2383.

Plutner H, Saraste J, Balch WE (manuscript submitted) Permeabilized cells reconstitute the vectorial flow of protein from the endoplasmic reticulum (ER) through pre-Golgi intermediates. J. Cell Biol

Plutner H, Cox AD, Pind S,Khosravi-Far R, Bourne JR, Schwaninger R, Der CJ, Balch WE (1991) Rab1b regulates vesicular transport between the endoplasmic reticulum and successive Golgi compartments. J Cell Biol 115: 31-43

Plutner H, Saraste J, Davidson H, Balch WE (1992a)Permeabilized cells reconstitute the vectorial flow of protein form the endoplasmic reticulum (ER) through pre-Golgi intermediates. J Cell Biol (manuscript submitted)

Plutner H, Balch WE (1992b) Multiple GTP-binding proteins regulate endoplasmic reticulum to Golgi transport. J. Cell Biol (manuscript submitted)

Robinson MS, Kreis TE (1992) Recruitment of coat proteins onto Golgi membranes in intact and permeabilized cells: effects of brefeldin A and G protein activators. Cell 69: 129-138

Ross EM (1989) Signal sorting and amplification through G protein-coupled receptors. Neuron 3: 141-152

Rothman JE, Orci L (1992) Molecular dissection of the secretory pathway. Nature 355: 409-415

Schwaninger R, Beckers CJM, Balch WE (1991) Sequential transport of protein between the endoplasmic reticulum and successive Golgi compartments in semi-intact cells. J Biol Chem 266: 13055-13063

Schwaninger R, Plutner H, Bokoch GM, Balch WE (manscript submitted) G protein gated export from the ER. J. Cell Biol.

Serafini T, Stenbeck G, Brecht A, Lottspeich F, Orci L, Rothman JE, and Wieland, FT. 1991. A coat subunit of Golgi-derived non-clathrin-coated vesicles with homology to the clathrin-coated vesicle coat protein B-adaptin. Nature. 349: 215-219.

Simon MI, Strathmann MP, Gautam N (1991) Diversity of G proteins in signal transduction. Science 252: 802-808

Stow JL, de Almeida JB, Narula N, Holtzman KJ, Ercolani L, Ausiello DA (1991) A heterotrimeric G protein, $G_{\alpha i-3}$, on Golgi membranes regulates the secretion of a heparan sulfate proteoglycan in LLC-PK epithelial Cells. J Cell Biol 114: 1113-1124

Tisdale EJ, Bourne JR, Khosravi-Far R, Davidson HW, Der CJ, Balch WE (manuscript submitted) GTP-binding mutants of rab1 and rab2 are potent inhibitors of endoplasmic reticulum (ER) to Golgi transport. J Cell Biol

Waters MG, Serafini T, Rothman JE (1991) 'Coatomer': a cytosolic protein complex containing subunits of non-clathrin-coated Golgi transport vesicles. Nature 349: 248-251.

Weingarten R, Ransnas L, Mueller H, Sklar LA, Bokoch GM (1990) Mastoparan nteracts with the carboxyl terminus of the $\alpha$ subunit of Gi. J Biol Chem 265: 11044-11049

Yang L, Baffy G, Rhee SG, Manning D, Hansen CA, Williamson JR (1991) Pertussis toxin-sensitive Gi protein involvement in epidermal growth factor-induced activation of phospholipase $C\gamma$ in rat hepatocytes. J Biol Chem 266: 22451-22458

# FUSION RAPIDLY FOLLOWS VESICLE TRANSPORT TO THE TARGET MEMBRANE IN PROTEIN TRANSPORT THROUGH THE GOLGI APPARATUS *IN VITRO*. A RE-EVALUATION OF TRANSPORT KINETICS BASED ON THE FINDING THAT THE GLYCOSYLATION USED TO MARK TRANSPORT, AND NOT TRANSPORT ITSELF, IS RATE LIMITING IN THE ASSAY.

R.R. Hiebsch and B.W. Wattenberg
Department of Cell Biology
The Upjohn Company
Kalamazoo, MI 49001

## Abstract

A well characterized assay measuring protein transport between compartments of the Golgi apparatus (W.E. Balch et al, 1984, Cell 39:405-416) utilizes glycosylation of a membrane protein to mark transit. Previously, kinetic analysis of the assay relied on the assumption that the glycosylation that is used to mark transport occurs very rapidly upon fusion of transport vesicles with their target membranes. Here we report that, in fact, glycosylation is rate limiting in this assay and therefore that fusion proceeds substantially faster than previously thought. This finding alters the previous model of transport which held that there was a lengthy interval after vesicles attach to their target before fusion occurs. Our findings indicate that a presumptive pre-fusion intermediate, the "NEM resistant intermediate" is an artifact of the kinetic experiments used to define it.

A cell free assay reconstituting transport between compartments of the Golgi apparatus has been used to explore the biochemical mechanisms underlying the formation, targeting, and fusion of transport vesicles (Balch et al., 1984a). To mark transport, this assay relies upon the glycosylation of vesicular stomatitis virus (VSV) G protein upon its arrival in the medial cisternae of an "acceptor" Golgi population from a "donor" population. Previously, analysis of the kinetics of this transport have been used to construct models of the events in vesicle formation and fusion. One intermediate in the transport process occurs relatively early in transport, and was thought to represent the attachment of transport vesicles to the acceptor membranes prior to their fusion. Subsequent glycosylation, thought to represent vesicle fusion, requires relatively low cytosol concentrations once this intermediate is attained and so it was given the name the "Low Cytosol Requiring Intermediate" (LCRI). This low cytosol requirement is a

NATO ASI Series, Vol. H 74
Molecular Mechanisms of Membrane Traffic
Edited by D. J. Morré, K. E. Howell, and J. J. M. Bergeron
© Springer-Verlag Berlin Heidelberg 1993

reflection of the fact that a single protein, termed POP (Prefusion Operating Protein), is the sole cytosolic protein required in the transport assay after the formation of the LCRI. We previously purified POP and produced monoclonal antibodies to it (Wattenberg, et al., 1989).

**Cloning of the POP gene reveals POP to be Uridine Monophosphokinase.**

POP was cloned by a well established expression cloning technique (Young and Davis, 1983). It was confirmed that the cloned sequences encoded POP by preparing lysates from E. coli harboring the cloned sequences and measuring POP activity in those lysates (Figure 1). Because E. coli lysates contain substances that inhibit the transport assay it was necessary to fractionate the lysates by gel filtration chromatography to detect POP activity. It can be seen that lysates derived from E. coli containing the putative POP sequence exhibited POP activity, whereas lysates from cells containing empty plasmid did not.

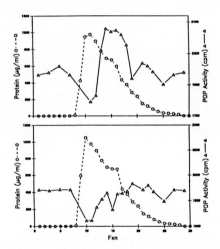

**Figure 1. E. coli carrying a plasmid encoding POP produce POP activity.** E. coli transformed with a pUC19 plasmid containing the putative POP coding sequence or with pUC19 alone were grown and lysates prepared and chromatographed on Sephadex G-75. Column fractions were assayed for POP activity (6) (triangles). Protein content is indicated by open circles.

Restriction analysis (not shown) defined a minimal sequence which contained a single open reading frame. Sequencing of the DNA was performed and the derived sequence was compared to sequences in the database. An exact match was found to the sequence of the yeast URA6 gene which encodes uridine monophosphokinase (UMP kinase) (Liljelund et al., 1989).

## POP/UMP Kinase Stimulates the Glycosylation Reaction used to Mark Transport and Not Transport Itself.

In this assay transport is marked by a glycosylation reaction in which the substrate is UDP-N-acetylglucosamine (UDP-GlcNAc). The uptake of UDP-GlcNAc into the lumen of the Golgi, where the glycosyltransferase resides, is dependent on an anti-port system which imports UDP-GlcNAC and exports UMP (Perez et al., 1985). Thus UMP on the cytosolic face of the Golgi will act as an inhibitor of UDP-GlcNAc uptake, and therefore of the glycosylation reaction used to mark transport. This indicated that POP/UMP-kinase might be scavenging cytosolic UMP and promoting UDP-GlcNAc uptake into the Golgi. This was directly tested by measuring UDP-GlcNAc transport into the Golgi using an assay developed by Hirschberg and colleagues (Perez et al., 1985) (Figure 2). Indeed the presence of POP/UMP-kinase enhanced the uptake of UDP-GlcNAc into Golgi vesicles markedly. This demonstrates that instead of enhancing fusion reactions, as previously supposed, POP is stimulating the transport assay by enhancing the glycosylation that is used as an indicator of transport in this assay.

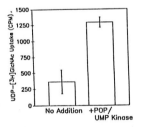

**Figure 2. POP/UMPK enhances UDP-GlcNAc uptake into Golgi membranes under assay conditions.** Acceptor membranes were incubated under conditions which generate the LCRI (6) and then assayed for UDP(6-$^3$H)GlcNAc uptake (5). One aliquot was incubated without any further additions and a second was incubated with approximately 90 μg/ml (final concentration) of a partially purified POP fraction. Shown is the mean plus and minus standard deviation of triplicate assays.

**Glycosylation and not Protein Transport is Rate Limiting in the Golgi Transport Assay.**

The kinetics of glycosylation were previously thought to follow the kinetics of vesicle formation, attachment, and fusion, in the Golgi transport assay. On this basis the kinetics of transport have been used to construct a model of intermediates in the transport process. However the demonstration that what was thought to be a transport factor in fact enhances glycosylation suggested that glycosylation, and not reactions in vesicular transport, might be rate limiting. To address this, the presumptive late intermediate, the LCRI, was formed by pre-incubation of donor and acceptor membranes for 25 minutes. UDP-[$^3$H]-GlcNAc was added either immediately, or after a further incubation for 20 or 40 minutes (Figure 3). If fusion of vesicles is rate limiting, and glycosylation is not, it would be expected that addition of the glycosylation substrate at later times would lead to a burst of glycosylation, as the G protein from vesicles that had already fused would be rapidly glycosylated. Instead it was found that the kinetics of glycosylation were independent of when the substrate was added after incubation of the LCRI. This indicates that glycosylation is slower than any fusion reactions which occur after the formation of the LCRI.

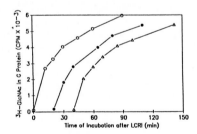

**Figure 3. Glycosylation is rate limiting for incorporation of [$^3$H]-GlcNAc into G protein.** The kinetics of incorporation of [$^3$H]-GlcNAc into G protein was measured when UDP-[$^3$H]-GlcNAc was added either at t=0 (closed circles), t=20 (open triangles), or t=40 (closed triangles) minutes after initiation of the post-LCRI reaction (6).

Because the concentration of sugar nucleotide used in these assays (0.4 μM) is below the reported levels which saturate the sugar nucleotide transporter (Perez et al., 1985), it seemed reasonable that the rates of glycosylation might be increased by increasing the concentration of sugar nucleotide. Again, the LCRI was formed, and the effect of increasing the sugar nucleotide concentration on the rate of G protein glycosylation in a subsequent incubation was measured (Figure 4). The rate of glycosylation was doubled by increasing the sugar nucleotide concentration. Because it is extremely unlikely that the sugar nucleotide concentration would have an effect on vesicle fusion reactions, this experiment directly shows that it is the rate of glycosylation, and not fusion, that is being measured in these assays.

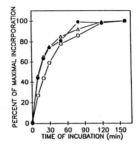

**Figure 4. Increased concentrations of UDP-[³H]-GlcNAc increases the rate of incorporation of label into G protein after the formation of the LCRI.** A time course of incorporation of [³H]-GlcNAc into G protein was measured after the formation of LCRI. UDP-[³H]-GlcNAc was added at either 0.4 μM (closed circles), 4.0 μM (closed triangles) or 20 μM (open triangles). For ease of comparison [³H]-GlcNAC incorporation in each time course is expressed as the percentage of incorporation at t=160 minutes. That incorporation was 8465 cpm for 0.4 μM UDP-[³H]-GlcNAC, 15613 cpm for 4.0 μM UDP-[³H]-GlcNAc, and 18336 for 20 μM UDP-[³H]-GlcNAC.

## Conclusions

Kinetics can be an invaluable tool in placing the biochemical components of a system in a mechanistic context. Since glycosylation is used as the signal that a G protein containing vesicle has fused with the acceptor cisternae in the Golgi transport assay, it is important to understand the temporal relationship between glycosylation and vesicle fusion when applying a kinetic analysis to transport. The kinetics of glycosylation of G protein in this assay are characterized by a 7-10 minute lag time followed by a 60-90 minute linear incorporation. The lag time has been shown not to be due to an effect of glycosylation (Balch et al., 1984b), and so can reliably be attributed to transport reactions. However it has been assumed, without support, that this was also true of the later kinetics. We show here that after the formation of the LCRI, an intermediate previously thought to represent vesicles attached but unfused to their target, the rate of glycosylation is limited by the glycosylation reaction itself, and not by vesicle fusion. How does this change the current model of intermediates in the transport process? One alteration involves a intermediate that was thought to represent the final step before vesicle fusion. When the thiol alkylating agent N-ethylmaleimide is added to transport reactions, glycosylation briefly continues, and then subsides (Balch et al., 1984b). This slight continued glycosylation was thought to represent G protein in a transport intermediate just preceding fusion, indicating that the last steps in fusion were not sensitive to NEM. Our results indicate that this interpretation is unlikely to be correct. We have found that NEM inhibits POP/UMP-kinase (data not shown). Therefore, the small amount of NEM "resistant" glycosylation probably results from a time dependent build-up of UMP because of the inactivation of POP/UMP-kinase by NEM. What the true rate of vesicle fusion is in this assay cannot yet be determined. The limiting case would be that formation of the LCRI in fact represents vesicle fusion. As the time-course of formation of the LCRI is relatively rapid ($T_{1/2}$=12 minutes) this would place the rate of transport observed in vitro close to that observed in vivo.

The reconstitution of various protein transport processes has been successfully attained in a wide variety of systems. Many of these use glycosylation as a marker for transport (Brandli, 1991). This work underscores the caution that must be taken when interpreting results from these systems. However, it also demonstrates that careful analysis can dissect the components of the system required in the transport processes from those involved in glycosylation.

# REFERENCES

Balch WE, Dunphy WG, Braell WA, Rothman JE (1984a) Reconstitution of the transport of protein between successive compartments of the Golgi measured by the coupled incorporation of N-acetylglucosamine. Cell 39:405-416.

Balch WE, Glick BS, Rothman JE (1984b) Sequential intermediates in the pathway of intercompartmental transport in a cell-free system. Cell 39:525-536.

Brandli AW (1991) Mammalian glycosylation mutants as tools for the analysis and reconstitution of protein transport. Biochem J 276:1-12.

Liljelund P, Sanni A, Friesen JD, Lacroute F (1989) Primary structure of the S. cerevesiae gene encoding uridine monophosphokinase. Biochem Biophys Res Commun 165:464-473.

Perez, M, Hirschberg CB (1985) Translocation of UDP-N-Acetylglucosamine into vesicles derived from rat liver endoplasmic reticulum and Golgi apparatus. J Biol Chem 260:4671-4678.

Wattenberg BW, Hiebsch RR, LeCureux, LW and White MP (1989) Identification of a 25-kD protein from yeast cytosol that operates in a pre-fusion step of vesicular transport between compartments of the Golgi. J Cell Biol 110:947-954.

Young RA and Davis RW (1983) Efficient isolation of genes by using antibody probes. Proc Natl Acad Sci 80:1194-1198.

# Immunocytochemical analysis of the transfer of vesicular stomatitis virus G glycoprotein from the intermediate compartment to the Golgi complex.

S. Bonatti,[1]L.V. Lotti,[1]M.R. Torrisi, M.C. Pascale
Department of Biochemistry and Medical Biotechnology
University of Naples
via S. Pansini 5
80131 Naples
Italy

## Introduction

Several independent lines of evidence suggest that the traffic between the ER and the Golgi complex involves other membrane-bound structure that may form distinct compartment(s). These structures have been variously named, but their morphology has yet to be defined, and the mechanism of the transport between the ER and Golgi complex is still largely unknown. To address an aspect of this problem we used G glycoprotein synthesized by the VSV ts-045 mutant strain. At the non-permissive temperature (39 °C) G glycoprotein accumulates in the ER as a multimeric aggregate and there is very little export. When the temperature is lowered to 31 °C, the aggregated protein trimerizes, leaves the ER and reachs the Golgi complex; if the temperature is lowered to 15 °C, the protein correctly trimerizes but it is arrested in an intermediate location before reaching the Golgi complex. Here we describe the morphology of this intermediate structure and the transfer of G glycoprotein from this compartment to the Golgi complex.

## Results

The intermediate compartment seemed to consist of about 30-40 separate units of clustered small vesicles and short tubules (Lotti et al. 1992). The clusters occupied a circular area of average diameter less than 1 $\mu$m, and vesicles and tubules showed a constant diameter of about 80 nm. No membrane continuity was detected between the units and either the ER or the Golgi complex. The units contained Rab2 protein and were spread throughout the cytoplasm, with a ratio of about 6:4 in the peripheral versus perinuclear site (Table I). Time-course experiments revealed a progressive

---

[1] Department of Experimental Medicine, University of Rome" La Sapienza",Rome,Italy

NATO ASI Series, Vol. H 74
Molecular Mechanisms of Membrane Traffic
Edited by D. J. Morré, K. E. Howell, and J. J. M. Bergeron
© Springer-Verlag Berlin Heidelberg 1993

transfer of G glycoprotein from the intermediate compartment to the Golgi stacks, while the tubulo-vesicular units did not appear to change their intracellular distribution (Table I). Moreover, the labeling density of peripheral and perinuclear units decreased in parallel during the transfer (Table I).

## Conclusion

These results support the notion that the intermediate compartment is a station in the secretory pathway, and that a vesicular transport connects this station to the Golgi complex.

**Table I** Quantitation of the immunoelectron microscopical analysis of the time-course of G protein transfer from the intermediate compartment to the Golgi complex.

| | A | | B | |
| | Density of labeling $\pm$ S.E.M. | | Percent distribution of the tubulo-vesicular structure | |
| min at 31 °C | Golgi complex | Tubulo-vesicular structures | Peripheral location | Perinuclear location |
| --- | --- | --- | --- | --- |
| 0.0 | 13 $\pm$ 2.4 | 88 $\pm$ 5.0 | 60 (86.0) | 40 (91.1) |
| 2.5 | 49 $\pm$ 8.6 | 88 $\pm$ 11.0 | 55 (80.2) | 45 (99.8) |
| 5.0 | 96 $\pm$ 9.1 | 55 $\pm$ 5.0 | 58 (59.0) | 42 (49.9) |
| 10.0 | 158 $\pm$ 4.2 | 35 $\pm$ 5.3 | 53 (34.3) | 47 (35.7) |

Cells were fixed with glutaraldehyde, embedded in LR White resin, and immunolabeled as described previously (Lotti et al. 1992). **A**: average total number of colloidal gold particles on Golgi complexes and tubulo-vesicular structures after incubation with anti G glycoprotein antibody. In uninfected cells treated in parallel, 2.42$\pm$1.36 gold particles were detected on Golgi stacks and 1.95$\pm$1.52 on tubulo-vesicular structures. **B**: intracellular distribution of the tubulo-vesicular structure units during the transfer of G protein to the Golgi complex. Units found within 2 $\mu$m of the nuclear envelope were scored as "perinuclear", all others were defined "peripheral". Values are given as per cent of total; the corresponding density of labeling, calculated as in A, is indicated in parentheses. Each time-point value derives from the analysis of at least 15 profiles.

## References

Lotti L V , Torrisi M R , Pascale M C , and Bonatti S (1992) J Cell Biol 118: in the press.

# SYNTHESIS OF GLYCOSYL-PHOSPHATIDYLINOSITOL ANCHORS IS INITIATED IN THE ENDOPLASMIC RETICULUM

J. Vidugiriene and A.K. Menon
The Rockefeller University, 1230 York Avenue,
New York, NY 10021-6399, USA.

Numerous proteins from eucaryotic organisms are covalently modified by inositol-containing glycophospholipids (GPIs). Addition of a GPI anchor to protein occurs by cleavage of a carboxy-terminal signal sequence and attachment of a GPI precursor to the newly exposed α-carboxyl group of the polypeptide [1]. The experiments described here are aimed at defining the sub-cellular localization of GPI synthesis. Since recent data from other laboratories indicated that GPI synthesis could be easily assayed in T cell (BW5147.3 thymoma) lysates [3], we chose to determine the intracellular location of GPI assembly by analyzing sub-cellular fractions from T cells. After disruption of the cells by nitrogen cavitation and removal of nuclei, 70-90% of the lysosomes (β-hexosaminidase activity) and peroxisomes (catalase activity) by low speed centrifugation, the post nuclear supernatant (PNS) was layered on a series of sucrose steps and centrifuged [4]. Fractions were collected from the top of the tube and assayed for organelle-specific marker enzymes. As shown in Fig. 1, the endoplasmic reticulum (ER; dolichol-P-mannose synthase activity), Golgi (α-mannosidase II activity) and plasma membrane (PM; alkaline phosphodiesterase activity) were clearly separated.

GPI assembly is initiated by the transfer of N-acetylglucosamine (GlcNAc) fron UDP-GlcNAc to phosphatidylinositol (PI) to form GlcNAc-PI. GlcNAc-PI is then deacetylated to form GlcN-PI. The structure is completed by sequential addition of three mannose residues and phosphoethanolamine [1, 2]. In order to define the sub-cellular localization of the GPI biosynthetic pathway we concentrated on the first step of GPI biosynthesis, *i.e.*, formation of GlcNAc-PI and GlcN-PI. Fractions were incubated with UDP-[³H]GlcNAc, and lipid products were analyzed by thin layer chromatography (as in Ref. 3).

NATO ASI Series, Vol. H 74
Molecular Mechanisms of Membrane Traffic
Edited by D. J. Morré, K. E. Howell, and J. J. M. Bergeron
© Springer-Verlag Berlin Heidelberg 1993

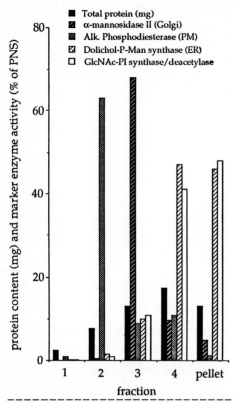

Legend:
- ■ Total protein (mg)
- ▨ α-mannosidase II (Golgi)
- ▩ Alk. Phosphodiesterase (PM)
- ▨ Dolichol-P-Man synthase (ER)
- ☐ GlcNAc-PI synthase/deacetylase

y-axis: protein content (mg) and marker enzyme activity (% of PNS)

x-axis: fraction — 1, 2, 3, 4, pellet

**Figure 1. Distribution of organelle-specific marker enzymes and GPI biosynthesis (GlcNAc-PI synthase and deacetylase) in PNS fractions.** BW5147.3 thymoma cells (4 x $10^8$) were harvested, washed, then resuspended in 10 ml of buffer A (0.25M sucrose in 10mM Tris/HCl buffer (pH 7.5) containing 1 mM DTT and 1 mM PMSF). The cells were broken by nitrogen cavitation and the resulting lysate was clarified by centrifugation (10,000$g$; 15min). The supernatant (PNS) was layered over a series of sucrose steps (38%, 30% and 20%) and centrifuged (100,000$g$; 2hr). Fractions (1-4 and pellet; pellet resuspended in 3 ml buffer A) were collected from the top of the tube and assayed for organelle-specific marker enzymes [4, 5].

activities and GlcNAc/GlcN-PI biosynthesis in the various PNS fractions showed that the GlcNAc-PI/GlcN-PI biosynthetic activity co-fractionated with the endoplasmic reticulum marker, dolichol-P-mannose synthase. The results are consistent with the proposal that GPI synthesis is initiated in the endoplasmic reticulum.

1. Cross, G.A.M. (1990) Glycolipid anchoring of plasma membrane proteins. *Annu. Rev. Cell Biol.* **6**, 1-34.
2. Doering, T.L., Masterson, W.J., Hart, G.W., and Englund, P.T. (1990) Biosynthesis of glycosylphosphatidylinositol membrane anchors. *J. Biol. Chem.* **265**, 611-614.
3. Stevens, V.L., and Raetz, C.R.H. (1991) Defective glycosylphosphatidylinositol biosynthesis in extracts of three Thy-1 negative lymphoma cell mutants. *J. Biol. Chem.* **266**, 10039-10042.
4. Storrie, B., and Madden, E. (1990) Isolation of subcellular organelles. *Methods in Enzymol.* **182**, 203-235.
5. Braell, W.A. (1988) Two sensitive, convenient, and widely applicable assays for marker enzyme activities specific to endoplasmic reticulum. *Anal. Biochem.* **170**, 328-334.

# PROTEIN TRAFFICKING ALONG THE EXOCYTOTIC PATHWAY

Wanjin Hong
Institute of Molecular and Cell Biology
National University of Singapore
Singapore 0511, Singapore

The exocytotic (secretory) pathway is composed of the endoplasmic reticulum (ER), the Golgi apparatus, the endosomal/lysosomal/lysosomal system, the plasma membrane, which is separated into apical and basolateral domains in epithelial cells, and various vesicular and tubular intermediates that connect dynamically these membrane compartments. The signals and molecular mechanisms that govern the selective targeting of proteins to various compartments of the exocytotic pathway are the major focus of the laboratory.

**Retention of a type II integral membrane protein in the ER by carboxyl-terminal KEDL sequence:** Carboxyl-terminal KDEL is a well-defined signal that mediates localization of ER luminal proteins. To test if this signal could function as a ER localization signal for type II membrane proteins, we appended either KDEL or KDEV sequence to the carboxyl terminus of a type II cell surface protein, dipeptidyl peptidase IV (DPPIV). Both biochemical and morphological data demonstrated that DPPIV-KDEL, unlike DDPIV and DPPIV-KDEV, is mainly localized to the ER (Tang et al. 1992a).

**The putative KDEL receptor is preferentially localized to the cis-Golgi and can recycle between the ER and the Golgi:** Antibodies against the C-terminus of the mammalian homology of yeast ERD2 gene product demonstrated that the protein (p23) is expressed in all cell types examined and is mainly localized on the cis-Golgi. Furthermore, the p23 is able to recycle between the Golgi and the ER in NRK cells. These data suggest that the retrieval of KDEL-containing luminal ER proteins occurs probably in the cis-Golgi.

**The transmembrane domain of Golgi glycosyltransferases is sufficient for Golgi retention:** Cellular localization of chimeric proteins constructed between Golgi α2,6-sialytransferase (ST) and surface DPPIV demonstrated that the 17-residue transmembrane domain of ST is sufficient for Golgi retention (Wong et al. 1992). Similarly, the Golgi localization sequence of N-

NATO ASI Series, Vol. H 74
Molecular Mechanisms of Membrane Traffic
Edited by D. J. Morré, K. E. Howell, and J. J. M. Bergeron
© Springer-Verlag Berlin Heidelberg 1993

acetylglucosaminyltransferase I has been mapped to the membrane spanning region (Tang et al. 1992b).

**Brefeldin A selectively inhibits apical trafficking in MDCK cells:** In tight monolayer of MDCK cells, brefeldin A (BFA) at 0.5-2 $\mu$g/ml was found to inhibit apical protein secretion with concomitant enhanced secretion from the basolateral surface, while total protein secretion and Golgi morphology were unaffected (Low et al. 1991b). Furthermore, apical targeting of a membrane protein (DPPIV) (Low et al. 1991a) was similarly abolished by BFA under a condition (about 1 $\mu$/ml BFA) where total surface expression and ER-Golgi transport of DPPIV, as well as basolateral targeting of uvomorulin, were not affected. Further experiments demonstrated that TGN to apical transport was selectively inhibited (Low et al. 1992).

### References

Low SH, Wong SH, Tang BL, Subramaniam VN, Hong W (1991a) Apical cell surface expression of rat dipeptidyl peptidase IV in transfected MDCK cells. J Biol Chem 266: 13391-13396

Low SH, Wong SH, Tang, BL, Tan P, Subramaniam VN, Hong W (1991b) Inhibition by brefeldin A of protein secretion from the apical cell surface of Madin-Darby canine kidney cells. J Biol Chem 266: 17729-17732

Low SH, Tang BL, Wong SH, Hong W (1992) Selective inhibition of protein targeting to the apical domain of MDCK cells by brefeldin A. J Cell Biol, in press

Tang BL, Wong SH, Low SH, Hong W (1992a) Retention of a type II surface membrane protein in the endoplasmic reticulum by the Lys-Asp-Glu-Leu sequence. J Biol Chem 267: 7072-7076

Tang BL, Wong SH, Low SH, Hong W (1992b) The transmembrane domain of N-acetylglucosaminyltransferase I contains a Golgi retention signal. J Biol Chem 267: 10122-10126

Wong SH, Low SH, Hong W (1992) The 17-residue transmembrane domain of $\beta$-galactoside $\alpha$2,6-sialytransferase is sufficient for Golgi retention. J Cell Biol 117: 245-258

# A SHORT N-TERMINAL SEQUENCE IS RESPONSIBLE FOR THE RETENTION OF INVARIANT CHAIN (Ii) IN THE ENDOPLASMIC RETICULUM.

Marie-Paule SCHUTZE, Michael R. JACKSON and Per A. PETERSON.
The Scripps Research Institute,
Dept. Immunology,
10666 N. Torrey Pines Road,
La Jolla, CA 92037.
USA.

Proteins destined for the exocytic pathway are directed into the endoplasmic reticulum (ER) by an N-terminal signal peptide. Current views support the idea that transport occurs by default such that proteins inserted into the ER are carried passively to the cell surface unless retained by structural motifs (Pfeffer and Rothman 1987). In the case of ER resident proteins, such motifs have been identified for both lumenal and type I transmembrane proteins (Munro and Pelham 1987, Jackson et al 1990). When transplanted onto reporter proteins, these motifs maintain the chimeric proteins in the ER.

In the present study, we have investigated the putative existence of structural motifs for the retention of type II transmembrane protein in the ER. In particular, we have focused our efforts on the invariant chain (Ii) , a type II transmembrane protein which associates with MHC class II molecules in the ER. In contrast to type I transmembrane protein, it is the N-terminal extremity which is exposed on the cytoplasmic side for type II proteins. Therefore it is this region which should contain the signal for retention. Two forms of Ii exist, Iip31 and Iip33, which are the result of initiation from two ATGs in the Ii mRNA , i.e initiation at the first codon produces Iip33 and at the second codon, Iip31. The two forms of Ii differ in their N-terminal tails, Iip33 contains a 45 residue cytoplasmic domain preceeding the transmembrane segment, whereas the cytoplasmic portion of Iip31 is only 30 residues. By eliminating one of the two initiation codons by point mutation, constructs encoding either Iip33 or Iip31 were produced. When expressed in HeLa cells, it was shown that Iip31 was exported out of the ER whereas Iip33 was localized to the endoplasmic reticulum (Lotteau et al. 1990). Based on N-terminal deletions of the Iip31 cytoplasmic domain, it has been suggested that endosomal targetting information resides

NATO ASI Series, Vol. H 74
Molecular Mechanisms of Membrane Traffic
Edited by D. J. Morré, K. E. Howell, and J. J. M. Bergeron
© Springer-Verlag Berlin Heidelberg 1993

within this domain (Lotteau et al 1990, Bakke and Dobberstein 1990). As Iip33 differs from Iip31 by 15 residues in the N-terminal region, the retention signal should be contained in this region. To identify the putative retention motif of Iip33, a series of truncated mutant molecules were expressed. The subcellular localisation of these mutated forms of Iip33 were analysed by immunofluorescence microscopy and biosynthetic pulse chase experiments. Following such analyses, we found that deletions of more than four residues in the N-terminal region of Iip33 resulted in the transport of the truncated molecules out of the ER. This clearly indicated that the first four residues at the N-terminal extremity contained information important for the retention of Iip33 in the ER. Futhermore, in contrast to the transported forms of Ii which were degraded with half times less than two hours, the ER retained forms of Ii were very stable with a half life of more than five hours. In order to more precisely define the ER retention motif of Iip33, we have made an extensive series of point mutations in the amino-terminal sequence of Iip33. Preliminary results from these analyses suggest that three arginine residues located close to the amino terminus are crucial for the ER retention motif. This raises the possibility that amino terminal arginine residues maintain type II transmembrane protein in the ER, in much the same way as lysine residues located close to the C-terminus of type I transmembrane protein do.

## REFERENCES

Bakke, O., and Dobberstein, B. (1990). MHC class II-associated invariant chain contains a signal sorting for endosomal compartments. Cell 63: 707-716.

Jackson, M. R., Nilsson T. and P.A.Peterson. (1990). Identification of a consensus motif for retention of transmembrane proteins in the endoplasmic reticulum. EMBO J. 9: 3153-3162.

Lotteau, V., Teyton, L., Peleraux, A., Nilsson, T., Karlsson, L., Schmid, S., Quaranta, V., and P.A. Peterson. (1990). Intracellular transport of class II MHC molecules directed by invariant chain. Nature 348: 600-605.

Munro, S. and Pelham, H. R. B. (1987). A C-terminal signal prevents secretion of luminal ER proteins. Cell 48: 899-907.

Pfeffer, S.R. and Rothman, J.E. (1987). Biosynthetic protein transport and sorting by the endoplasmic reticulum and Golgi. Annu. Rev. Biochem. 56: 829-852.

Isolation and characterization of the principal ATPase of transitional elements of the endoplasmic reticulum of rat liver

L. Zhang and D. J. Morré
Department of Medicinal Chemistry
Purdue University, West Lafayette, IN 47906

Cell -free transfer of radiolabeled membrane protein from part-rough, part-smooth transitional elements of the endoplasmic reticulum to Golgi apparatus requires ATP (Balch *et al* ., 1986). One ATP - dependent step is that of the budding of the transition vesicles involved in membrane transport ( Paulik *et al* ., 1988).

Transitional endoplasmic reticulum of rat liver contains an ATPase activity (TER ATPase). This ATPase shows a $Mg^{2+}$ optimum at 2mM and pH optimum at 8.5. TER ATPase shows a pattern of response to inhibitors that differs from other known membrane ATPases (Table 1) ( for review see Nelson *et al.*, 1989).

**Table 1.  Properties of the proton ATPases and TER ATPase**

|  | P | F | V | TER ATPase |
|---|---|---|---|---|
| pH optimum | 6.5 | 8.0 | 7.0 | 8.5 |
| Ion stimlulation | $K^+$ | - | $CL^-$ | - |
| inhibitors | $VO_4^-$ DCCD | Azide DCCD Oligomycin | $NO_3^-$ DCCD NEM | $Co^{2+}$ Azide |

From the results summarized above, TER ATPase is a unique ATPase and is distinguished from other known ATPases on the basis of inhibitor specificities, pH optima and ion requirements.

To isolate and sequence the TER ATPase, the ATPase of detergent-solubilized transitional ER from rat liver was partially purified from DEAE-52 ion-exchange column. Pooled fractions from the DEAE-52 column were resolved on a 6% native gel and  ATPase activity was developed using an

NATO ASI Series, Vol. H 74
Molecular Mechanisms of Membrane Traffic
Edited by D. J. Morré, K. E. Howell, and J. J. M. Bergeron
© Springer-Verlag Berlin Heidelberg 1993

activity stain. The region containing the ATPase was then cut from the gel and analyzed by SDS-PAGE. A major band with ca. of molecular weight of a 100kDa protein was identified by silver staining.

A 25 amino acid sequence obtained by digestion of the 100kDa with cyanogen bromide exhibited 80% identity to p97-ATPase ( Peters *et al.*, 1990). This 25 amino acid sequence shared 55% identity with cell division control protein( CDC 48) ( Fröhlich *et al.*, 1991). Both p97-ATPase and CDC 48 display homology to the mammalian N-ethylmaleimide sensitive fusion protein (NSF) and yeast Sec 18p which are essential for fusion in secretory process ( Wilson *et al.*, 1989).

References

Balch W E, Elliott M M, Keller D S (1986) ATP-coupled transport of Vesicular Stomatitis Virus G protein between the endoplasmic reticulum and the Golgi. J. Biol. Chem. 261: 14681-14689.

Fröhliich K-U, Fries H-W, Rüdiger M, Erdmann R, Botatein D (1991) Yeast cell cycle protein CDC 48p shows full-length homology to the mammalian protein VCP and is a membrane of a protein family involved in secretion, peroxisome formation, and gene expression. J. Cell. Biol. 114: 443-453

Nelson N and Taiz L (1989) The evolution of $H^+$- ATPases. TIBS 114: 113-116

Paulik M, Nowack D D, and Morr'e D J ( 1988) Isolation of a Vesicular intermediate in the cell-free transfer of membrane from transitional elements of the endoplasmic reticulum to Golgi apparatus cisternae of rat liver. J. Biol. Chem. 263: 17738-17748

Peters J-M, Walsh M J and Franke W W (1990) An abundant and ubiquitous homo-oligomeric ring-shapeed ATPase particle related to the putative vesicle fusion protein Sec18p and NSF. EMBO J. 9: 1757-1767

Wilson D W, Wilcox C A, Flynn G C, Chen E, Kuang W-J, Henzel W J, Block M R, Ullrich A and Rothman J E ( 1989) A fusion protein required for vesicle-mediated transport in both mammalian cells and yeast. Nature 339: 355-359

We thank Dr. Gerald Becker of Eli Lilly Research laboratories, Indianapolis for the generating the amino acid sequence.

# A GTP HYDROLASE ACTIVITY PURIFIED FROM TRANSITIONAL ENDOPLASMIC RETICULUM OF RAT LIVER BINDS RETINOL

J. Zhao and D. M. Morré
Department of Foods and Nutrition
Purdue University
West Lafayette, Indiana 47907

## INTRODUCTION

GTP hydrolysis by an endoplasmic reticulum fraction from rat liver enriched in part-rough, part-smooth transition elements (transitional endoplasmic reticulum is inhibited by all-_trans_ retinol (Zhao et al. 1990). The inhibition is non-competitive and the $K_m$ for GTP is 0.3 mM. The inhibitory effect is most apparent with GTP as substrate. Retinol did not significantly inhibit hydrolysis of the nucleoside diphosphates or of ATP or UTP. A GTP hydrolase enriched fraction from transitional endoplasmic reticulum of rat liver binds retinol. This protein may fulfill some role in mediating intracellular interactions, such as regulation of membrane traffic. GTP-binding proteins appear to be required for most vesicle formation-fusion events in eukaryotic secretory pathways (Bourne 1988). Within the cell, the transfer of materials from the endoplasmic reticulum to the Golgi apparatus occurs via transition vesicles. Retinol stimulates formation of transition vesicles, but not their fusion with the cis Golgi apparatus in a cell-free system from rat liver (Nowack et al. 1990). This study further characterizes the GTP hydrolase that is inhibited by retinol.

## METHODS

Transitional endoplasmic reticulum was isolated from livers of male Sprague-Dawley rats (Morré 1971). The transitional endoplasmic reticulum was detergent-solubilized with 1% $C_{12}E_8$. After centrifugation, the supernatant was applied to a DE-52 column followed by G-200 gel filtration chromatography, HPLC and native gel electrophoresis. The protein was transferred by electroblotting to Immobilon-P, excised and fragmented by chemical cleavage at the methionine residues using cyanogen bromide. GTP hydrolase activity was monitored at all stages of purification. The ability

NATO ASI Series, Vol. H 74
Molecular Mechanisms of Membrane Traffic
Edited by D. J. Morré, K. E. Howell, and J. J. M. Bergeron
© Springer-Verlag Berlin Heidelberg 1993

of the purified protein to bind [$^3$H]retinol was determined by equilibrium dialysis.

## RESULTS

A GTP hydrolase, purified from transitional endoplasmic reticulum by DEAE-cellulose column chromatography, gel filtration chromatography, HPLC and native gel electrophoresis, was inhibited by all-<u>trans</u> retinol with a $K_i$ of 0.03 mM. The hydrolase was tightly membrane-associated and required detergent for solubilization. The purified protein bound retinol with a kD of ca. 0.1 nM and with a stoichiometry of ca. 1 mole retinol/bound per mole of purified protein. Cleavage by cyanogen bromide at methionine residues produced three peptides. The amino acid composition of the three peptides has been determined by microsequencing (courtesy of Dr. G. Becker, Eli Lilly Research Laboratories, Indianapolis, IN) and none exhibit any obvious sequence homology with previously described retinoid binding proteins. The GTP hydrolase activity was also inhibited by $AlF_4^-$, GTP$\gamma$S and pertussis toxin but not by cholera toxin.

### References

Bourne HR (1988) Do GTPases direct membrane traffic in secretion? Cell 53: 669-671
Morré DJ (1971) Isolation of Golgi apparatus. Method Enzymol 22: 130-148
Nowack DD, Paulik M, Morré DJ, Morré DM (1990) Retinoid modulation of cell-free membrane transfer between endoplasmic reticulum and Golgi apparatus. Biochem Biophys Acta 1051: 250-258
Zhao J, Morré DJ, Paulik M, Yim J, Morré DM (1990) GTP hydrolysis by transitional endoplasmic reticulum from rat liver inhibited by all-<u>trans</u> retinol. Biochim Biophys Acta 1055: 230-233

# A 38 kDa PROTEIN RESIDENT TO CIS GOLGI APPARATUS CISTERNAE OF RAT LIVER IS RECOGNIZED BY AN ANTIBODY DIRECTED AGAINST α SUBUNITS OF TRIMERIC G-PROTEINS

A. O. Brightman[1*], M. Paulik[1], J. B. Lawrence[1], T. Reust[1], C. C. Geilen[2], K. Spicher[3], W. Reutter[2], D. M. Morré[4] and D. J. Morré[1]

[1]Department of Medicinal Chemistry & Pharmacognosy and [4]Department of Foods and Nutrition
Purdue University
West Lafayette, IN

Using a completely cell-free system from rat liver, we have demonstrated the formation of 50 - 70 nm ATP-dependent transfer vesicles and the transfer of both lipids and proteins to Golgi apparatus immobilized on nitrocellulose. The ATP-dependent transfer is specific for cis, and to a lesser degree, medial Golgi apparatus elements and not exhibited by trans Golgi apparatus elements. Therefore, evidence was sought for a cis-Golgi apparatus "docking" protein capable of specifically binding transition vesicles.

When isolated Golgi apparatus cisternae were fractioned into cis-, medial- and trans-derived fractions using free-flow electrophoresis, a 38 kDa protein present in cis Golgi apparatus, but not trans Golgi apparatus, was detected when the fractions were analyzed by SDS-PAGE. The 38 kDa protein also was present in transition vesicles, but not in the transitional endoplasmic reticulum (ER). The 38 kDa protein was extracted from cis Golgi apparatus membranes by high salt or high pH.

Western Blot analysis revealed that anti-p38 recognized p38 and a protein of higher molecular weight (p72) in cis and medial subfractions of isolated Golgi apparatus. To determine the interaction of these proteins with transition vesicles, extrinsic proteins, including p38 and p72, were extracted with high salt from cis Golgi apparatus and labeled with $^{125}$I. Labeled p38 and p72, incubated with isolated membranes for 30 min at 37°C, bound to transition vesicles significantly more than to plasma membrane. Furthermore, binding of the p38 and p72 extrinsic cis Golgi apparatus proteins by transition vesicles was inhibited by immunoprecipitation with anti-p38 antibodies. These results indicated that the anti-p38 antibody

*Present address, Eli Lilly Company, Indianapolis, IN, USA, [2]Institut fur Molekulargiologie und Biochemie der Freien Universitat Berlin, D-100 Berlin 33, Germany, [3]Institut fur Pharmakologie der Freien Universitat Berlin, D-1000 Berlin 33, Germany

could recognize the native form of p38 and p72.

The effect of the anti-p38 antibodies on membrane transfer from the ER to cis Golgi apparatus was determined since the antibodies could recognize the native forms of both p38 and p72. The presence of the anti-p38 antibodies in the cell-free transfer system inhibited 50% of the vesicular transfer between the ER and cis Golgi at 37°C with no effect at 4°C using a cell-free ER/cis-Golgi transfer system. The inhibition of transfer comparable to transfer in the absence of ATP occurred only when cis-Golgi apparatus membranes were preincubated for 15 min with the antisera. Preincubation of the ER with antibody had no effect. Furthermore, the antisera specifically inhibited transfer to cis Golgi apparatus with isolated transition vesicles as donor.

An involvement of small GTP binding proteins (Goud and McCaffrey 1991) and trimeric G-proteins (Burgoyne 1992) in the regulation of intracellular vesicle traffic is already established from various lines of investigation. Evidence is most complete for intra-Golgi traffic and for transfer between the trans-Golgi-network (TGN) and the plasma membrane. Less data are available for vesicular transport between the ER and the cis-Golgi. Interestingly, p38 was recognized by a peptide-specific antibody directed against the α subunit of trimeric G-proteins. When p38 was immunoprecipitated by the anti-p38 antibodies and separated by SDS-PAGE, subsequent western blotting showed that p38 also was recognized by the peptide antibody α common which is directed against the amino acid sequence GAGES (the putative nucleotide binding site of the α subunit of trimeric G-proteins).

Taken together, the findings suggest that p38 may be a target of GTP binding as well as a possible candidate for a targeting or docking protein. The existence of such a protein would help explain why GTP-γ-S inhibits specific binding of transition vesicles to cis-Golgi apparatus elements.

## References

Goud B, McCaffrey M (1991) Small GTP-binding proteins and their role in transport. Curr Opinion Cell Biol 3: 626-633

Burgoyne RD (1992) Trimeric G proteins in Golgi transport. Trends Biol Sci 17: 87-88

# ACYL TRANSFER REACTIONS ASSOCIATED WITH CIS GOLGI APPARATUS

J. B. Lawrence, T. W. Keenan, P. Moreau and D. J. Morré
Department of Medicinal Chemistry
109 HANS
Purdue University
West Lafayette, IN 47907

## INTRODUCTION

Recent work in our laboratory involving cell-free lipid transfer from transitional endoplasmic reticulum (ER) to Golgi apparatus has shown a latent phospholipase A activity localized in the cis Golgi of rat liver (Moreau 1991). Subsequently, phosphatidylethanolamine (PE) was shown to be an important component of ER to Golgi transport vesicles. Because PE has a small hydrophilic head group and a large hydrophobic tail, it has a conical shape which is very conducive to Hexagonal II phase structure formation. The possibility that we examined was that the lysophosphatidylcholine (lysoPC), formed by the phospholipase A activity, could interact with the PE to destabilize the Hexagonal II phase and form a bilayer. However, because levels of PE decrease through the Golgi stacks, the presence of lysoPC will then destabilize the bilayer structure. To retain a stabile bilayer structure, the lysoPC must be removed. This can be accomplished in two ways: 1) degradation by a Golgi lysophospholipid phospholipase or 2) reacylation by a Golgi acyl-CoA:lysoPC acyltransferase activity. The reacylation of Golgi apparatus lysoPC via the Lands pathway (Lands 1976) was investigated.

## METHODS

Golgi apparatus were isolated from rat livers as described (Morre 1972) . The Golgi apparatus were used either directly or subfractionated into cis, medial and trans compartments. To assay acyl-CoA:lysoPC acyltransferase, Golgi apparatus fractions (100 mg of protein) were incubated with [14C] lysoPC in 50 mM Hepes pH 7.0 at 37° C for 60 minutes, unless otherwise noted. The activity measured was the conversion of [14C] lysoPC to [14C] PC as determined by thin layer chromatography. The thin layer chromatography plates were developed in a

NATO ASI Series, Vol. H 74
Molecular Mechanisms of Membrane Traffic
Edited by D. J. Morré, K. E. Howell, and J. J. M. Bergeron
© Springer-Verlag Berlin Heidelberg 1993

chloroform:methanol:water (65:35:8) solvent system and analyzed by autoradiography. Zones corresponding to PC were scraped and radioactivity determined.

RESULTS AND DISCUSION

Golgi apparatus acyl-CoA:lysoPC acyltransferase activity was located primarily in cis and medial Golgi apparatus fractions. This activity was optimal between pH 6.0 and 7.5 and was stimulated by various acyl-CoA derivatives but not by fatty acids plus ATP. Malonyl-CoA stimulated acyltransferase activity much more strongly than did palmitoyl-CoA, oleolyl-CoA, eicosanoyl-CoA or arachadonyl-CoA. Activity was unaffected by EGTA and was inhibited by high concentrations of the phospholipase inhibitor manoalide. Acyltransferase activity was temperature dependent but showed no definite transition temperature over the range of 15 to 37º C.

The lysoPC formed at the cis face of the Golgi apparatus may be in response to the fusion of ER to Golgi transport vesicles which have a high PE content. Because the PE levels in the Golgi apparatus decrease from cis to trans, the lysoPC must be removed to retain a stable lipid bilayer structure. The Golgi apparatus acyl-CoA:lysoPC acyltransferase activity presented here may be involved in this stabilization process.

T. W. Keenan, Department of Biochemistry and Nutrition
   VPI   Blacksburg, VA  24061

P. Moreau, CNRS   1 rue Camille-Saint-Saens
   33077 Bordeaux Cedex  France

Lands, W. E. M. and Crawford, C. G. (1976) Enzymes of Membrane Phospholipid Metabolism in Animals, In The Enzymes of Biological Membranes (Edited by Martouosi, A.), Vol. 2, pp. 3-85  Plenum Press, New York.

Moreau, P. and Morré, D. J. (1991) Cell-Free Transfer of Membrane Lipids. Evidence For Lipid Processing, J. Biol. Chem. 266, 4329-4333

Morré, D. J., Cheetham, R. D., Nyquist, S. E. and Ovtracht, L. (1972) A Simplified Procedure For Isolation of Golgi Apparatus From Rat Liver, Prep. Biochem. 2, 61-69

# HEPATIC BILE FORMATION: INTRACELLULAR TRAFFICKING OF BILE SALTS

James M. Crawford, Stephen Barnes, Rebecca C. Stearns, Cynthia L. Hastings, Deborah C.J. Strahs and John J. Godleski
Department of Pathology
Brigham and Women's Hospital
75 Francis Street
Boston, MA 02192, USA

Bile formation facilitates the digestion and absorption of lipids from the gut, and provides a mechanism for the elimination of a wide variety of potentially toxic compounds from the body. The primary organic solutes in bile are amphiphilic bile salts and phospholipids, with the substantial flux of bile salts through the liver as part of the enterohepatic circulation providing the major driving force for bile formation. Although biliary phospholipid appears to be derived from intracellular pools, the interactions of bile salts with intracellular membranes and their transport through the hepatocyte remain poorly understood. We have examined the role of bile salt hydrophilicity in regulating bile salt interactions with microtubule-dependent membrane-based transport mechanisms, and have developed a novel methodological approach for the ultrastructural localization of bile salts within hepatocytes.

Bile Salt Hydrophilicity: Intact rats were subjected to overnight biliary drainage to deplete the endogenous bile salt pool, and reinfused with selected bile salts (taurodehydrocholate, TDHC; tauroursodeoxycholate, TUDC; and taurocholate, TC; in order of decreasing hydrophilicity; 200 nmol/min.100g). Biliary excretion of i.v.-injected tracer [$^3$H]taurocholate was monitored, using colchicine-sensitivity as an indicator of microtubule-dependent movement through the hepatocyte (Crawford et al., 1988). The colchicine-insensitive component of bulk bile salt excretion remained constant at ~130 nmol/min. 100g, and the colchicine-sensitive component increased from ~0 in TDHC-reinfused animals to 35 and 60 nmol/min.100g in TUDC- and TC-reinfused animals, respectively. In control animals, peak biliary excretion of tracer

NATO ASI Series, Vol. H 74
Molecular Mechanisms of Membrane Traffic
Edited by D. J. Morré, K. E. Howell, and J. J. M. Bergeron
© Springer-Verlag Berlin Heidelberg 1993

[³H]taurocholate (occuring 2.5 min after i.v. injection) increased linearly from 15.3 (TDHC-infused) to 18.0 (TC-infused) %/min with decreasing hydrophilicity of the reinfused bile salt ($P<0.002$). Peak excretion decreased linearly from 13.8 (TDHC-infused) to 9.2 (TC-infused) %/min in colchicine-pretreated animals ($P<0.001$). We conclude that utilization of an intracellular microtubule-dependent pathway increases with decreasing bile salt hydrophilicity. This pathway permits more efficient excretion of bile salts, but increases the susceptibility of bile salt excretion to microtubule disruption.

<u>Ultrastructural studies</u>: Isolated hepatocytes were incubated 5 or 20 min with the 2-fluoro-$\beta$-alanine conjugate of cholic acid (FBAL-cholate; 50 $\mu$M), a bile salt analogue which is taken up and excreted into bile in a manner similar to native bile salts (Sweeny et al., 1990). Hepatocytes were cryofixed, freeze-dried and vapor fixed (paraformaldehyde followed by $OsO_4$) using the Life-Cell® system. Resin-embedded sections were examined, without further staining, in a Zeiss CEM902 electron microscope equipped for Electron Energy Loss Spectroscopy. Using the imaging capabilities of the instrument, fluorine was detected primarily in association with intracellular membranes, particularly membranes of the endoplasmic reticulum ($P<0.05$ by morphometric analysis). The fluorine signal was confirmed by serial energy loss spectra of cell regions containing this organelle. Fluorine also was detected in association with membranes of the Golgi appratus. We conclude that cryofixation and freeze-dry processing followed by electron microscopy with EELS is a valuable technique for examining intracellular transport of bile salts. Our results with FBAL-cholate support the concept that bile salts interact with intracellular membranes during transport through hepatocytes.

Crawford JM, Berken CA, Gollan JL (1988) Role of the hepatocyte microtubular system in the excretion of bile salts and biliary lipid: Implications for intracellular vesicular transport. J Lipid Research 29:144-156

Sweeny DJ, Daher G, Barnes S, Diasio RB (1990) Biological properties of the 2-fluoro-$\beta$-alanine conjugates of cholic acid and chenodeoxycholic acid in the isolated perfused rat liver. Biochim Biophys Acta 1054:21-25

# Wheat storage proteins as a model system to study the mechanism of protein sorting within the endoplasmic reticulum.

Gad Galili and Yoram Altschuler
Department of Plant Genetics
The Weizmann Institute of Science
Rehovot 76100
Israel

Although cellular factors that control the sorting of secretory proteins within the endoplasmic reticulum have been identified, very little is known about the signals and mechanisms by which secretory proteins interact with these factors. Wheat storage proteins can serve as an excellent model system for identification of such signals. These proteins are synthesized in developing wheat grains and then accumulate in protein bodies where they survive the period of grain dessication and later serve as a source of nitrogen and energy for the germinating seedlings. The wheat storage proteins are synthesized on membrane bound polysomes and then sequester into the endoplasmic reticulum (ER). Following sequestration, these proteins are either being retained within the ER and form dense protein bodies inside this organelle or are transported via the Golgi to vacuoles and condense into protein bodies at a post-ER location. The signals on these proteins that determine whether they will be retained within the ER or be exported to the Golgi are not known, although they are not of the K/HDEL type as these proteins lack such signals.

The majority of the wheat storage proteins (termed gliadins) appear naturally as chimeric proteins containing two separately folded autonomous regions which were evolved independently as separate proteins and were apparently fused during their evolution (Shewry and Ththam, 1990). A schematic diagram of these proteins is illustrated in Fig. 1. The N-terminal region is composed of seven to 16 tandem repeats rich in glutamine and proline that are apparently arranged in β-turn configurations (Shewry and Tatham, 1990). The tandem repeats vary slightly between the different types of gliadins, but they are essentially based on either PQQPFPQ, PQQQPPFS or PQQPQ concensus sequences (Shewry and Tatham, 1990). The C-terminal regions of the gliadins are apparently arranged predominantly in α-helix and β-sheet configurations (Shewry and

NATO ASI Series, Vol. H 74
Molecular Mechanisms of Membrane Traffic
Edited by D. J. Morré, K. E. Howell, and J. J. M. Bergeron
© Springer-Verlag Berlin Heidelberg 1993

Tatham., 1990). These regions also contain six to eight cysteine residues, which form three to four intramolecular disulfide bonds (Shewry and Tatham, 1990).

In the present report, we have studied the role of the N- and C-terminal regions of a wheat γ type gliadin in its sorting within the ER. This was addressed by expression in *Xenopus* oocytes of a wild type γ-gliadin as well as two deletion mutants containing only one of its autonomous regions. *Xenopus* oocytes is a very powerful system for this study inasmuch as these cells lack vacuoles and hence storage proteins exporting from the ER to the Golgi are secreted into the medium by the default pathway (Simon *et al.*, 1990) and thus can be easily traced. Our results demonstrate that the N-terminal glutamine and proline-rich autonomous region of the gliadin was retained entirely within the ER, indicating that this sequence harbors an ER retention signal. In contrast, the C-terminal autonomous region was efficiently exported from the ER. The wild type γ-gliadin, containing both regions, was exported from the endoplasmic reticulum at a lower rate and efficiency than its C-terminal region. Correct formation of the intramolecular disulfide bonds within the C-terminal unique region of the γ–gliadin was essential for the export from the ER. These results suggest that the rate and efficiency of export of wheat gliadins from the endoplasmic reticulum is determined by a balance between the effectiveness of their signals for retention within this organelle and the competence of other protein regions for export.

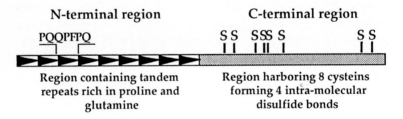

**Fig. 1** Schematic structure of a wheat gliadin storage protein. POOPFPQ- a consensus repeat. S- location of cysteine residues.

## REFERENCES

Shewry, P R and Tatham, A S (1990) The prolamin storage proteins of cereal seeds: structure and evolution. Biochem. J. 267: 1-12.

Simon, R , Altschuler, Y , Rubin, R and Galili, G (1990) Two closely related wheat storage proteins follow a markedly different subcellular route in Xenopus laevis oocytes. The Plant Cell 2: 941-950.

# DETERMINANTS OF OUTER MEMBRANE PROTEIN SORTING AND TOPOLOGY IN MITOCHONDRIA

G.C. Shore, H.M. McBride, D.G. Millar, and J.-M. Li
Department of Biochemistry
McGill University
Montreal, Quebec H3G 1Y6
Canada

The mitochondrial outer and inner membrane each contains a distinct import machinery for translocating proteins into or across their respective lipid bilayers. These two translocators may function either independently of each other or they may align in tandem and cooperate during translocation such that an incoming polypeptide can span both membranes simultaneously (Glick, et al., 1992; Pfanner et al., 1992). The latter situation is particularly relevant to proteins that are translocated, in whole or in part, to the matrix. Among this latter group of proteins may be polypeptides that are subsequently exported to the inner membrane or intermembrane space, via the conservative sorting pathway (Hartl and Neupert, 1990). For proteins that are sorted to the outer membrane, intermembrane space, or inner membrane by a unidirectional import pathway, they contain signals that are independently recognized and interpreted by the translocation machineries of the two membranes. This is well illustrated by the inherent problem of sorting to the inner membrane, in which the membrane anchor domain that results in arrest of the imported protein at the inner membrane, must be allowed to pass freely through the outer membrane.

To better understand some of these problems with respect to sorting to the outer and inner membrane, we have focused on a simple bitopic protein of the outer membrane in yeast mitochondria, OMM70 (Hase et al., 1984) (also called MAS70, Hines et al., 1990). Earlier studies revealed that all of the requisite topogenic information to target and insert OMM70 into the outer membrane in the correct orientation ($N_{in}$-$C_{cyto}$) resides in the N-terminal 29 amino acids of the polypeptide (Hase et al., 1984). This region is comprised of two

NATO ASI Series, Vol. H 74
Molecular Mechanisms of Membrane Traffic
Edited by D. J. Morré, K. E. Howell, and J. J. M. Bergeron
© Springer-Verlag Berlin Heidelberg 1993

structural domains (Fig. 1): a hydrophilic, positively-charged sequence (amino acids 1-10) and an apolar sequence (amino acids 11-29) which is the predicted transmembrane segment. Earlier work suggested that amino acids 1-12 function as a matrix-targeting signal while the transmembrane segment functions as a stop-transfer sequence. When amino acids 1-12 of OMM70 were fused to a reporter protein (dihydrofolate reductase), the hybrid protein was imported to the matrix (Hurt et al., 1985). However, the rate and extent of this import was very low and might have arisen from the ability of random, positively-charged sequences at the N-terminus of a protein to permit low-efficiency import into mitochondria.

**pOMD29 (OMM70)**

MLSNLRILLNKAALRKAHTSMVRNFRYGKPVQSQVQGL̲I̲L̲A̲A̲V̲A̲A̲T̲G̲T̲A̲I̲G̲A̲Y̲Y̲Y̲GPLNC - DHFR
1            15                30

**pO-OMD**

Figure 1. N-terminal amino acid sequence of pOMD29 and pO-OMD. The single-letter amino acid code has been used. The identical region of the predicted transmembrane segment in the two proteins is underlined.•,substitution of alanine for threonine; +, positively-charged residues. See text for additional details.

To further analyze this question, we created the hybrid protein, pOMD29, in which the N-terminal 29 amino acids of OMM70 were fused to dihydrofolate reductase (DHFR) (Fig. 1) (Li and Shore, 1992a). The hybrid protein was efficiently targeted and inserted into the outer membrane of intact mitochondria from rat heart, in the expected orientation ($N_{in}$-$C_{cyto}$). Targeting and insertion was specified by the transmembrane segment (amino acids 11-29), whereas the hydrophilic N-terminus (amino acids 1-10) enhanced the overall rate of import and insertion, due largely to the presence of the basic amino acids at positions 2, 7 and 9 (McBride et al., 1992).

We suggest that amino acids 1-10 and 11-29 of pOMD29 (OMM70) cooperate to form the functional equivalent of the signal-anchor sequence found in type II and type III proteins (von Heijen, 1988) that are inserted into the membrane of the endoplasmic reticulum (Wickner and Lodish, 1985), in which the targeting domain is coincident with, or overlaps, the membrane anchor domain. A hallmark feature of a signal-anchor function is that the sequence that triggers translocation across the membrane is also the sequence that abrogates translocation and results in lateral release of the hydrophobic core of the signal-anchor into the surrounding lipid bilayer (Blobel, 1980; Singer, 1990). In the context of pOMD29, an important consequence of a signal-anchor sequence, therefore, may be that it selects the outer membrane for insertion, simply because this is the first membrane encountered by the incoming precursor protein. Similarly, the combination of a matrix-targeting signal followed immediately by a stop-transfer domain may also select the outer membrane for insertion if, again, the stop-transfer segment enters the outer membrane translocation machinery and abrogates translocation before the protein is committed for import into the interior of the organelle (Nguyen et al., 1988).

Does the protein translocation machinery of the inner membrane also recognize and interpret the topogenic information of the pOMD29 (OMM70) signal-anchor sequence, a signal which is otherwise never encountered by the inner membrane? This question can be addressed by selectively rupturing the mitochondrial outer membrane while leaving the inner membrane intact (Ohba and Schatz, 1989), thereby allowing a precursor protein direct access to the import machinery of the inner membrane without first passing through the outer membrane. When such an approach was applied to pOMD29, the protein was efficiently inserted into the inner membrane, with the same orientation ($N_{in}$-$C_{cyto}$) that the protein adopts in the outer membrane. In contrast to insertion into the outer membrane, however, insertion into the inner membrane was dependent on the electrochemical potential (Li and Shore, 1992a). These results suggest that the import machineries of the outer and inner membrane interpret topogenic sequences and insert membrane proteins in a remarkably similar way. This is consistent with the notion that the signal-anchor sequence of pOMD29 selects the outer membrane because this is the first membrane encountered by the incoming protein. Moreover, it implies that if the outer membrane translocator is negatively regulated with respect to lateral release of the signal-anchor into the surrounding lipid bilayer, then pOMD29 by default may be recognized by the inner membrane translocator.

An unexpected finding with regard to the pOMD29 (OMM70) signal-anchor sequence is that, once inserted into the lipid bilayer of the outer membrane, this domain forms homo-oligomers (at the minimum, dimers), as judged by the ability of a bifunctional reagent to cross-link pOMD29 polypeptides. This observation may have functional implications for OMM70 since there is evidence that this protein is a component of the protein import machinery (Hines et al., 1990). An interesting structural feature of the signal-anchor sequence is an enrichment of alanine residues on one side of the predicted transmembrane helix (amino acids 11-29). Whether or not the presence on one side of the helix of amino acids bearing a small side chain is required for oligomerization of the signal-anchor sequence in the membrane, is currently under study.

Again by analogy to type II and type III proteins that are inserted into the ER membrane, the orientation of pOMD29 in the outer membrane might be expected to be conferred by structural determinants within the signal-anchor sequence. In the case of the ER, orientation is determined by sequences that flank the hydrophobic core of the signal-anchor, especially as this pertains to the presence or absence of a retention signal located on the N-terminal side of the hydrophobic core, which is made up, in part, of positively-charged residues.

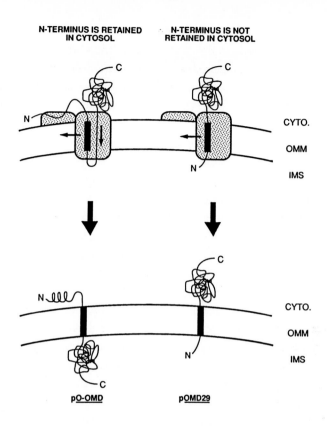

Figure 2. Working model to describe how pOMD29 AND pO-OMD assume their respective topology in the OMM. The transmembrane segment of the two proteins is shown as a black bar. CYTO, cytoplasm; OMM, outer mitochondrial membrane; IMS, intermembrane space.

As illustrated in Fig. 1, we replaced amino acids 1-10 of pOMD29 with a stretch of 38 amino acids that contains the potent matrix-targeting signal of preornithine carbamyl transferase (pOCT). The new protein, designated pO-OMD, was inserted into the outer membrane, but in an orientation opposite to that of pOMD29 (i.e., $N_{cyto}$-$C_{in}$) (Li and Shore, 1991b). A model to explain these finding is given in Fig.2. Whether or not the determinants that either cause retention of the pOCT signal of pO-OMD on the cytosolic side of the membrane or allow translocation of the N-terminus of pOMD29 across the membrane are related to their strengths as matrix-targeting signals remains to be determined. It also remains to be determined if retention of the N-terminus of pO-OMD on the cytosolic side of the membrane are the result of interactions of the pOCT signal with a specific membrane protein (receptor) or with the surface of the lipid bilayer or a result of some other process.

Finally, if the hydrophobic core of a signal-anchor sequence forms the transmembrane segment of pOMD29 in the outer membrane, how is it that hydrophobic sequences that anchor proteins in the inner membrane pass through the outer membrane translocator without arresting and inserting the protein in the bilayer? As an illustration, if the VSV G stop-transfer domain is engineered into a matrix precursor protein, pOCT, at a position immediately downstream of the matrix-targeting signal, the protein is inserted into the outer membrane. When placed toward the C-terminus of pOCT, the VSV-G stop-transfer sequence results in insertion of pOCT into the inner membrane (Nguyen et al., 1988). Presumably, the outer membrane translocator must interpret the context of the membrane anchor domain within the polypeptide (Nguyen et al., 1988 Shore et al., 1992). One possibility is that, when the matrix-targeting signal of a protein passes through the outer membrane and engages the inner membrane translocator at contact sites, interactions between the inner membrane and outer membrane translocators prevent the latter from responding to a stop-transfer segment (Nguyen et al., 1988). Such a model is consistent with the effect of position of the stop-transfer segment on sorting between the two membranes. Alternatively, sequences flanking a potential transmembrane segment may prevent this segment from being recognized as a stop-transfer signal by the outer membrane translocator; or perhaps proteins destined for the inner membrane recruit unique cytosolic factors or receptors that, upon interaction with the outer membrane translocator, prevent it from arresting and inserting transmembrane segments into the outer membrane bilayer. A comprehensive analysis of this problem will require the isolation of a functional outer membrane translocator free of the inner membrane translocator.

Blobel G (1980) Intracellular protein topogenesis. Proc Natl Acad Sci USA 77:1496-1499

Glick B, Wachter C, Schatz G (1991) Protein import into mitochondria: Two systems acting in tandem? Trends Cell Biol 1:99-103

Hartl F-U, Neupert W (1990) Protein sorting to mitochondria. Evolutionary considerations of folding and assembly. Science 247:930-938

Hase T, Müller U, Riezman H, Schatz G (1984) A 70-kD protein of the yeast mitochondrial membrane is targeted and anchored via its extreme amino terminus. EMBO J 3:3157-3164

Hines V, Brandt A, Griffiths G, Horstmann H, Brütsch H, Schatz G (1990) Protein import into yeast mitochondria is accelerated by the outer membrane protein MAS70. EMBO J 9:3191-3200

Hurt EC, Müller U, Schatz G (1985) The first twelve amino acids of a yeast outer mitochondrial membrane protein can direct a nuclear-encoded cytochrome oxidase subunit to the mitochondrial inner membrane. EMBO J 4:3509-3518

Li J-M, Shore GC (1992a) Protein sorting between mitochondrial outer and inner membranes. Insertion of an outer membrane protein into the inner membrane. Biochim Biophys Acta 1106:233-241

Li J-M, Shore GC (1992b) Reversal of the orientation of an integral protein of the mitochondrial outer membrane. Science, 256:1815-1817

McBride HM, Millar DG, Li J-M, Shore GC (1992) A signal-anchor sequence selective for the mitochondrial outer membrane. Submitted for publication

Nguyen M, Bell AW, Shore GC (1988) Protein sorting between mitochondrial membranes specified by position of the stop-transfer domain. J Cell Biol 106:1499-1505

Ohba M, Schatz G (1989) Disruption of the outer membrane restores protein import to trypsin-treated yeast mitochondria. EMBO J 6:2117-2122

Pfanner N, Rassow J, van der Klei IJ, Neupert W (1992) A dynamic model of the mitochondrial protein import machinery. Cell 68:999-1002

Shore GC, Millar DG, Li J-M (1992) Protein insertion into mitochondrial outer and inner membranes via the stop-transfer sorting pathway. In: Neupert W, Lill R (eds) Membrane biogenesis and protein targeting, New Comprehensive Biology 22. Elsevier, Amsterdam, in press

Singer SJ (1990) The structure and insertion of integral proteins in membranes. Annu Rev Cell Biol 6:247-296

von Heijne G (1988) Transcending the impenetrable: How proteins come to terms with membranes. Biochim Biophys Acta 947:307-333

Wickner WT, Lodish HF (1985) Multiple mechanisms of protein insertion into and across membranes. Science 230:400-407

# MOLECULAR CHAPERONES HSP70 AND HSP60 IN PROTEIN FOLDING AND MEMBRANE TRANSLOCATION

Jörg Martin and F.-Ulrich Hartl
Program of Cellular Biochemistry & Biophysics
Sloan-Kettering Institute
1275 York Avenue
New York, NY 10021 USA

Given the difficulties protein chemists may encounter when attempting to renature unfolded proteins *in vitro*, it is noteworthy that the acquisition of the correctly folded structure seems to be much less of a traumatic experience for a nascent polypeptide chain *in vivo*. Generally, unfolded polypeptides have the tendency to aggregate. The cellular environment with its extremely high concentration of total protein (~0.3 g/ml) and of newly-synthesized, folding chains (30-50 µM in *Escherichia coli*) may result in even further reduction of solubility and thus should strongly favor misfolding and aggregation of a folding protein (Zimmerman and Trach, 1991). Nevertheless, the yield of folded protein *in vivo* can reach almost 100% (Gething et al., 1986). It has become clear over recent years that the action of molecular chaperones, helper proteins which interact with folding intermediates and prevent unproductive off-pathway reactions (Ellis, 1987; Rothman, 1989; Gething and Sambrook, 1992), is essential in accomplishing this high efficiency of physiological protein folding.

The members of the Hsp70 and Hsp60 families of molecular chaperones are well characterized with respect to their role in cellular protein folding. These proteins are ubiquitous, occurring in almost every cell type, in the cytosol and within subcellular compartments including mitochondria, chloroplasts and the endoplasmic reticulum (Hartl et al., 1992). Several lines of evidence suggest that in addition to the pathway of protein folding, directed by the amino acid sequence of the folding protein, there may be a chaperone pathway in which Hsp70s and Hsp60s cooperate in mediating protein folding. Hsp70s appear to interact first with a nascent polypeptide chain emerging from the ribosome or at the trans-side of a membrane during translocation. This leads to the stabilization of an early folding intermediate, perhaps resembling the 'molten globule', which is subsequently made available to Hsp60 for mediation of folding to the native state. Such a pathway has been described for proteins which fold following import into

NATO ASI Series, Vol. H 74
Molecular Mechanisms of Membrane Traffic
Edited by D. J. Morré, K. E. Howell, and J. J. M. Bergeron
© Springer-Verlag Berlin Heidelberg 1993

mitochondria and is likely to exist in the bacterial cytosol as well. In the following sections we will summarize the functional properties established for the sequential action of Hsp70 and Hsp60 in protein folding.

## Molecular chaperones and mitochondrial biogenesis

The analysis of the folding of proteins imported into mitochondria from the cytosol has contributed considerably to our present understanding of chaperone-mediated protein folding. In particular, this experimental system served to establish the potential of Hsp70 and Hsp60 to act in a sequential folding pathway.

*Stabilisation of an unfolded state by cytosolic Hsp70:* Following their synthesis in the cytosol, most mitochondrial precursor proteins have to traverse the outer and inner mitochondrial membranes at so-called translocation contact sites (Hartl and Neupert, 1990). Regions of the polypeptide chain engaged in translocation appear to be in an extended conformation (Rassow et al., 1990). Consequently, the maintenance of an unfolded state in the cytosol is a prerequisite for efficient translocation. The unfolded state is maintained by Hsp70 and probably additional components such as homologues of the *E. coli* stress protein DnaJ (see Fig. 1). For example, in a yeast mutant defective in cytosolic Hsp70, the precursor of subunit β of the $F_1$-ATPase has been shown to accumulate in the cytosol (Deshaies et al., 1988). It appears that the majority of newly-synthesized proteins interact with Hsp70 as nascent chains emerging from the ribosome (Beckman et al., 1990), except perhaps for those proteins which undergo co-translational transport into the endoplasmic reticulum. It is thought that precursor proteins reach the receptors at the outer mitochondrial membrane in a complex with the chaperone. As described below, the release of the Hsp70-bound protein or its transfer to another component in the chaperone cascade is probably a highly regulated process, depending on two other stress proteins cooperating with Hsp70.

*Requirement of mitochondrial Hsp70 for translocation:* Once a mitochondrial precursor protein has docked at specific receptors of the outer membrane, it is released from the cytosolic chaperone in an ATP-dependent manner and can then be translocated into the matrix compartment through a hydrophilic environment (Pfanner et al., 1987), which is most likely a proteinaceous pore. Initially, this process is driven by the membrane potential, $\Delta\Psi$, which allows insertion of the positively charged N-terminal targeting sequence into the inner membrane (Martin et al., 1991a). The translocation of the mature protein part, however, depends mainly on other driving forces. In a yeast mutant defective in the mitochondrial Hsp70 (mHsp70) Ssc1p, precursor proteins accumulate as intermediates spanning both membranes in a permanent association with mHsp70 (Kang et al., 1990). It was concluded that binding of mHsp70 is of importance in pulling the protein into the matrix in a tug of war with the cytosolic chaperones. This process would be defective in the mutant. If the unfolded state is a prerequisite for translocation, then import of proteins following their artificial unfolding in 8 M urea and the removal of cytosolic chaperones should be less affected in the mutant mitochondria. Indeed, precursor proteins which are diluted from denaturant into the import reaction are able to traverse the membranes of the mutant mitochondria, but are then unable to refold. They remain associated with the mutant mHsp70 as trapped folding intermediates. The direct physical interaction of a translocating polypeptide chain with the wild-type form of mHsp70 at the matrix side of the inner membrane has been demonstrated (Scherer et al. 1990, Ostermann et al., 1990). This interaction may resemble that of cytosolic Hsp70 with nascent chains emerging from ribosomes. The organellar mHsp70 would thus have extended its function after the endosymbiotic event which led to the development of mitochondria from prokaryotic ancestors. Import into mitochondria (and membrane translocation of proteins in general) is unidirectional. It is currently unknown how this is accomplished, but posttranslocational events including proteolytic processing (and glycosylation in case of secretory proteins) and the interaction with molecular chaperones following translocation are probably important in preventing a protein from returning to the cytosol.

*Transfer to Hsp60 and folding:* In order to fold to the native state, at least some of the proteins imported into mitochondria have to be transferred from mHsp70 to Hsp60 (Kang et al., 1990; Neupert et al., 1990; Manning-Krieg et al., 1991) (see Fig. 1). Hsp60 mediates the folding of proteins to their native state or into monomers that subsequently undergo oligomeric assembly (Cheng et al., 1989). This reaction requires ATP-hydrolysis (Ostermann et al., 1989) and regulation by Hsp10, the mitochondrial GroES homologue (see below). The importance of Hsp60 for the biogenesis of mitochondrial proteins has again been established using a specific temperature sensitive yeast mutant. The defect of Hsp60 in the mutant strain *mif4* results in the misfolding of a variety of imported proteins such as the β subunit of the $F_1$-ATPase or the trimeric matrix enzyme ornithine transcarbamoylase (Cheng et al., 1989). These proteins are imported into the mutant mitochondria but then form misfolded aggregates. Among the proteins affected by the defect in Hsp60 are precursor proteins destined for the intermembrane space, such as cytochrome b2, which are first imported into the matrix and then re-exported across the inner membrane (Hartl and Neupert, 1990). A hydrophobic signal-sequence mediates their re-export step into the intermembrane space. This hydrophobic sequence, either by interfering with folding of the mature protein part and/or by direct interaction with Hsp60, results in a tight association of the protein with Hsp60 preventing folding prior to export (Koll et al., 1992). Release from Hsp60 then apparently requires the interaction of the signal sequence with a component of the export machinery, for example a receptor at the inner surface of the inner membrane.

## Molecular mechanism of chaperone action: Interplay between Hsp70 and Hsp60

Recently, various reactions of chaperone-mediated protein folding have been reconstituted *in vitro* using pure chaperone components (mostly from *E. coli*) and artificially unfolded substrate proteins. Considerable insight has been gained through these studies into the molecular mechanisms of Hsp70 and Hsp60 action and into how they may cooperate in the ordered protein folding pathway observed in mitochondria. It appears likely that homologues of two additional *E. coli* stress proteins, DnaJ and GrpE, are involved in this reaction whose distinct steps can now be defined:

*Stabilisation of a conformational intermediate:* Members of the Hsp70 family consist of two domains, an N-terminal ATP-hydrolysing domain of about 44 kD (Flaherty et al., 1990), and a C-terminal putative 'substrate' binding domain. ATP-hydrolysis is generally required for the release of bound substrate proteins and the binding of peptides or unfolded proteins stimulates the ATP-ase of Hsp70 (Flynn et al., 1991). DnaK, the *E. coli* homologue of Hsp70, has been shown biochemically and genetically to interact with two other heat shock proteins, DnaJ and GrpE. All three genes were originally discovered through mutations blocking DNA-replication of bacteriophage $\lambda$ (Friedman et al., 1984). The function of the chaperones in this process is to disassemble a preprimsosmal protein complex at ori$\lambda$. More recently, DnaK, DnaJ and GrpE have also been shown to drive the disassembly of repA dimers into active monomers (Wickner et al., 1991a). Insight into the functional cooperation of DnaK, DnaJ and GrpE in protein folding came

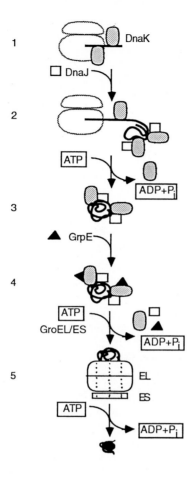

Fig. 1 Model for the chaperone-mediated folding pathway of newly synthesized proteins in the bacterial cytosol. The reaction scheme may correspond to the steps involved in the folding of mitochondrial proteins if synthesis on ribosomes is replaced by membrane-translocation, DnaK by mitochondrial Hsp70 and GroEL/ES by Hsp60/Hsp10. Taken from Langer et al., 1992 (see text for details).

from studies using the monomeric mitochondrial protein rhodanese as a model substrate. Dilution of rhodanese from urea into buffer solution results in rapid aggregation. The presence of DnaK can suppress this process by formation of a complex between DnaK and an early folding intermediate of rhodanese (Langer et al., 1992). The efficiency of complex formation is rather low however, and a high molar excess of DnaK is required. DnaJ alone is also able to bind a folding intermediate of rhodanese, thereby preventing its aggregation. Interestingly, both Dnak and DnaJ act synergistically in stabilizing rhodanese in a tight complex. This complex of rhodanese-DnaK-DnaJ is not disrupted by ATP-hydrolysis but rather is stabilized by the presence of hydrolysable ATP (Fig. 1, stage 3).

Why do DnaK and DnaJ form a much tighter complex with rhodanese than each chaperone alone? An attractive possibility is that they jointly stabilize a folding intermediate by recognizing different conformational features of the polypeptide. Such differential recognition by DnaJ and DnaK is inferred from binding studies employing defined folding intermediates of α-lactalbunin ( α-LA). Whereas DnaK is able to bind efficiently to a carboxymethylated form of α-LA, which is extended and lacks stable secondary structure, DnaJ has only a weak affinity for this state. Its binding affinity appears to increase α-LA intermediates which contain structure and possess features of the molten globule. The preference of DnaK for extended conformations has also been indicated in a recent study using nuclear magnetic resonance (NMR) to analyse the conformation of a Hsp70-bound peptide (Landry et al., 1992).

*GrpE-dependent transfer.* To allow intermediates in protein folding to progress to the native state, the chaperone bound forms then have to be released from DnaK and DnaJ. This step is mediated by the heat shock protein GrpE. Its interaction with a preformed rhodanese-DnaK-DnaJ protein complex leads to the dissociation of the complex and to the release of the folding intermediates. GrpE thus can be regarded as a coupling factor of the transfer step: It allows takeover of a conformational intermediate by other chaperones from DnaJ-DnaK. GrpE may fulfill this function by directly interacting with DnaK and stimulating nucleotide exchange from an ADP- to an ATP-bound form of DnaK (Fig. 1, stage 4) (Liberek et al., 1991). This type of reaction may also be involved in the release of chaperone-bound precursor proteins for membrane translocation (see above). In *Escherichia coli* , for example, the chaperone SecB may serve as the acceptor for at least a subset of proteins destined for export to the periplasm (Wickner et al., 1991b).

*GroEL/ES-mediated folding.* The rhodanese folding intermediate can be transferred *in vitro* from DnaJ-DnaK to GroEL, the *E. coli* Hsp60 homologue, for folding to the native state (Fig. 1, stage 5). Unlike DnaK/DnaJ, GroEL appears to be able to mediate folding to more compact intermediates in the folding pathway in an ATP-dependent process. The remarkable quarternary structure of GroEL, consisting of two stacked heptameric rings of identical ~60 kDa subunits (Hohn et al.,1979;Hendrix,1979), may signal responsibility for this capacity. Only one or two folding polypeptides bind per GroEL tetradecamer (Martin et al., 1991b; Viitanen et al., 1991). It is a possibility that hydrophobic motifs are recognized by the chaperone which are buried in the native state of the substrate protein but are exposed in early folding intermediates. Based on an analysis by NMR, it was proposed that peptides interacting with GroEL adopt an α-helical conformation upon binding to the chaperone (Landry et al., 1992). In order to execute its full function in folding, GroEL depends on yet another heat shock protein, GroES (Hsp10). This smaller protein, a heptamer of 10 kDa subunits (Chandrasekhar et al., 1986), serves as a 'pacemaker' by regulating the ATP-ase activity of GroEL (Martin.et al., 1991b). Once a folding intermediate is bound to GroEL, rounds of ATP driven release and rebinding may take place. As in the case of Hsp70, binding of the protein stimulates the ATP-ase of GroEL. In the absence of GroES, ATP-hydrolysis does not result in a permanent release however, except for certain proteins such as dihydrofolate reductase, which fold spontaneously with high efficiency *in vitro*. As demonstrated with rhodanese, the substrate protein rather undergoes rapid cycles of release and rebinding which can be interrupted by adding an unfolded competitor molecule such as casein, a protein which exposes hydrophobic surfaces. Casein competes with rhodanese for binding sites at GroEL and displaces the former when hydrolysable ATP is present. The labile folding intermediate of rhodanese will then aggregate. Apparently, the rapid 'whole sale' release of a substrate protein is not productive for folding.

In the presence of GroES, however, the substrate protein is released from GroEL in a conformation which is able to reach the native state having circumvented the danger of aggregation (Fig. 1, stage 5). Regulation by GroES may prevent the complete release of the polypeptide in a single step, resulting in a step-wise release of the protein. Released segments of the protein substrate would thereby be free to fold, whereas other parts of the protein would remain associated with GroEL. This mechanism would serve a dual function: Firstly, the folding protein would be shielded from the cellular environment thus preventing its interaction with other folding proteins or sticky membranes. Secondly, it may allow parts of a protein to fold without interference from other, more distal regions of the polypeptide chain, thus avoiding unproductive intramolecular contacts. If folding at the surface of GroEL turns out to be unsuccessful, the protein segment could then rebind, triggering a new cycle of ATP-hydrolysis and partial release. As expected, the competitor casein has no influence on the folding process of rhodanese in the presence of GroES. No displacement occurs because rhodanese constantly remains in close association with GroEL during its progressive folding at the surface of the chaperone. Using fluorescence spectroscopy, the presence of intermediates closer to the native state than the initially bound form could be observed. These conformational intermediates were still in contact with GroEL. There is no evidence that GroEL alters the folding pathway of a protein. Its function seems to be restricted to provide an optimal intracellular microenvironment where aggregation and unproductive side steps are prevented. The folding protein follows its intrinsically determined route to the native state. Partial release from GroEL would simply allow a folding protein to explore this route. The exact molecular mechanism of GroEL/ES action remains to be established.

## Future aspects

Is there a general pathway of chaperone-mediated protein folding? Given the role of Hsp60/GroEL in protein folding in mitochondria and the bacterial cytosol, it is intriguing that a structural or functional equivalent of the chaperonins has not yet been identified in the eukaryotic cytosol, which is known to contain both Hsp70 and DnaJ homologues (Caplan and Douglas, 1991; Luke et al., 1991; Atencio and Yaffe, 1992). However, a chaperonin candidate for the cytosol, t-complex polypeptide-1 (TCP-1), has been proposed (Ellis, 1990; Trent et al., 1991). Sequence analysis has revealed that TCP-1 has a 40% sequence identity with TF55, an archaebacterial stress protein, which forms a double-ring complex containing 8-9 membered rings of 55 kDa subunits and which has some chaperonin properties (Phipps et al., 1991; Trent et al., 1991). TCP-1 is indeed part of a similar ring-complex which appears to be present in the cytosol of most eukaryotic cells. Based on recent findings (J. Frydman, E. Nimmesgern and F.U. Hartl, unpublished) it seems likely that this complex indeed has a chaperone function in protein folding and its potential to cooperate with Hsp70 in the folding of newly-synthesized polypeptide chains is at present being explored.

## Acknowledgments

We thank Dr. R. Hlodan for critically reading the manuscript. J. M. is a recipient of a post-doctoral fellowship from the 'Dr. Mildred Scheel Stiftung für Krebsforschung'.

# References:

Atencio DP, Yaffe MP (1992) MAS5, a yeast homolog of DnaJ involved in
mitochondrial protein import. Mol Cell Biol 12:283-291

Beckman RP, Mizzen LA, Welch WJ (1990) Interaction of hsp70 with newly
synthesized proteins: implications for protein folding and assembly. Science
248:850-854

Caplan AJ, Douglas MG (1991) Characterization of YDJ1: a yeast
homologue of the bacterial dnaJ protein. J Cell Biol 114:609-621

Chandrasekhar GN, Tilly K, Woolford C, Hendrix R, Georgopoulos C (1986)
Purification and properties of the groES morphogenetic protein of *Escherichia
coli*. J Biol Chem 21:12414-12419

Cheng MY, Hartl FU, Martin J, Pollock RA, Kalousek F, Neupert W, Hallberg EM,
Hallberg RL, Horwich AL (1989) Mitochondrial heat-shock protein hsp60 is
essential for assembly of proteins imported in yeast mitochondria. Nature
337:620-625

Deshaies RJ, Koch BD, Werner-Washburne M, Craig EA, Schekman R (1988) A
subfamily of stress proteins facilitates translocation of secretory and
mitochondrial precursor polypeptides. Nature 332:800-805

Ellis RJ (1987) Proteins as molecular chaperones. Nature 328:378-379

Ellis RJ (1990) Molecular chaperones: the plant connection. Science 250:954-959

Flaherty KM, DeLuca-Flaherty, McKay DB (1990) Three-dimensional structure of
the ATPase fragment of a 70K heat-shock cognate protein. Nature 346:623-
628

Flynn GC, Pohl J, Flocco MT, Rothman JE (1991) Peptide-binding specificity of the
molecular chaperone BiP. Nature 353:726-730

Friedman, DE, Olson ER, Georgopoulos C, Tilly K, Herskowitz I, Banuett F (1984)
Interactions of bacteriophage and host macromolecules in the growth of
bacteriophage λ.Microbiol Rev 48:299-325

Gething MJ, McCammon K, Sambrook J (1986) Expression of wild type and mutant
forms of influenza hemagglutinin: the role of folding in intracellular transport.
Cell 46:939-950

Gething MJ, Sambrook J (1992) Protein folding in the cell. Nature 355:33-45

Hartl FU, Neupert W (1990) Protein sorting to mitochondria: evolutionary
conservations of folding and assembly. Science 247:930-938

Hartl FU, Martin J, Neupert W (1992) Protein folding in the cell: the role of
molecular chaperones hsp70 and hsp60. Annu Rev Biophys Biomol Struct
21:293-322

Hendrix RW (1979) Purification and properties of groE, a host protein in
bacteriophage assembly. J Mol Biol 129:375-392

Hohn T, Hohn B, Engel A, Wurtz M (1979) Isolation and characterization of the host
protein groE involved in bacteriophage lambda assembly. J Mol Biol 129:359-
373

Kang PJ, Ostermann J, Shilling J, Neupert W, Craig EA, Pfanner N (1990)
Requirement of hsp70 in the mitochondrial matrix for translocation and folding
of precursor proteins. Nature 348:137-143

Koll H, Guiard B, Rassow J, Ostermann J, Horwich AL, Neupert W, Hartl FU (1992)
Antifolding activity of hsp60 couples protein import into the mitochondrial matrix
with export to the intermembrane space. Cell 68:1163-1175

Landry SJ, Jordan R, McMacken R, Gierasch LM (1992) Different conformations for
the same polypeptide bound to chaperones DnaK and GroEL. Nature 355:455-
457

Langer T, Lu C, Echols H, Flanagan J, Hayer MK, Hartl FU (1992) Successive action of molecular chaperones DnaK, DnaJ and GroEL along the pathway of assisted protein folding. Nature 356:683-689

Liberek K, Marszalek J, Ang D, Georgopoulos C, Zylicz M (1991) *Escherichia coli* DnaJ and GrpE heat shock proteins jointly stimulate ATPase activity of DnaK. Proc Natl Acad Sci USA 88:2874-2878

Luke MM, Sutton A, Arndt KT (1991) Characterization of SIS1, a *Saccharomyces cerevisiae* homologue of bacterial dnaJ proteins. J Cell Biol 114:623-638

Manning-Krieg UC, Scherer PE, Schatz G (1991) Sequential action of mitochondrial chaperones in protein import into the matrix. EMBO J 10:3273-3280

Martin J, Mahlke K, Pfanner N (1991a) Role of an energized inner membrane in mitochondrial protein import. J Biol Chem 266:18051-18057

Martin J, Langer T, Boteva R, Schramel A, Horwich AL, Hartl FU (1991b) Chaperonin-mediated protein folding at the surface of groEL through a 'molten globule'-like intermediate. Nature 352:36-42

Neupert W, Hartl FU, Craig EA, Pfanner N (1990) How do polypeptides cross the mitochondrial membranes? Cell 63:447-450

Ostermann J, Horwich AL, Neupert W, Hartl FU (1989) Protein folding in mitochondria requires complex formation with hsp60 and ATP hydrolysis. Nature 341:125-130

Ostermann J, Voos W, Kang PJ, Craig EA, Neupert W, Pfanner N (1990) Precursor proteins in transit through mitochondrial contact sites interact with hsp70 in the matrix. FEBS Lett 277:281-284

Pfanner N, Hartl FU, Guiard B, Neupert W (1987) Mitochondrial precursor proteins are imported through a hydrophilic membrane environment. Eur J Biochem 169:289-293

Phipps BM, Hoffman A, Stetter KO, Baumeister W (1991) A novel ATPase complex selectively accumulated upon heat shock is a major cellular component of thermophilic archaebacteria. EMBO J 10:1711-1722

Rassow J, Hartl FU, Guiard B, Pfanner N, Neupert W (1990) Polypeptides traverse the mitochondrial envelope in an extended state. FEBS Lett 275:190-194

Rothman JE (1989) Polypeptide chain binding proteins: catalysts of protein folding and related processes in cells. Cell 59:591-601

Scherer PE, Krieg UC, Hwang ST, Vestweber D, Schatz G (1990) A precursor protein partly translocated into yeast mitochondria is bound to a 70 kd mitochondrial stress protein. EMBO J 9:4315-4322

Trent JD, Nimmesgern E, Wall JS, Hartl FU, Horwich AL (1991) A molecular chaperone from a thermophilic archaebacterium is related to the eukaryotic protein t-complex polypeptide-1. Nature 354:490-493

Viitanen PV, Donaldson GK, Lorimer GH, Lubben TH, Gatenby AA (1991) Complex interactions between chaperonin 60 molecular chaperone and dihydrofolate reductase. Biochemistry 30:9716-9723

Wickner S, Hoskins J, McKenney K (1991a) Function of DnaJ and DnaK as chaperones in origin-specific DNA binding by RepA. Nature 350:165-167

Wickner W, Driessen AJM, Hartl FU (1991b) The enzymology of protein translocation across the *Escherichia coli* plasma membrane. Annu Rev Biochem 60:101-124

Zimmerman SB, Trach SO (1991) Estimation of macromolecule concentrations and excluded volume effects for the cytoplasm of *Escherichia coli*. J Mol Biol 222:599-620

# CYTOSOLIC REACTIONS IN MITOCHONDRIAL PROTEIN IMPORT

Douglas M. Cyr and Michael G. Douglas
Department of Biochemistry and Biophysics
University of North Carolina
Chapel Hill,  North Carolina, 27599-7260
USA.

## Abstract

Cytosolic hsp70 molecules are involved in maintenance of pre-proteins in transport competent conformations.  Genetic and biochemical studies in procaryotes indicate that hsp70 (DnaK) functionally interacts with two other cytosolic proteins , DnaJ and GrpE (Ang et al., 1991).  Recently, several eukaryotic DnaJ homologs have been identified. We have purified cytosolic hsp70 (SSA1p) and DnaJ (YDJ1p) homologs from S. cerevisiae  and  characterized interactions between them. SSA1p exhibited a weak ATPase activity which was stimulated about 10 fold by YDJ1p. Stable complex formation  between SSA1p  and  a mitochondrial presequence peptide was demonstrated.  Significant  reductions in complex formation were only observed in the presence of  both ATP and YDJ1p.  Thus, an eukaryotic dnaJ homolog functionally interacts an eukaryotic hsp70 family member to regulate affinity of the chaperone for a polypeptide substrate.

## Introduction

Many mitochondrial proteins are encoded by nuclear genes, synthesized on cytosolic ribosomes and must be imported.  The import machinery requires that mitochondrial precursor proteins have a loosely folded conformation in order for membrane translocation to occur.   The precise mechanism for stabilization of mitochondrial  precursors in import  competent conformations is unclear.   However, the molecular chaperone hsp70, which binds nascent polypeptide chains as they emerge from ribosomes (Beckman et al., 1990), participates in this process (Deshaies et al., 1988).  It is proposed that ATP dependent cycling of hsp70 on and off  precursor proteins maintains their conformation competent for import  (reviewed in Gething and

NATO ASI Series, Vol. H 74
Molecular Mechanisms of Membrane Traffic
Edited by D. J. Morré, K. E. Howell, and J. J. M. Bergeron
© Springer-Verlag Berlin Heidelberg 1993

Sambrook, 1992). There is also evidence that cytosolic factors in addition to hsp70 act in this process (Deshaies et al., 1988 and Chrico et al., 1988). Insight into the identity of these cytosolic factors comes from recent studies with E. coli hsp70 (dnaK). DnaK functionally interacts with two other heat shock proteins, dnaJ and grpE in bacteriophage λ and P1 replication (Ang et al., 1991). Purified dnaJ and grpE synergistically stimulate the ATPase activity of dnaK up to 50 fold; dnaJ stimulates ATPase activity, while grpE stimulates adenine nucleotide exchange (Liberek et al.,1991a). DnaJ can also bind directly to protein substrates independent of dnaK and may be a chaperone molecule (Ang et al., 1991, Gamer et al., 1992, Langer et al.,1992, Wickner et al.,1991). Several eukaryotic DnaJ homologs have recently been identified (Bork et al., 1992, Caplan and Douglas ,1991). Thus, dnaJ homologs might act with eukaryotic hsp70 homologs in some biological reactions.

YDJ1p is an abundant cytosolic dnaJ homolog found in S. cerevisiae that is 32% identical to E. coli dnaJ, a heat shock protein, and is modified by farnesyl at the C-terminus (Caplan and Douglas, 1991, Atencio and Yaffe, 1992, Caplan et al., 1992). To test for interactions between eukaryotic dnaJ homologs and hsp70 family members, YDJ1p and a cytosolic hsp70 homolog of S. cerevisiae , SSA1p, were purified. The influence of YDJ1p on SSA1p ATPase activity and polypeptide binding were then tested. The results of such experiments are reported below.

Results and Discussion

The proteins used in this study were purified from two sources (see Cyr et al., 1992 for details). YDJ1p was overexpressed and purified from E. coli . SSA1p was purified from S. cerevisiae strain MW141 which is genetically engineered to constitutively express SSA1 and not the other three SSA genes that encode cytosolic hsp70 homologs (Deshaies et al., 1988).

To test for interactions between YDJ1p and hsp70 the influence of YDJ1p on SSA1p ATPase activity was determined. SSA1p hydrolyzed ATP at 2-5 nmol/ mg/ min. YDJ1p exhibited no detectable ATPase activity. However, addition of YDJ1p to reaction mixtures stimulated SSA1p ATPase $9.6 \pm 1$ fold (n=6, $\pm$.standard error).

Since ATP hydrolysis by hsp70 has been correlated with conformational changes in hsp70 and release of bound polypeptide substrates (Liberek et al., 1991b), the influence of YDJ1p on polypeptide binding to SSA1p was next determined. For binding experiments, a peptide, F1β1-51, corresponding to the first 51 amino acids of the mitochondrial F1-ATPase β-subunit precursor was employed.   This peptide resembles a nascent polypeptide chain as it emerges from a ribosome and is specifically recognized by the mitochondrial import apparatus (Cyr and Douglas, 1991). $^{125}I$ -F1β 1-51+3 and SSA1p were incubated in buffer containing 50mM hepes, 7.4, 50 mM NaCl, 0.1 mM EDTA and 0.4% BSA for 20 min at 30 °C.  Assay of stable complex formation between peptide and SSA1p was accomplished by native gel electrophoresis (Cyr et al., 1992 ).  Quantitation of such an assay is shown in Figure 1. YDJ1p or ATP alone had little effect on complex-formation.  However, the combination of YDJ1p and ATP reduced complex formation by 60%.   A 100  fold molar excess of unlabeled peptide blocked binding of $^{125}I$-peptide, indicating complex formation was specific.

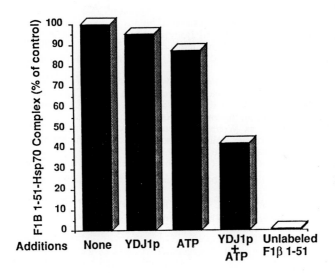

Figure 1.   YDJ1p and ATP reduce F1β  peptide-SSA1p complex formation.

These data provide the first example of the manner in which polypeptide binding by a eukaryotic hsp70 family member is regulated by an associated protein. Tight presequence peptide binding in the presence of ATP implies that interaction of hsp70 with eukaryotic dnaJ homologs at different locations within the cell provides a trigger for release of hsp70 bound polypeptides. Regulation of hsp70 by dnaJ homologs could be a mechanism for coupling the release of polypeptide substrates to their entrance into protein folding or transport pathways.

# References

Ang, D , Liberek, K , Skowyra, D , Zylicz, M and Georgopoulos, C (1991). Biological role and regulation of the universally conserved heat shock proteins. J Biol Chem **266**: 24233-24236.

Atencio, D P and Yaffe, M P (1992). MAS5, a yeast homolog of dnaJ involved in mitochondrial protein import. Mol Cell Biol **12**: 283-291.

Bork, P , Sander, C and Valencia, A (1992). A module of the dnaJ heat sock proteins found in malaria parasites. TIBS **17**: 29.2.

Beckman, R P , Mizzen, L A and Welch, W J (1990) Interaction of hsp70 with newly synthesized proteins: Implications for protein folding and assembly. Science **248**:850-854.

Caplan, A J and Douglas, M G (1991). Characterization of YDJ1: A yeast homologue of the E.coli dnaJ gene. J Cell Biol **114**:609-622.

Caplan, A J , Tsai, J , Casey, P J and Douglas, M G (1992). Farnesylation of YDJ1p is required for function at elevated temperatures in S.cerevisiae. J Biol Chem In press.

Cyr, D M , Lu, X L and Douglas M G (1992) Regulation of eukaryotic hsp70 by a dnaJ homolog (submitted).

Cyr, D M and Douglas M G (1991) Early events in the transport of proteins into mitochondria: Import competition by a mitochondrial presequence J Biol Chem **266**:21700-21708.

Chirico, W J , Waters, M G and Blobel. 1988. 70K heat shock related proteins stimulate protein translocation into microsomes. Nature **332**:805-810.

Deshaies, R J , B D Koch, M Werner-Washburne, E A Craig, and R Schekman. (1988). A subfamily of stress proteins facilitates translocation of secretory and mitochondrial precursor polypeptides. Nature **332**:800-805.

Gamer, J , Bujard, H and Bukau, B (1992). Physical interactions between heat shock proteins dnaK, dnaJ and grpE and the bacterial heat shock transcription factor σ32. Cell **69**:833-842.

Gething, M-J, and Sambrook, J (1992). Protein folding in the cell. Nature **355**:33-45.

Langer, T , Lu, C , Echols, H ,Flanagan, J , Hayer, M K and Hartl, F-U (1992). Successive action of dnaK, dnaJ and GroEL along the pathway of chaperone-mediated protein folding. Nature **356**:6883-6889.

Liberek, K , Marszalek, J , Ang, D , Georgopoulos, C and Zylicz, M (1991a) Escherichia coli dnaJ and grpE heat shock proteins jointly stimulate ATPase acitivity of dnaK. Proc Nat Acad Sci (U.S.A.). **88**:2874-2878.

Liberek,K , Skowyra,D , Zylic, M, Johnson, C and Georgopoulos (1991b) The E. coli DnaK chaperone, the 70-Kda heat shoch protein eukaryotic equivalent, changes conformation upon ATP hydrolysis, thus triggering its dissociation from bound traget protein. J Biol Chem. **266**:14491-96.

Wickner, S , Hoskins, J and McKenny, K (1991) Function of dnaJ and dnaK as chaperones in origin-specific DNA binding by RepA. Nature **350**:165-167

# HUMAN BIFUNCTIONAL ENZYME AND ITS IMPORT INTO PEROXISOMES

G. L. Chen, M. C. McGuinness,
G. Hoefler and A. B. Moser
Kennedy Krieger Institute,
Department of Neurology,
School of Mediciene,
Johns Hopkins University

Peroxisomes are present in almost all mammalian cells other than mature erythrocytes, but their size and number vary considerably. In the nervous system they are more abundant during development, specifically in the period of active myelination, and the early postnatal period. Peroxisomes contain more than 40 enzymes and play an important role in intermediary metabolism. In addition to the production of $H_2O_2$ (by oxidase) and its degradation (by catalase), peroxisomes are involved in a large variety of catabolic (oxidation of fatty acids, ethanol, L-pipecolic acid, polyamine and purines), and anabolic reactions (biosynthesis of bile acids, cholesterol and ether-linked phospholipids, such as plasmalogens, an essential myelin constituent) (Lazarow and Moser, 1989).

Patients with Zellweger syndrome lack morphologicl recognizable peroxisomes or greatly decrease in numbers. Peroxisomal deficiency leads to a wide range of severe phenotypic defects and show multiple biochemical abnormalities (Moser, 1992). Newborn infants with classic Zellweger syndrome rarely live more than a few months. Similar biochemical abnormalities in patients with milder clinical manifestations have been described. Of particular interest are the recently described patients characterized by an impairment in only a single peroxisomal enzyme (Theda et al, 1991). In pseudo-Zellweger syndrome patients, bifunctional enzyme proteins are either absent or dysfunctional.

Most eukaryotic cells have two fatty acid $\beta$-oxidation systems, one in mitochondria, and another in peroxisomes. The $\beta$-oxidation in peroxisomes requires four enzymes (fatty acyl-CoA synthetase, acyl-CoA oxidase, enoyl-CoA hydratase; 3-hydroxyacyl-CoA dehydrogenase (bifunctional enzyme), and 3-ketoacyl-CoA thiolase. They are entirely distinct from their mitochondrial counterparts. Synthesis of bile acids from cholesterol requires a series of reactions involving modifications to the steroid nucleus. $\beta$-oxidation of the side chain commences with a $3\alpha,7\alpha,12\alpha$-trihydroxy-5$\beta$-cholestanoyl-CoA oxidase which is

NATO ASI Series, Vol. H 74
Molecular Mechanisms of Membrane Traffic
Edited by D. J. Morré, K. E. Howell, and J. J. M. Bergeron
© Springer-Verlag Berlin Heidelberg 1993

distinct from its counterpart for very long chain fatty acids. However, the same thiolase and bifunctional enzyme are responsible for the oxidation of very long chain fatty acids and the side chain of cholesterol (Clayton, 1991). Inborn errors of metabolism with the bifunctional enzyme deficiency will lead to accumulation of bile acid intermediates as well as very long chain fatty acids. Patients with isolated peroxisomal β-oxidation enzyme deficiency therefore offer ideal subjects with which to study in detail the mechanism of peroxisomal protein import. In this report we present the results of cDNA cloning of bifunctional enzyme and its import into human cells.

## *Molecular biology of human bifunctional enzyme genes - cDNA*

By screening $5 \times 10^5$ recombinant phage of a human cDNA library with rat bifunctional enzyme cDNA as a probe, we obtained four positive clones (1/120,000). Clone HB5 contained a 2.3 kilobase insert (Chen et al, 1991). Sequence analysis of HB5 revealed a 669 bp ORF (Open Reading Frame) that showed 90% nucleotide homology to the 3' end coding region of the rat bifunctional enzyme. This sequence verified the identity of human bifunctional enzyme clones and was used to probe the human cDNA library. Four positive clones were isolated. The longest clone, BC93 of 3.4 Kb, covers 90% of human bifunctional enzyme including the 3' noncoding region and 1.8 kb ORF. Clone BC93 has a nucleotide sequence corresponding to the rat bifunctional enzyme cDNA extending from +400 to 3390 bp (400-1788 ORF). Our preliminary sequence data cover the entire pBC93 clone and show 77 % nucleotide homology in the coding region and <50 % homology in non-coding region. The deduced amino acid sequence is 75 % identical with the rat bifunctional enzyme protein sequence.

## In vitro import assay

The schematic diagram of in vitro assay for protein import into human peroxisomes is shown in Fig. 1. [$^{35}$S]-labeled bifunctional enzyme was synthesized by in vitro transcription and translation from normal or mutated cDNA. The labeled bifunctional enzyme protein was then mixed with the postnuclear supernatant fraction of cells in the presence of ATP. Part of the mixture was treated with protease under conditions that digest the nonspecific binding protein, but maintain the integrity of organelle. The optimum concentration of protease is confirmed by the complete degradation of cytosolic labeled bifunctional enzyme, but not the bifunctional enzyme already inside peroxisomes.

The entire mixture was then subjected to subcellular fractionation in Nycodenz gradient (Chen et al, 1991). The localization of peroxisomes was determined by the marker enzyme assay. Finally, by SDS-PAGE and fluorography analysis, the import of bifunctional enzymes was established by its colocalization with catalase.

Fig. 1   In vitro assay for protein import into peroxisomes

1. *In vitro import*
   *in vitro transcription*
   *in vitro translation [35S]*
   *PNS fraction*
   *Buffer (ATP)*
        ⇓
   *incubation 60 min, 26°C*

2. *Protease  protecyion assay*

|          | a | b | c |
|----------|---|---|---|
| Protease | - | + | + |
| Detergent| - | - | + |

        ⇓
   *incubation 15-30 min, 0°C*

3. *Subcellular fractination*
   *Nycodenz/sucrose gradient*
   *Centrifugation: 35min, 74kg*
   *Marker enzyme assay*

4. *Analysis*
   *SDS-PAGE*
   *Fluorography*

The bifunctional enzyme (HB5) was imported into peroxisomes of hepatoma cells as shown in Fig. 2, a & b. The 24 kDa bands cosedimented with catalase. Deletion of the last nine amino acids (HBH) abolished its ability to undergo import as shown in Fig. 2, c & d, indicating that Ser-Lys-Leu also served as targeting signal in human cells as in insect, plant, and rat (Gould et al, 1989). A similar result was found in human skin fibroblasts (data not shown). Recently a patient lacking bifunctional protein has been identified. Messenger RNA of normal size and amount was present in cultured cells, and peroxisomal targeting signal Ser-Lys-Leu was found in the cDNA. One possible mechanism for bifunctional protein deficiency is at peroxisomal import level. Mutations in internal region of bifunctional gene and defect of the import machinery may be responsible for the impairment of protein in the organelle. The in vitro import assay provides a specific means to answer these questions.

98

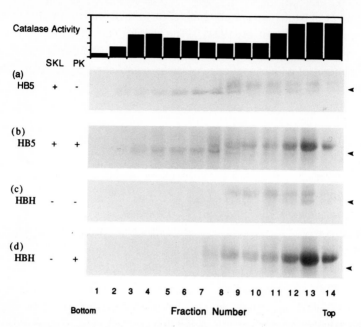

Catalase Activity

SKL PK

**(a)**
HB5  +  −

**(b)**
HB5  +  +

**(c)**
HBH  −  −

**(d)**
HBH  −  +

1  2  3  4  5  6  7  8  9  10  11  12  13  14

Bottom          Fraction Number          Top

Fig. 2 In vitro import of human bifunctional enzyme into human hepatoma cells.
[$^{35}$S]-labeled polypeptide from HB5 (a, b; with SKL) and HBH (c, d; without SKL) were incubated with HEP G2 cells and analysis as described in text. The expected 24 kDa protein bands are indicated.

## Literature Cited

Chen GL, Balfe A, Erwa W, Hoefler G, Gaertner J, Aikawa J, and Chen WW "Import of human bifunctional enzyme into peroxisome of human hepatoma cells in vitro" (1991) *Biochem and Biophy Res Comm* 178: 1084-1091

Clayton PT "Inborn errors of bile acid metabolism" (1991) *J Inher Metab Dis* 14:478-496

Gould SJ, Keller G, Hosken N, Wilkinson J, and Subramani S, (1989) "A conserved tripeptidesort proteins to peroxisomes" J Cell biol 108, 1657-1664

Lazarow P, and Moser HW, (1989) Disorders of Peroxisome Biogenesis. In: The Metabolic Basis of Inherited Disease. Scriver CR, Beaudet AL, Sly WS, Valle D (Eds). New York: McGraw Hill, Sixth Edition, Chapter 57, pp. 1479-1509.

Moser HW, (1992) "Peroxisomal disorders" In The Molecular and Genetic Basis of Neurological Disease. Rosenberg RN, Prusiner SB, DiMauro S, Barchi RL, and Kunkel LM, (Eds). Stoneham, MA: Butterworth Publishers, in press.

Theda C, McGuinness MC, and Moser HW, (1991) "Peroxisomal bifunctional enzyme deficiency - Biochemical and clinical findings. *Pediatr Res* 29:364a.

# TRANSPORT OF MICROINJECTED PROTEINS INTO THE PEROXISOMES OF MAMMALIAN CELLS.

P.A. Walton
Dept. of Anatomy
McGill University
Montreal, Quebec
Canada H3A 2B2

S.J. Gould
Kennedy Institute
Johns Hopkins Univ.
Baltimore, Maryland
U.S.A. 21205

S. Subramani
Dept. of Biology
University of California
La Jolla, California
U.S.A. 92093-0322

Peroxisomes are single-membrane-bound organelles found in virtually all eukaryotic cells. Proteins destined for the peroxisomes are synthesized on free polysomes in the cytoplasm and are transported into the peroxisome post-translationally (Fujiki et al., 1984). Although at least two proteins, thiolase and acyl-CoA oxidase, undergo proteolytic processing after transport (Fujiki et al., 1985), most proteins are synthesized at their mature size. Import of proteins into the peroxisome is dependent upon, among other factors, the presence of a peroxisomal targeting signal on the newly synthesized protein. A C-terminal tripeptide peroxisomal targeting signal (PTS) with the sequence serine-lysine-leucine-COOH (or a conservative variant) was identified initially in firefly luciferase (Gould et al., 1987), and has subsequently been found in most peroxisomal proteins (reviewed by Subramani, 1992).

Initially we microinjected two proteins: luciferase, and albumin conjugated to a peptide ending in the sequence ser-lys-leu (albumin-SKL) into the cytoplasm of BALB/c 3T3 cells grown in tissue culture. Following microinjection, incubation of the cells at 37°C for 16 hours resulted in import of these exogenous proteins into vesicles containing endogenous peroxisomal proteins (Figure 1). The translocation was both time and temperature dependent. No import was observed in cells incubated at 4°C. Digitonin-permeabilization experiments confirmed that the import of the albumin-SKL was into the matrix of the peroxisomes, and not merely surface aggregation. The import could be inhibited by coinjection of synthetic peptides bearing various peroxisomal targeting signal motifs. These proteins could be imported into peroxisomes in normal human fibroblast cell lines, but not in cell lines derived from patients with Zellweger syndrome. These results demonstrated that microinjection of peroxisomal proteins yielded an authentic *in vivo* system with which to study peroxisomal transport. Furthermore, these results confirmed that the process of peroxisomal transport does not involve irreversible modification of the protein; artificial hybrid substrates can be transported and employed as tools to study peroxisomal transport; and that the defect in Zellweger syndrome is indeed the inability to transport proteins containing the ser-lys-leu targeting signal into the peroxisomal lumen.

NATO ASI Series, Vol. H 74
Molecular Mechanisms of Membrane Traffic
Edited by D. J. Morré, K. E. Howell, and J. J. M. Bergeron
© Springer-Verlag Berlin Heidelberg 1993

Microinjection of a purified peroxisomal protein, alcohol oxidase, from *Pichia pastoris* into tissue culture cells resulted in import of this protein into vesicular structures. Transport was into membrane-enclosed vesicles as judged by digitonin-permeabilization experiments. The transport was time and temperature dependent. Vesicles containing alcohol oxidase could be detected as long as six days after injection. Coinjection of synthetic peptides containing a consensus carboxy-terminal tripeptide peroxisomal targeting signal resulted in abolition of alcohol oxidase transport into vesicles in all cell lines examined. Double-label experiments indicated that, although some of the alcohol oxidase was imported into vesicles containing other peroxisomal proteins, the bulk of the alcohol oxidase did not appear to be

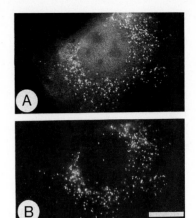

Figure 1. Colocalization of micro-injected albumin-SKL (A) with the endogenous peroxisomal protein CAT-PMP-20 (B). Bar, 10μm.

transported to pre-existing peroxisomes. While the inhibition of transport of alcohol oxidase by peptides containing the peroxisomal targeting signal suggests a competition for some limiting component of the machinery involved in the sorting of proteins into peroxisomes, the organelles into which the majority of the protein is targeted appear to be unusual and distinct from endogenous peroxisomes. Microinjected alcohol oxidase was transported into vesicles in normal fibroblasts and also in cell lines derived from patients with Zellweger syndrome, which are unable to transport proteins containing the ser-lys-leu-COOH peroxisomal targeting signal into peroxisomes. The intriguing result that peptides containing the tripeptide PTS inhibit transport of alcohol oxidase into peroxisome-like vesicles would then imply that although alcohol oxidase and luciferase contain different peroxisomal targeting signals, they share a common downstream component of the recognition and/or translocation machinery.

Fujiki, Y., R.A. Rachubinski, and P.B. Lazarow. 1984. Synthesis of a major integral membrane polypeptide of rat liver peroxisomes on free polysomes. *Proc. Nat. Acad. Sci. USA.* 81,7127-7131.

Fujiki, Y., R.A. Rachubinski, R.M. Mortensen, and P.B. Lazarow. 1985. Synthesis of 3-ketoacyl-CoA thiolase of rat liver peroxisomes on free polysomes as a larger precursor. Induction of thiolase mRNA activity by clofibrate. *Biochem. J.* 226,697-704.

Gould, S.J., G.-A. Keller, and S. Subramani. 1987. Identification of a peroxisomal targeting signal at the carboxy terminus of firefly luciferase. *J. Cell Biol.* 105,2923-2931.

Subramani, S (1992) Targeting of Proteins into the Peroxisomal Matrix. *J. Membrane Biol.* 125, 99-106.

# Isolation And Characterization Of A Functionally Active Protein Translocation Apparatus From Chloroplast Envelopes.

K. Waegemann, J. Soll
Botanisches Institut, Universität Kiel, Olshausenstraße 40,
D-2300 Kiel, Germany

Chloroplast structure and function depends vitally on the import of nuclear coded and cytosolically synthesized polypeptide constituents (de Boer and Weisbeek 1991). Proteins of the outer and inner envelope from chloroplasts collaborate to form an import machinery which is responsible for the specific recognition of chloroplast destined precursor proteins and their translocation through the two membrane barrier. Outer envelope membrane vesicles are purified from pea chloroplasts in a right side-out orientation, i.e. like in the intact organelle (Waegemann et al. 1992). Precursor proteins are bound to the membrane vesicles in an ATP, receptor and transitpeptide dependent manner (Waegemann and Soll 1991, Soll and Waegemann 1992). The translocation process of a precursor proceeds via distinct steps which can be detected <u>in vitro</u> as translocation intermediates, named deg 1-4 in chloroplasts (Fig 1). The outer envelope localized import apparatus yields deg 1 and 2 while deg 3 and 4 are translocation intermediates which occur in connection with the inner envelope import machinery (Fig 1).

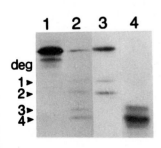

Fig 1) Localisation of pSSu translocation intermediates. pSSu binds to intact chloroplasts in the presence of 100 $\mu$M ATP (lane 1). Protease treatment results in the occurence of deg 1-4 (lane 2). Envelope membranes isolated from these plastids show that deg 3 and 4 are localized in the inner envelope (lane 4). Outer envelopes incubated with pSSu as above yield deg 1 and 2 after protease treatment (lane 3).

NATO ASI Series, Vol. H 74
Molecular Mechanisms of Membrane Traffic
Edited by D. J. Morré, K. E. Howell, and J. J. M. Bergeron
© Springer-Verlag Berlin Heidelberg 1993

Using a solubilisation protocol of outer envelope membranes as outlined in Fig 2A (Kiebler et al. 1990) a protein complex could be enriched by sucrose density centrifugation which contained bound precursor protein (import complex I) (Fig 2B). Detection of this import complex I depended on the presence of ATP, a transit sequence and protease sensitive components, e.g. receptor molecules during the incubation of the membrane vesicle with the precursor (Waegemann and Soll 1991). The protein composition of import complex I is distinctively different from total outer envelope membrane proteins and shows an enrichment of a number of proteins as judged by gel electrophoresis followed by silver staining (Fig 2C). Using immunological technics we have identified four outer envelope proteins (OEP) of 86, 70, 75, and 34 kDa (Fig 2D). The 70 kDa polypeptide crossreacts with an antiserum against heatshock protein 70 (hsp 70). Furthermore it is possible to immunoprecipitate the precursor protein by antibodies against hsp 70 (Waegemann and Soll 1991) indicating a close interaction between these two components. Envelope localized hsc 70 might thus act as a chaperone to facilitate the translocation of a precursor protein by interacting with the transport competent, i.e. partially folded, conformation of the polypeptide chain (Marshall et al. 1990, von Heijne and Nishikawa 1991, Waegemann and Soll 1991). The role and function of other import complex I constituents remains to be established.

After solubilisation of outer envelope membranes and centrifugation through sucrose-gradients the fractions which contained import complex I were pooled and subsequently incubated with precursor protein (see Fig. 2A, right panel). This isolated membrane complex recognized specifically the precursor polypeptide in relation to the mature form and interacted with it in an ATP dependent manner. Following protease treatment translocation intermediates deg 1 and 2, identical to those described above (Fig 1) for the organellar system were detected. Transit peptide dependent interaction of the isolated import complex with the precursor points to the

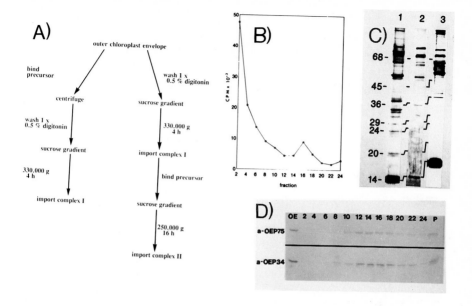

Fig 2) Characterization of the protein import complex from chloroplast outer envelopes. A) Solubilisation and isolation protocol of import complex I and II. B) Distribution of radioactivity in the sucrose density gradient as determined by liquid scintillation counting (line drawing). C) Polypeptide composition of outer chloroplast envelope (lane 1), complex I (lane 2) and complex II (lane 3). D) Immunoblot analysis of fractions obtained from a sucrose density gradient to enrich complex I using antisera against OEP 75 and OEP 34. Numbers on top indicate fraction numbers.

presence of a receptor in this complex. Protease sensitive components in import complex II are OEP 86 (Fig 3) and OEP 34 (see Fig 2C, D), however, direct evidence for their function as receptor is still missing. We were unable to detect a putative receptor of 30 kDa molecular (Pain et al. 1988, Schnell et al. 1990) in import complex II and conclude from this and from other results that the receptor for chloroplastic protein import is not yet conclusively identified (Flügge et al. 1991). The presence of hsc 70 in import complex II indicates again the involvement of this protein in the translocation mechanism (Fig 3). The protein composition of import complex I and II is similar (Fig 2C). To our knowledge the results described above demonstrate for the first time that it is possible to isolate a protein translocation apparatus as a functional active unit from chloroplasts. This will enable us to study the mechanism of protein translocation in greater detail.

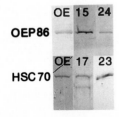

Fig 3) OEP 86 and a hsc 70 homologue are present in import complex II. Outer envelope membranes (OE) and samples of the sucrose density gradient (figures indicate fraction numbers) were tested using the respective antiserum.

## Acknowledgement

This work was supported by the Deutsche Forschungsgemein-schaft.

## References

de Boer, A D & Weisbeek, P J (1992) Biochim Biophys Acta, in press.

Flügge, U -I , Weber, A , Fischer, K , Lottspeich, F , Eckerskorn, Ch , Waegemann, K & Soll, J (1991) Nature 353, 364-367.

Kiebler, M , Pfaller, R , Söllner, T , Griffiths, G , Hartmann, H , Pfanner, N & Neupert, W (1990) Nature 348, 610-616.

Marshall, J S , DeRocher, A E , Keegstra, K & Vierling, E (1990) Proc Natl Acad Sci USA 87, 374-378.

Pain, D , Kanwar, Y S & Blobel, G (1988) Nature 331, 232-237.

Schnell, D J , Blobel, G & Pain, D (1990) J Cell Biol 111, 1825-1838.

Soll, J & Waegemann, K (1992) Plant J. 2, 253-256.

v.Heijne, G & Nishikawa, K (1991) FEBS Lett 278, 1-3.

Waegemann, K & Soll, J (1991) Plant J 1, 149-158.

Waegemann, K , Eichacker, S & Soll, J (1992) Planta 187, 89-94.

# Protein Insertion Into The Outer Mitochondrial Membrane

D.G. Millar and G.C. Shore
Department of Biochemistry
McGill University
Montreal, Quebec, H3G 1Y6
Canada

Biogenesis of the outer mitochondrial membrane (OMM) requires that cytoplasmically synthesized proteins be targeted to the organelle and selectively inserted into the outer membrane. Unlike most proteins of the mitochondrial matrix, inner mitochondrial membrane (IMM), and the intermembrane space (IMS), outer membrane proteins do not have cleavable N-terminal signal sequences and membrane insertion is independent of a transmembrane potential. In order to be correctly sorted, the protein must signal termination of translocation across the OMM and become stably integrated into the hydrophobic core of the phospholipid bilayer. Hydrophobic proteins of the IMM or matrix must be allowed to slip freely through the OMM. Protein receptors and components of the insertion machinery which interact specifically with OMM proteins remain largely uncharacterized. We are interested in the extent of overlap between the pathway of OMM insertion and import to the matrix or IMM and how this relates to proper sorting.

Separate membrane insertion machineries exist in both the OMM and IMM which are not permanently coupled and can independently translocate proteins across each membrane. This was demonstrated with the IMS protein cytochrome c heme lyase which is released into the IMS after crossing the OMM and by the observation that translocation intermediates of matrix targeted proteins arrested at the IMM can still cross the OMM to become partially exposed to the IMS (reviewed by Glick et al., 1991; Pfanner et al., 1992). In addition, the IMM can import matrix targeted proteins directly, bypassing the OMM and contact sites. Furthermore, a hybrid OMM targeted protein, when given access to the IMS by rupturing the OMM with osmotic shock, is inserted into the IMM

NATO ASI Series, Vol. H 74
Molecular Mechanisms of Membrane Traffic
Edited by D. J. Morré, K. E. Howell, and J. J. M. Bergeron
© Springer-Verlag Berlin Heidelberg 1993

(Li and Shore, 1992). These results suggest that sorting may be achieved by the action of dynamic translocation systems which can function in tandem or independently in response to different protein sorting signals.

To address this question, we have employed chimeric proteins of mouse cytosolic dihydrofolate reductase (DHFR) fused to the N-terminal targeting sequences of either the matrix enzyme pre-ornithine carbamyl transferase, termed pO-DHFR, or the yeast 70kDa outer mitochondrial membrane protein OMM70, termed pOMD29. Excess pO-DHFR effectively competed for import of in vitro translated pO-DHFR into the matrix of purified rat heart mitochondria but not for outer membrane insertion of pOMD29 (McBride et al., 1992). Thus it appears that saturation of some point on the matrix import pathway is compatible with continued OMM insertion. We are currently assessing whether or not translocation of pO-DHFR across the OMM and insertion of POMD29 share a common OMM translocation apparatus.

Inserted pOMD29 and an import intermediate of pO-DHFR arrested at the OMM were used to examine putative components of the insertion machinery by chemical cross-linking. Several cross-linked products are observed which do not form if the mitochondria are pre-treated with protease (Millar and Shore, unpublished data). Work is in progress to assess whether OMM proteins in contact with translocation intermediates are general or exclusive for the differently sorted proteins.

Glick B, Wachter C, Schatz G (1991) Protein import into mitochondria: two systems acting in tandem? Trends Cell Biol 1:99-103

Li J-M, Shore GC (1992) Protein sorting between mitochondrial outer and inner membranes. Insertion of an outer membrane protein into the inner membrane. Biochim Biophys Acta 1106:233-241

McBride HM, Millar DG, Li J-M, Shore GC (1992) A signal-anchor sequence selective for the mitochondrial outer membrane. Submitted for publication

Pfanner N, Rassow J, van der Klei IJ, Neupert W (1992) A dynamic model of the mitochondrial import machinery. Cell 68:999-1002

# THE MEMBRANE-BOUND 95 KDA SUBUNIT OF THE YEAST VACUOLAR PROTON-PUMPING ATPase IS REQUIRED FOR ENZYME ASSEMBLY AND ACTIVITY

[1]M.F. Manolson, [1]D. Proteau, [1]R.A. Preston, [1]M.E. Colosimo, [2]B.T. Roberts, [2]M.A. Hoyt, and [1]E.W. Jones, [1]Dept. of Biol. Sci., Carnegie Mellon Univ., Pittsburgh, PA, [2]Dept. of Biol., The Johns Hopkins Univ., Baltimore, MD

Yeast vacuoles, like mammalian lysosomes, are maintained at an acidic pH by a vacuolar-type proton-pumping ATPase (V-ATPase). A genetic screen using the pH sensitive fluorescence of 6-carboxyfluorescein diacetate was developed to identity yeast defective in vacuolar acidification (Preston *et al.*, 1992). Yeast bearing the *vph1-1* (Vacuolar pH1-1) mutation have neutral vacuoles *in vivo* (Preston *et al.*, 1989) and have no detectable bafilomycin-sensitive ATPase activity or ATP dependent proton-pumping associated with purified vacuoles. The peripherally bound nucleotide-binding subunits (Vma1p and Vma2p) are present in wild type levels in yeast whole cell extract yet are not associated with the vacuolar membrane. The *VPH1* gene was cloned by screening a λgt11 expression library with antibodies directed against a 95 kDa vacuolar integral membrane protein and independently cloned by complementation of the *vph1-1* mutation. Disruption of the gene revealed that the *VPH1* gene product is required for V-ATPase assembly and vacuolar acidification but is not essential for cell viability or for targeting and maturation of vacuolar proteases. Through cell fractionation and immunobloting, the Vph1p antigen was shown to co-purify with alkaline phosphatase activity, a specific marker for vacuolar membranes. Furthermore, the Vph1p antigen was enriched and co-purified with vacuolar $H^+$-ATPase activity. The inability of alkaline $Na_2CO_3$

NATO ASI Series, Vol. H 74
Molecular Mechanisms of Membrane Traffic
Edited by D. J. Morré, K. E. Howell, and J. J. M. Bergeron
© Springer-Verlag Berlin Heidelberg 1993

extraction to remove the Vph1p antigen from the membrane suggest that the *VPH1* gene product is an integral membrane protein. The *VPH1* gene was sequenced to completion and found to encode a putative polypeptide of 840 amino acids with a predicted molecular mass of 95.4 kDa. Computer algorithms predict 6 to 8 membrane spanning regions in the Carboxyl-terminus half of the polypeptide. The *VPH1* gene product has over its entire open reading frame 42% identity to the 116-kDa polypeptide of the rat clathrin-coated vesicles/synaptic vesicle proton pump, 42% identity to the TJ6 mouse immune suppressor factor, 42% identity to the *C. elegan* proton pump homologue and 54% identity to the putative polypeptide encoded by the yeast gene *STV1* (Similar To VPH1). *STV1* was identified as an open reading frame next to the *BUB2* gene. The details of this work will be presented in Manolson *et al.*, 1992. Disruption of the *STV1* gene did not result in any observable vacuolar acidification defect. Over expression of the *STV1* gene did partially complement phenotypes resulting from the disruption of the *VPH1* gene suggesting that Vph1p and Stv1p are functional homologs. Work is in progress to determine whether Stv1p and Vph1p are differentially expressed subunits of the same V-ATPase enzyme or equivalent subunits for specific V-ATPases located on different organelles.

REFERENCES

Manolson MF, Proteau D, Preston RA, Stenbit A, Roberts BT, Hoyt MA, Preuss D, Mulholland J, Botstein D, Jones EW (1992) The *VPH1* gene encodes a 95-kDa integral membrane polypeptide required for *in vivo* assembly and activity of the yeast vacuolar H$^+$-ATPase. J. Biol. Chem. in press

Preston RA, Murphy RF, Jones EW, (1989) Assay of vacuolar pH in yeast and identification of acidification-defective mutants. Proc. Natu. Acad. Sci. U.S.A. 86:7027-7031.

Preston RA, Reinagel PS, Jones EW, (1992) Genes required for vacuolar acidity in *Saccharomyces cerevisiae*. Genetics, 131:551-558.

# THE TRAFFIC OF MOLECULES ACROSS THE TONOPLAST OF PLANT CELLS

B. P. Marin

Laboratoire de Biotechnologie : Physiologie et Métabolisme Cellulaires

ORSTOM

911, Avenue Agropolis

34.032-Montpellier-Cedex 1

France

Studies on the compartmentation of alkaloids in a cell suspension of *Catharanthus roseus* demonstrate that, while unspecific ion trapping in the vacuolar compartment occurs for such such molecules, other accumulation mechanisms also exist in cells. In any cases, a protonmotive force is involved with $H^+$ recycling, in co-transport of solutes, and, with net active $H^+$ influx. To illustrate such a point of view, the latex of *Hevea brasiliensis* could be examplified.

## - The nature of *Hevea* latex

The latex of *Hevea brasiliensis* constitutes the fluid cytoplasm expelled from wounded laticifers. The lutoids form a polydisperse vacuo-lysosomal system limited by a single membrane which is of particular importance in the exchange of metabolites between the cytoplasm and the intravacuolar medium.

## - The compartmentation of citrate

Usually, in *Hevea* latex, citrate is compartmentalized inside the lutoids. Nevertheless, this molecule is an effector of several enzymes important in the rubber biosynthesis taken place in the cytoplasm. Any traffic of this molecule across the tonoplast is important for such biosynthesis.

## - The bioenergetic machinery of lutoids

Actually, several activities present at the interface beween the vacuo-lysosomal compartment and the cytoplasm are capable to transport actively protons. The proton fluxes, the cytoplasmic as well as the vacuolar pH changes, are shown to be under the control of at least two opposing proton translocating systems : a magnesium-dependent ATPase system and an electron-transporting system, an NADH-cytochrome-oxido-reductase. The first one catalyzes an influx of protons acidifying the internal space of vacuoles. The second one induces an efflux of protons collapsing the transtonoplastic pH gradient. Recently, a pyrophosphatase bound to the tonoplast is evidenced as a third proton pumping system.

In each case , by different methods of analysis, these proton pumps create a ΔpH lowering the intravacuolar pH by at least 0.5 pH unit and make the lutoid interior more positive by 60 mV, reversing the preexisting Δpsi.

NATO ASI Series, Vol. H 74
Molecular Mechanisms of Membrane Traffic
Edited by D. J. Morré, K. E. Howell, and J. J. M. Bergeron
© Springer-Verlag Berlin Heidelberg 1993

Methylamine uptake and different acridine and quinacrine dyes quenching confirm such acidification. Parallely, TPP$^+$ and TPMP$^+$ uptake and oxonol VI quenching demonstrate the generation of a $\Delta$psi (interior negative). Protonophores reversed all these movements of protons. Consequently, ATPase, PPase and NADH-cytochrome-oxidoreductase pump electrogenically the protons creating a $\Delta$psi.

The protonmotive force generated by such proton-pumping systems provides the driving force for the transport of most solutes accumulated inside the vacuolar compartment.

## - The mechanism of tonoplastic citrate translocator

From data reported on vacuolar vesicles and those described with native lutoids, it will be suggested the occurrence of an internal compartmentation of the vacuolar citrate. As no morphology is visible inside the lutoidic space, a kinetic process is proposed. In each case, magnesium is present in stoichiometric concentrations with citrate. Consequently, a part of vacuolar citrate could chelate this cation. Acid-base equilibrium calculations and associations of citrate with Mg$^{2+}$ predict an accumulation of the impermeant complex (citrate)$^{3-}$-Mg$^{2+}$. Parallely, a pool of permeant (citrate)$^{2-}$ is formed. Such conclusions come from the shape of the curves of citrate uptake vs pH but also the variation of K$_m$ vs pH. Kinetic simulations based on such an hypothesis are quite reconcilable with the experimental data.Such a model involves the working of citrate translocator antiportly with a proton, one molecule of magnesium being transferred from the cytoplasm to the internal lutoidal space. By this way, the dynamic nature of the exchange of citrate across the tonoplast will be finely controlled by the tonoplast energization level, related to the transtonoplast proton fluxes catalyzed by the ATPase system and the redox pump.

The vacuolar storage avoids the development of too high a concentration in the cytoplasm, which would cause a specific feedback inhibition of the enzymes involved in the rubber synthesis.

## - Conclusions

The plant cell is characterized by the compartmentation of its metabolism. The cytoplasm could be regarded as a very complicated network of several metabolic pathways of molecules involving the different cell organelles. Nevertheless, some important enzymatic reactions involved some cofactors not present in the cytoplasm. They could be accumulated in one cell compartment (the vacuole or an other acid compartment) and excluded from another one (where they play an important role). In most cases, in terms of cell dynamics, they are actively transported across the membranes limiting the different compartments involved. The compartmentalization is one of the essential mechanisms of metabolic regulation in plants. Parallely, in order to compartmentalize metabolism, each newly-synthesized enzyme must be delivered specifically to the organelles that require it. From these studies, it become possible to understand how a great lot of metabolites was transported across the tonoplast of plant cells.

# FUSICOCCIN BINDING AND INTERNALIZATION BY SOYBEAN PROTOPLASTS

M. A. Villanueva, R. Stout* and L. R. Griffing
Department of Biology
Texas A & M University
College Station, TX 77843-3258

## INTRODUCTION

Fusicoccin (FC) is a fungal toxin which binds to a receptor on the plasmalemma of higher plants causing hyperpolarization of the membrane potential and acidification of the apoplastic space (Feyerabend and Weiler 1989). In order to understand the fate of the toxin after its binding, we have developed an FC-containing probe which can be seen by light- or electron microscopy and can be quantified biochemically. The colloidal gold conjugate of bovine serum albumin covalently linked to FC (FC-BSA-Gold) binds to the plasmalemma of intact protoplasts and is partially internalized. Cells treated with FC-BSA-Gold are compared to control cells treated with the gold conjugate of BSA only (BSA-Gold), a marker for fluid-phase uptake. The amount and location of the gold label differ between FC-BSA-Gold and BSA-Gold treated protoplasts. When internalization of FC-BSA-Gold takes place, it may do so, in part, via receptor-mediated endocytosis. The technique described here will allow us to measure and visualize receptor-mediated endocytosis in plant cells.

FC-BSA (Pini *et al.* 1979) and BSA were conjugated to 10 nm colloidal gold as described (Slot and Geuze 1985). Protoplasts were suspended in solutions containing the gold probes at various concentrations and incubated 15 min at 25°C followed by two washes in wash medium (10 min, 25°C, 150 g). The protoplasts were fixed and silver-enhanced *en bloc* (Villanueva *et al.* J. Exp. Bot. 1992, submitted). Silver-enhanced gold was exam-

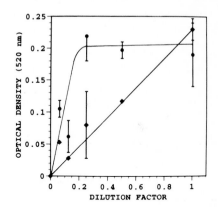

*Figure 1. FC-BSA-Gold ( ● ) and BSA-Gold ( ◆ ) association with soybean protoplasts after fixation and silver enhancement of the gold (read as optical density of silver-enhanced cell suspension, untreated cell turbidity subtracted). Dilution factor is the factor by which the initial gold colloid concentration ($0.5 A_{520}$) is multiplied to achieve the final incubation concentration (45 min total incubation time).*

## METHODS

Protoplasts from liquid cultured soybean cells (*Glycine max* L., line SB-1) were floated on a 10% (w/v) dextran cushion (Mersey *et al.* 1985) to remove cell debris and dead cells. Intact protoplasts were collected from the interface.

* Department of Biology, Montana State University, Bozeman, MT

NATO ASI Series, Vol. H 74
Molecular Mechanisms of Membrane Traffic
Edited by D. J. Morré, K. E. Howell, and J. J. M. Bergeron
© Springer-Verlag Berlin Heidelberg 1993

ined with epipolarization confocal laser scanning optics (CLSM, Zeiss). Nomarski and epipolarization optical sections were superimposed using NIH Image software on a Macintosh computer. For spectrophotometric analysis of probe binding and/or uptake, silver-enhanced cells were suspended in 80 % (v/v) glycerol and the $A_{520}$ of the suspension read (Taylor et al. The Plant Journal 1992, in preparation).

## RESULTS AND DISCUSSION

FC-BSA-Gold bound in a saturable manner to soybean protoplasts (Figure 1, ● ), indicating that it bound to its plasma membrane receptor. BSA-Gold, the control for fluid phase endocytosis (Villanueva et al. 1992), showed linear, non-saturable uptake with concentration (Figure 1,◆). When protoplasts were observed with Nomarski and CSLM epipolarization optics, the FC-BSA-Gold was found at and within the periphery (Figure 2d-f). In contrast, protoplasts treated with the same concentration of BSA-Gold showed labeling throughout the cell in organelles in the perinuclear region and transvacuolar strands (Figure 2 a-c). Neither BSA-Gold nor FC-BSA-Gold accumulated in the vacuole even after long incubation times (up to 4 h), whereas a gold conjugate of a polysaccharide elicitor entered the vacuole after 45 min incubation (Taylor et al. 1992). These different patterns of localization may indicate that different pathways are taken by different ligands, with possible down regulation of the elicitor receptors (Taylor et al. 1992). To distinguish between the fluid phase

and receptor-mediated components of uptake of FC-BSA-Gold in further experiments it will be necessary to block the fluid phase component during the initial binding stage.

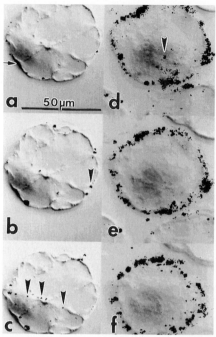

*Figure 2. Distribution of the BSA-Gold (a-c) and FC-BSA-Gold (d-f) in representative soybean protoplasts after 45 min. incubation with identical probe concentrations. The epipolarization signal is colored black. In one position, indicated by arrow in (a), the density of the nomarski image is also very dark, but does not indicate the presence of probe. Label is in the cell cortex, arrowhead in (b), and in a transvacuolar strand, arrowheads in (c). Perinuclear label is indicated by arrowhead in (d). The images are optical sections (a&d=25 µm into the cell, b&e=20µm, and c&f=15µm).*

## REFERENCES

Mersey, B. G., Griffing, L. R., Rennie, P. J. and Fowke, L. C. (1985) The isolation of coated vesicles from protoplasts of soybean. Planta 163:317-327.

Feyerabend, M. and Weiler, E.W. (1989) Photoaffinity labeling and partial purification of the putative plant receptor for the fungal elicitor toxin, fusicoccin. Planta 178:282-290.

Pini, C., Vicari, G., Ballio, A., Federico, R., Evidente, A. and Randazzo, G. (1979) Antibodies specific for fusicoccins. Plant Sci. Lett. 16:343-353.

Slot, J. W. and Geuze, H. J. (1985) A new method for preparing gold probes for multiple-labeling cytochemistry. Eur. J. Cell Biol. 38:87-93.

# TGN38 AND SMALL GTP BINDING PROTEINS ARE PART OF A MACROMOLECULAR COMPLEX IN THE TRANS-GOLGI NETWORK

J. R. Crosby, S. M. Jones, M.S. Ladinsky and K.E. Howell

Department of Cellular and Structural Biology, Box B-111

University of Colorado Medical School

4200 East 9th Ave.

Denver, CO 80262

USA

TGN38 is a type 1 integral membrane protein which is predominantly localized to the trans-Golgi network [Luzio et al., 1990]. Indirect immunofluorescence in NRK cells using antibodies against TGN38 shows localization to cisternal structures representative of the Golgi complex. Neither cytoplasmic vesicles nor the plasma membrane are labeled (Fig. 1).

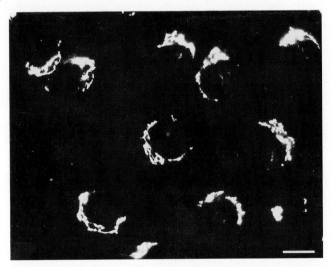

Fig. 1. Immunofluorescence localization of TGN38 in NRK cells using a mouse antisera raised against a peptide of the 16 amino-terminal amino acids of TGN38. Bar 3µm- 3,780x.

However, electron microscopy of NRK cryosections labelled with antibodies against TGN38 and protein-A colloidal gold show that localization of TGN38 is limited to the trans-most cisternae and the trans-Golgi network (Fig. 2).

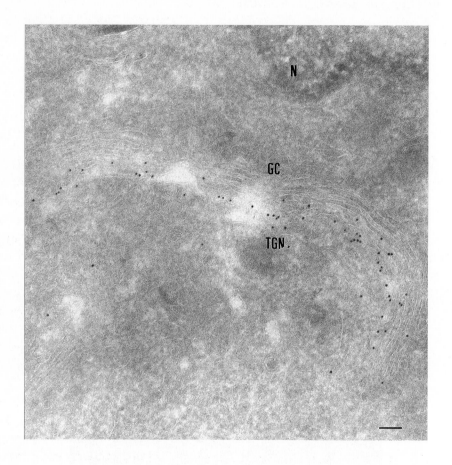

Fig. 2. Image from an ultrathin-cryosection immunolabeled with antibodies against TGN38 and protein-A colloidal gold (10 nm). This sample is unusual in that the majority of the label is in the trans-most cisterna. Nucleus ( N), Golgi cisternae (GC). trans-Golgi network (TGN) Bar 0.1 μm-79,750x.

However, in NRK cells TGN38 cycles to the plasma membrane and returns via the endocytic pathway within 30 min [Ladinsky & Howell, 1992]. Based on this data, we can hypothesize that TGN38 is involved in vesicle formation or targeting of vesicles from the trans-Golgi network to the plasma membrane. It therefore follows that the cytoplasmic domain of TGN38 would interact with small molecular weight GTP-binding proteins; molecules that are thought to determine the specificity of vesicular transport [Bourne, 1988; also see reviews, Balch, 1990; Gruenberg and Clague, 1992; Pfeffer, 1992].

To identify the proteins which associate with TGN38, mixed micelles were prepared by CHAPS solubilization of a carbonate washed stack Golgi fraction. Increasing stringencies of solubilization were used from 2.5 to 40 mM CHAPS. The resulting micelles were immunoprecipitated using either a pre-immune serum, antisera against TGN38 or the luminal domain of polymeric IgA receptor (pIgA-R). The pIgA-R was chosen as a control since it is a type 1 integral membrane protein present in Golgi membranes at concentrations similar to TGN38. However, it is not a resident Golgi protein, but is in transit to the plasma membrane [Sztul et al., 1983; Salamero et al., 1990]. The immunoprecipitates were resolved by SDS-PAGE, transfer to nitrocellulose and sequentially ligand blotted with [$\alpha^{32}$P]-GTP [Lapatina & Reep. 1986] and immunoblotted (Fig 3).

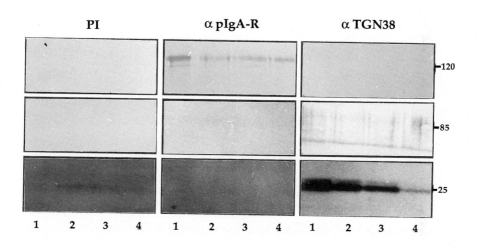

Fig. 3. Immunoprecipitations were carried out using either preimmune serum [left], antibodies against the pIgA-R [center] or TGN38 [right]. Top panels are immunoblots with αpIgA-R, middle panels immunoblots with αTGN38 and lower panels are [$\alpha^{32}$P]-GTP-ligand blots. Lanes: (1) 2.5 mM CHAPS, this is in the range for the Critical Micellar Concentrations for CHAPS; (2) 10 mM CHAPS, (3) 20 mM CHAPS and (4) 40 mM CHAPS.

Above the CMC for CHAPS, TGN38 antibodies immunoprecipitated a relatively constant amount of TGN38, yet no detectable pIgA-R [TGN38 is resolved as a 'smeary' doublet at ~85 kD, reflecting extensive glycosylation of the 38 kD protein backbone]. Notably, GTP-binding proteins at 25 kD were co-immunoprecipated at all detergent/lipid ratios, although the intensity of the bands on the ligand blot decreased with increasing stringency of the detergent extraction. At the highest stringency, a single band of GTP-binding proteins with reduced intensity remains associated with TGN38. Similarly, pIgA-R antibodies immunoprecipitated constant amount of pIgA-R without detectable amounts of TGN38. In contrast, GTP-binding proteins did not co-immunoprecipitate with pIgA-R antibodies. Pre-immune serum immunoprecipitated only background amounts of protein immunoreactive to either pIgA-R or TGN38 antibodies and background levels of GTP-binding proteins. This data demonstrates that above the CMC, CHAPS solubilization effectively produced small mixed detergent/lipid micelles, with TGN38 and the pIgA-R not being present in the same micelle. GTP-binding proteins are associated with the TGN38 micelles but are not associated with the micelles containing other integral membrane proteins. Since a high affinity complex exists between TGN38 and the GTP-binding proteins within CHAPS micelles, these molecules most likely form functional complexes *in vivo*.

The specificity of the interaction between TGN38 and small GTP-binding proteins in Golgi membranes has been confirmed by sedimentation in sucrose density gradients and fractionation on a Sephacryl S100 sizing column. The functions of these interactions are being studied using cell-free assays which reproduce the different steps of vesicular transport. Our goal is to understand the function of this complex in the process of membrane traffic.

REFERENCES
Balch WE (1990) Small GTP-binding proteins in vesicular transport. Trends Biochem. Sci. 15:473-477.
Bourne HR, Sanders DA, McCormick F (1990) The GTPase superfamily: a conserved switch for diverse cell functions. Nature 348:125-132.
Gruenberg J, Clague MG (1992) Regulation of intracellular membrane transport. Current Opin. in Cell Biol. 4:593-599.
Ladinsky MS, Howell KE (1992) The trans-Golgi network can be dissected structurally and functionally from the cisternae of the Golgi complex by brefeldin A. Eur. J. Cell Biol. 59:92-105
Lapatina E ,Reep B (1986) Specific binding of [a-$^{32}$P]GTP to cytosolic and membrane-bound proteins of human platelets correlates with the activation of phospolipase C. Proc. Nat. Acad. Sci. USA 84:2261-2265.
Luzio, P J, Brake B, Banting G, Howell KE,.Braghetta P, Stanley KK (1990) Identification, sequencing and expression of an integral membrane protein of the trans-Golgi network (TGN38). J. Biochem. 270:97-102.
Pfeffer S (1992) GTP-binding protiens in intracellular transport. Trends in Cell Biol. 2:41-46.
Salamero J Sztul ES Howell KE (1990) Exocytic transport vesicles generated in vitro from the trans-Golgi network carry secretory and plasma membane proteins. Proc. Natl. Acad. Sci. USA. 87:7717-7721.
Sztul ES Howell KE Palade GE (1983) Biogenesis of the polymeric IgA receptor in rat hepatocytes. J. Cell Biol. 100:1248-1254.
Sztul, E.S., P. Melançon,and K.E. Howell. 1992. Targetting and fusion in vesicular transport. Trends in Cell Biol. 2:381-386.

# β-COP, A COAT PROTEIN OF NONCLATHRIN-COATED VESICLES OF THE GOLGI COMPLEX, IS INVOLVED IN TRANSPORT OF VESICULAR STOMATITIS VIRUS GLYCOPROTEIN

R. Duden, B. Storrie, R. Pepperkok, J. Scheel, B. Joggerst-Thomalla, A. Sawyer, H. Horstmann, G. Griffiths and T. E. Kreis

European Molecular Biology Laboratory
Meyerhofstrasse 1
D-6900 Heidelberg
Germany

The Golgi complex is a polarized cytoplasmic organelle that is generally considered to be built up of at least three functionally distinct, membrane bounded subcompartments, the cis-Golgi network (CGN), the stacked Golgi cisternae and the trans-Golgi network (TGN) [Mellman and Simons, 1992]. It receives material from the endoplasmic reticulum (ER) and endosomes, transports cargo through its subcompartments and delivers it to the plasma membrane and endosomes. Vesicular carriers are believed to mediate the vectorial transport of this material from one membrane bounded subcompartment to the next. Recently, the idea that transport may also occur by transient tubular connections has also

NATO ASI Series, Vol. H 74
Molecular Mechanisms of Membrane Traffic
Edited by D. J. Morré, K. E. Howell, and J. J. M. Bergeron
© Springer-Verlag Berlin Heidelberg 1993

been discussed [Klausner et al., 1992; Kreis, 1992]; until now, however, the evidence for this mechanism of membrane transport is less convincing than for vesicular transport.

Two distinct classes of transport vesicles have been identified that are coated with different proteins on their cytoplasmic surfaces. The putative functions of these coat proteins include sorting of cargo, as well as regulation of budding, fusion  or targeting of the vesicles. Clathrin-coated vesicles are involved in signal mediated transport from the TGN to endosomes [Pearse and Robinson, 1990], and it has been postulated that nonclathrin-coated vesicles mediate transport through the Golgi complex (Rothman and Orci, 1992).

Some of the factors involved in regulating these processes of membrane transport have been identified and partially characterized by combining biochemical methods, cell-free assay systems and yeast genetics [Klausner et al., 1992; Kreis, 1992; Rothman and Orci, 1992; Schekman, 1985]. The coat proteins associated with the clathrin- and nonclathrin-coated vesicles play an essential function in the regulation of cytoplasmic membrane traffic, but their precise roles remain so far elusive.

COAT PROTEINS

The coats of clathrin- and nonclathrin-coated vesicles consist of specific complexes of proteins. Clathrin and adaptor proteins are associated with the former [Pearse and Robinson, 1990] and coat proteins (COPs) with the latter class of vesicles [Rothman and Orci, 1992]. The adaptor protein complex (HA1) of the TGN-derived clathrin-coated vesicles is a heterotetramer of $\gamma$-, $\beta'$-adaptin, a 47 kDa and a 20 kDa protein. The COP complex consists of $\alpha$-, $\beta$-, $\gamma$-COP (160, 110, 105 kDa, respectively), and a few smaller proteins of 30-40 kDa and ~20 kDa (ARF; Serafini et al., 1991). This COP complex can be purified using conventional biochemical methods [Waters et al., 1991]. Immunoprecipitation of rat liver extract or $^{35}$S-methionine labeled HeLa cytosol with a polyclonal antibody raised against a synthetic peptide of $\beta$-COP (EAGE; Duden et al., 1991) yields a native protein complex with virtually identical protein composition (Figure 1).

The proteins associated with the clathrin adaptor and the COP complex are remarkably similar in molecular weight [Duden et al., 1991] which led to the suggestion that clathrin- and COP-coated vesicles may perform similar functions. In fact homology between $\beta$-COP and $\beta$-adaptin has been demonstrated recently [Duden et al., 1991]. Whereas it is clear that clathrin-coated vesicles are involved in signal-mediated transport of cargo [Pearse and Robinson, 1990], it remains to be established whether the cargo of COP-coated vesicles is sorted or enters the carrier nonselectively. If the clathrin adaptor proteins and the COPs were indeed homologous, it would be tempting to assume that their functions are also homologous. Sequence comparison of the other "corresponding" subunits of the two vesicle coat complexes may help clarify this issue.

## β-COP IS PREFERENTIALLY ASSOCIATED WITH VESICLES CARRYING VSV-G TO THE GOLGI COMPLEX

The temperature sensitive glycoprotein of tsO45 mutant vesicular stomatitis virus (VSV-G) has been used as a model for analyzing transport of membrane components from the ER, through the Golgi complex to the cell surface. VSV-G can be arrested in the ER at nonpermisive temperature (39.5$^o$C), and it accumulates at 15$^o$C in the CGN [Schweizer et al., 1990] and at 20$^o$C in the TGN [Griffiths and Simons, 1986]. β-COP colocalizes with VSV-G in the CGN, the Golgi stack and the TGN by immunfluorescence labeling with specific antibodies, but not significantly with the viral glycoprotein that has been arrested in the ER (Figure 2).

Colocalization of VSV-G to putative carrier vesicles coated with the COP complex containing β-COP could be demonstrated by immunoelectron microscopy (Figure 3). The COP-coated vesicles carrying VSV-G are morphologically different from the clathrin-coated vesicles formed by the Golgi complex. They are smaller in diameter and have a thinner coat [Duden et al., 1991].

Transport of VSV-G from the ER to the Golgi complex occurs most likely in carrier vesicles. The appearance of vesicular structures containing β-COP can be visualized by immunfluorescence labeling with specific antibodies after a brief (5 min) shift to permissive temperature (31$^o$C). These probably newly formed vesicles contain VSV-G. Interestingly, β-COP colocalizes predominantly with the vesicles containing VSV-G that form from the ER and the CGN, but not with vesicles formed at the TGN

(Figure 4). Obviously, although VSV-G has been accumulated in the presence of cyclohexamide in the CGN and TGN at the respective temperatures, some vesicle formation may occur in the preceding compartments (ER, CGN) and may lead to a mixed population of vesicles of different origin carrying VSV-G.

CONCLUSIONS

Taken together, these data suggest that VSV-G, a model membrane protein following the constitutive membrane traffic route through the Golgi complex to the cell surface, is transported in COP-coated vesicles. It appears that β-COP is predominantly associated with vesicles transporting material to and through the Golgi complex and to a significantly smaller extent with the vesicles involved in transport from the TGN to the cell surface. This finding is consistent with the observation that β-COP is predominantly localized to the cis-side of the Golgi complex in rat pancreas cells [A. Oprins, R. Duden, T.E. Kreis, H.J. Geuze and J.W. Slot, submitted for publication]. If β-COP was not associated with the carrier vesicles involved in the constitutive transport of membrane components from the TGN to the cell surface, it could be speculated that closely related TGN-COPs (not TGN clathrin and adaptor complex)  exist, analogous to the related Golgi and plasma membrane clathrin adaptor complexes. Further work on the characterization of membrane associated coat proteins will help us understand better the general principles underlying the regulation of the formation of membrane transport intermediates.

## REFERENCES

Duden R, Griffiths G, Frank R, Argos P, Kreis TE (1991) β-COP, a 110kD protein associated with nonclathrin coated vesicles and cisternae of the Golgi complex shows homology to b-adaptin. Cell 46: 649-665

Duden R, Allan VJ, Kreis TE (1991) Involvement of β-COP in membrane traffic through the Golgi complex. Trends Cell Biol 1: 14-19

Griffiths G, Simons K (1986) The *trans* Golgi network: sorting at the exit site of the Golgi complex. Science 234: 438-443.

Ho WC, Allan VJ, van Meer G, Berger EG, Kreis TE (1989) Reclustering of scattered Golgi elements occurs along microtubules. Europ J Cell Biol 48: 250-263

Mellman I, Simons K (1992) The Golgi complex: in vitro veritas? Cell 8: 829-840

Klausner RD, Donaldson JG, Lippincott-Schwartz J (1992) Brefeldin A: insights into the control of membrane traffic and organelle structure. J Cell Biol 116: 1071-1080

Kreis TE (1992) Regulation of vesicular and tubular membrane traffic of the Golgi complex by coat proteins. Curr Op Cell Biol, in press

Pearse BMF, Robinson MS (1990) Clathrin, adaptors, and sorting. Annu Rev Cell Biol 6: 151-171

Pepperkok, R., Bré, M.-H., Davoust, J. and Kreis, T.E. (1990) Microtubules are stabilized in confluent epithelial cells but not in fibroblasts. J Cell Biol 111: 3003-3012

Rickard, J.E. and Kreis, T.E. (1990) Identification of a novel nucleotide-sensitive microtubule-binding protein in HeLa cells. J Cell Biol 110: 1623-1633

Rothman JE, Orci L (1992) Molecular dissection of the secretory pathway. Nature 355: 409-415

Scheel J, Kreis TE (1991) Motor protein independent binding of endocytic carrier vesicles to microtubules in vitro. J Biol Chem 266: 18141-18148

Schekman  R (1985) Protein localization and membrane traffic in yeast. Annu Rev Cell Biol 1: 115-144

Schweizer A, Fransen JAM, Matter K, Kreis TE, Ginsel L, Hauri HP (1990) Identification of an intermediate compartment involved in protein transport from the endoplasmic reticulum to Golgi apparatus. Eur J Cell Biol 53: 185-196

Serafini T, Orci L, Amherdt M, Brunner M, Kahn RA, Rothman JE (1991) ADP-ribosylation factor (ARF) is a subunit of the coat of Golgi-derived COP-coated vesicles: a novel role for a GTP-binding protein. Cell 67: 239-253

Waters MG, Serafini T, Rothman JE (1991) "Coatomer": a cytosolic protein complex containing subunits of non-clathrin-coated Golgi transport vesicles. Nature 349: 248-251

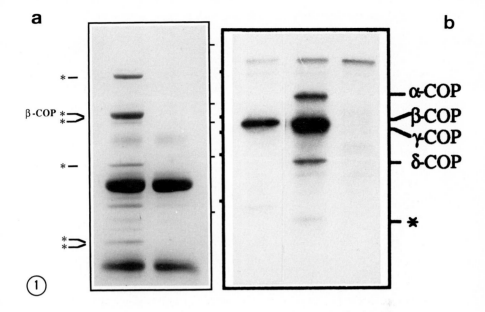

FIGURE 1: IMMUNOPRECIPITATION OF NATIVE COP COMPLEX

An affinity purified polyclonal antibody against a synthetic peptide of β-COP (EAGE) has been used to immunoprecipitate the native COP complex from a rat liver extract (a) or $^{35}$S-labeled HeLa cytosol (b). (a) Shows the Coomassie stained gel of COPs immunoisolated from rat liver (first lane) and the control (second lane), where the antibodies were preincubated with the EAGE peptide. (b) Shows the autoradiagraph of immunoisolated COPs from SDS denatured HeLa cytosol (first lane), and native HeLa cytosol immunoprecipitatied with untreated antibodies (second lane) or antibodies preincubated with EAGE peptide (third lane). (a) The strongly labeled bands at 50 and 25 kDa are immunoglobulin heavy and light chains. Asterisks indicate α-δ-COP and the two bands at ~30 kDa. (b) The asterisk indicates the COP complex associated proteins at ~30 kDa.

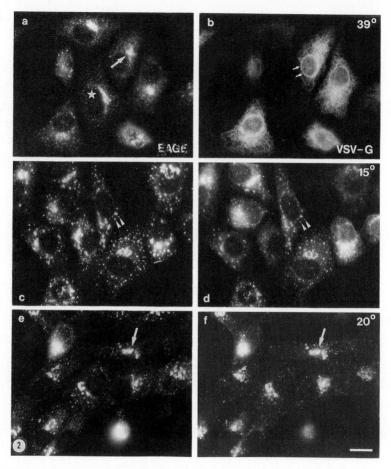

FIGURE 2: COLOCALIZATION OF β-COP AND VSV-G IN THE BIOSYNTHETIC PATHWAY

tsO45-VSV-infected Vero cells were kept for 2.5 hr at nonpermissive temperature (a,b), or were further incubated with 10μg/ml cyclohexamide for 3 hr at 15°C to accumulate VSV-G in the CGN (c,d), or at 20°C to accumulate VSV-G in the TGN (e,f). Cells were then fixed and double labeled with antibodies against β-COP and VSV-G as described [Duden et al., 1991]. The larger arrows indicate Golgi complex and the smaller arrows ER. Arrowheads indicate patches of CGN. The asterisk indicates an uninfected cell. Bar = 20μm.

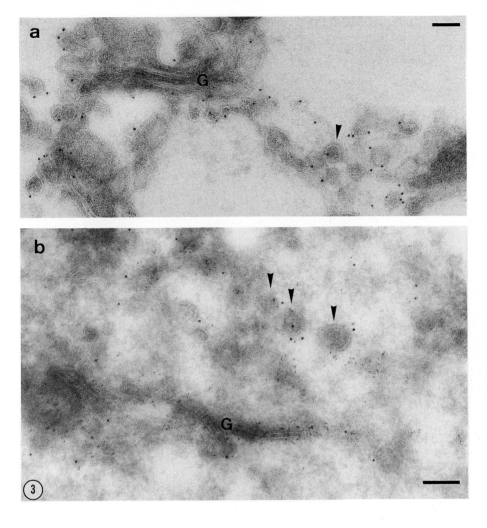

FIGURE 3: LOCALIZATION OF β-COP BY IMMUNOELECTRON MICROSCOPY

Control (a) and tsO45-VSV-infected (b) cells were single-labeled with antibodies against β-COP (a) and double-labeled with specific antibodies against β-COP and VSV-G (b) as described [Duden et al., 1991]. β-COP is associated with membranes of the Golgi complex and coated vesicular structures (a) that contain VSV-G in infected cells (b). Vesicular structures are indicated by arrowheads. Bar = 100 nm.

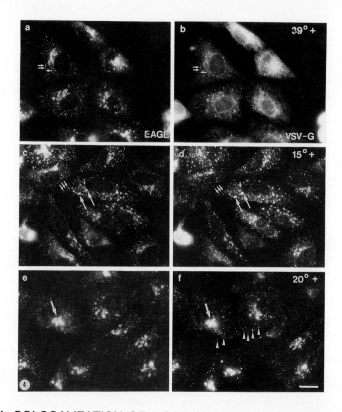

FIGURE 4: COLOCALIZATION OF β-COP AND VSV-G TO VESICLES THAT FORM FROM THE ENDOPLASMIC RETICULUM AND THE CIS-GOLGI NETWORK

tsO45-VSV-infected Vero cells were shifted for 5 min to permissive temperature after they were kept for 2.5 hr at the nonpermissive temperature to accumulate VSV-G in the ER (a,b), or were then further incubated with 10μg/ml cyclohexamide for 3 hr at 15°C to accumulate VSV-G in the CGN (c,d), or at 20°C to accumulate VSV-G in the TGN (e,f). Cells were then fixed and double labeled with antibodies against β-COP and VSV-G as described [Duden et al., 1991]. The larger arrows indicate in (c,d) patches of CGN and in (e,f) a Golgi complex. Smaller arrows (a-d) indicate vesicular structures that contain VSV-G and β-COP. Arrowheads in (e,f) indicated vesicles containing VSV-G but not β-COP. Bar = 20μm.

# DEFINING THE RETENTION SIGNAL IN A MODEL GOLGI MEMBRANE PROTEIN

C.E. Machamer, M.G. Grim, A. Esquela, K. Ryan, and A.M. Swift
Department of Cell Biology and Anatomy
Johns Hopkins University School of Medicine
725 North Wolfe Street
Baltimore, MD 21205 USA

## INTRODUCTION

Specific sequences found in resident proteins of the endoplasmic reticulum (ER) and Golgi complex are believed to direct retention of these proteins in the appropriate compartment of the exocytotic pathway. Such "retention signals" have been identified at the carboxy-termini of both lumenal and membrane-bound ER resident proteins (reviewed by Pelham, 1991). We have been studying the targeting of Golgi membrane proteins using the E1 glycoprotein from the avian infectious bronchitis virus (IBV) as a model protein. This protein is targeted to *cis* Golgi membranes when expressed from cDNA in animal cells (Machamer et al, 1990). We previously determined that the first of the three membrane-spanning domains of the E1 protein contained Golgi targeting information (Machamer and Rose, 1987; Swift and Machamer, 1991). This first membrane-spanning domain ("m1") can retain a plasma membrane protein (the vesicular stomatitis virus G protein) in the Golgi complex when inserted in place of the normal transmembrane domain. The chimeric protein ("Gm1") is retained in the early Golgi complex suggesting that the m1 domain is necessary and sufficient for the targeting of the IBV E1 protein to the *cis* Golgi complex (Swift and Machamer, 1991).

NATO ASI Series, Vol. H 74
Molecular Mechanisms of Membrane Traffic
Edited by D. J. Morré, K. E. Howell, and J. J. M. Bergeron
© Springer-Verlag Berlin Heidelberg 1993

Interestingly, others have recently reported that targeting information for several endogenous Golgi glycosyltransferases appears to be contained within transmembrane domains (Nilsson et al, 1991; Munro, 1991; Teasdale et al, 1992; Colley et al, 1992; Aoki et al, 1992, Russo et al, 1992; Wong et al, 1992; Tang et al, 1992). Recognition of specific sequence information within the lipid bilayer must therefore be important for proper targeting of resident Golgi proteins. As a first step in understanding the mechanism of targeting of proteins to the Golgi complex, we have begun to define the exact sequence requirements for retention of our model Golgi protein. The results presented here suggest that uncharged polar residues that line one face of a predicted alpha-helix in m1 are the important feature.

## RESULTS AND DISCUSSION

The chimeric Gm1 protein consists of the extracellular and cytoplasmic domains of the vesicular stomatitis virus G protein and the first membrane-spanning domain of the IBV E1 protein (Fig. 1).

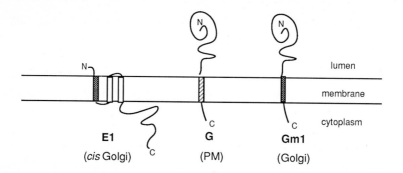

**Figure 1**. Diagrammatic representation of the IBV E1 protein, the VSV G protein, and the Gm1 protein indicating the cellular localization as determined by immunoelectron microscopy and/or indirect immunofluorescence microscopy.

The Gm1 protein is retained in the Golgi region of transfected cells and contains only high mannose asparagine-linked oligosaccharides (Swift and Machamer, 1991). This suggests that Gm1 may be retained in the *cis* Golgi like the E1 protein itself, although absolute confirmation requires immunoelectron microscopy, currently in progress.

To determine the amino acid residues in m1 required for Golgi targeting, we produced a number of point mutations within this sequence in the chimeric Gm1 protein by oligonucleotide-directed mutagenesis. Disruption of the retention signal would be expected to result in transport of the protein to the plasma membrane by default. After mutagenesis and sequencing, the mutant genes were expressed transiently in COS cells, and localization was determined by indirect immunofluorescence microscopy and oligosaccharide processing. The results are summarized in Fig. 2.

Nonconservative and conservative substitutions at three positions in m1 (Asn-465, Thr-476 and Gln-480; see Fig. 2) disrupted Golgi targeting. Most of the Gm1 proteins with mutations at any one of these positions were transported efficiently to the plasma membrane. Interestingly, two conservative replacements of Gln-480 (Asn and His) were tolerated. The three polar residues line up on one face of a predicted alpha-helix, suggesting this face may be involved in targeting. In contrast, substitutions at many other positions in m1 had no effect on Golgi retention of Gm1. We conclude that these residues are not directly involved in targeting. Consistent with the idea that one face of a helix might be required for targeting, insertion of two Ile residues in the middle of m1 also disrupted Golgi retention.

For the mutant Gm1 proteins that were not retained, we determined the rates of transport through the Golgi by measuring the rate of oligosaccharide processing. The two asparagine-linked oligosaccharides on the parent G protein are converted from the high mannose form to the complex, endoglycosidase H-resistant form in the *medial* Golgi with a half-time of about 20 minutes in transfected cells. The half-times for oligosaccharide processing of the mutant Gm1 proteins that were transported to the cell surface ranged from 25 to 55 minutes. Mutations at Gln-480 resulted in the fastest half-times (25-30 minutes), and those at Asn-465 the slowest (45-55 minutes).

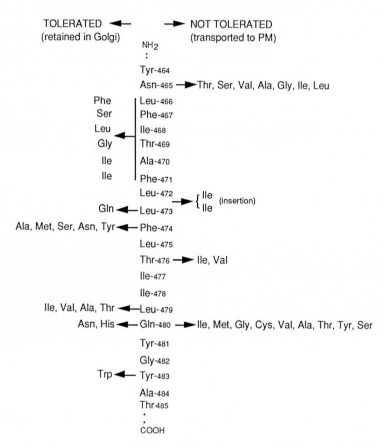

Figure 2. Summary of m1 mutations in Gm1. Single amino acid substitutions (or a block replacement or insertion as indicated) were introduced into Gm1 by site-directed mutagenesis. Cellular localization of mutant proteins was determined by indirect immunofluorescence microscopy and oligosaccharide processing. Only the sequence of the m1 transmembrane domain is shown (numbers pertain to position in the full-length protein). Mutations on the left were tolerated: no effects on Golgi localization were observed. Mutations on the right were not tolerated: mutant Gm1 proteins with any of these substitutions were efficiently transported to the plasma membrane.

The fact that mutations at Gln-480 generated Gm1 proteins that were transported nearly as fast as the parent G molecule indicated that Gm1 proteins with these substitutions were not significantly detained in the *cis* Golgi, and thus Gln-480 was essential for retention.

Our results suggest that at least some of the residues lining one face of a predicted alpha-helix are instrumental in targeting of the model Golgi membrane protein Gm1. Retention signals within transmembrane domains of resident Golgi proteins seem to be a common feature (Machamer, 1991). How might information buried in the lipid bilayer function in retaining resident proteins? Two general mechanisms can be considered (Fig. 3).

## A. RETRIEVAL

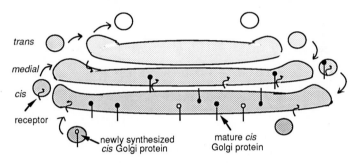

## B. RETENTION

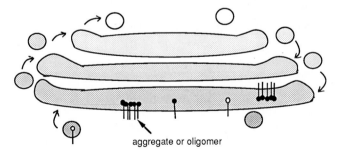

Figure 3. General models for targeting of a newly synthesized endogenous *cis* Golgi protein via sequence information in its transmembrane domain. Both mechanisms could operate together for efficient retention.

The retrieval model (Fig. 3A) implicates a constitutively recycling receptor with a ligand binding site within its own transmembrane domain. The receptor would bind any escaped proteins and return them to the appropriate Golgi subcompartment (or to the ER for another round of vesicular transport and chance to be properly retained). The retention model (Fig. 3B) suggests that in the appropriate subcompartment, the transmembrane domain induces some sort of oligomer or aggregate that cannot enter transport vesicles. These two mechanisms are not mutually exclusive, and may function together for efficient targeting. Both models require differences in microenvironment (such as lipid composition or divalent cation concentration) between the ER and Golgi, and within Golgi subcompartments, either for receptor binding and release, or for inducing oligomerization.

To date, we have not identified any potential receptor proteins that associate with our model Golgi protein. The single amino acid substitutions that prevent retention of the Gm1 protein in the Golgi do not result in a slow leak of molecules to the cell surface (as might be expected if retrieval was disrupted), but result in rapid and efficient transport. Thus, it seems likely that the polar residues we have identified must play a role in actually retaining the protein. We have preliminary evidence that the Gm1 protein forms large oligomers as it arrives in the Golgi complex. Mutant Gm1 proteins that are not retained do not form these oligomers. It is possible that the m1 domain interacts with itself, resulting in oligomers. Such aggregates could be retained for steric reasons, or specifically via interactions with other proteins. The Gm1 protein should continue to be a useful model for investigating the mechanism of retention of resident Golgi proteins, and the generation and maintenance of Golgi subcompartments.

ACKNOWLEDGEMENTS

This work was supported by NIH grant GM42522 and the Pew Charitable Trusts (Pew Scholars in the Biomedical Sciences Program).

REFERENCES

Aoki D, Lee N, Yamaguchi N, Dubois C, Fukuda MN (1992) Proc Natl Acad Sci USA 89:4319-4323
Colley KJ, Lee EU, Paulson JC (1992) J Biol Chem 267:7784-7793
Machamer CE (1991) Trends Cell Biol 1:141-144
Machamer CE, Mentone SA, Rose JK, Farquhar MG (1990) Proc Natl Acad Sci USA 87:6944-6948
Machamer CE, Rose JK (1987) J Cell Biol 105:1205-1214
Munro S (1991) EMBO J 10:3577-3588
Nilsson T, Lucocq JM, Mackay D, Warren G (1991) EMBO J 10:3567-3575
Pelham HRB (1991) Curr Opin Cell Biol 3:585-591
Russo RN, Shaper NL, Taatjes DJ, Shaper JH (1992) J Biol Chem 267:9241-9247
Swift AM, Machamer CE (1991) J Cell Biol 115:19-30
Tang BL, Wong SH, Low SH, Hong W (1992) J Biol Chem 267:10122-10126
Teasdale RD, D'Agostaro G, Gleeson PA (1992) J Biol Chem 267:4084-4096
Wong SH, Low SH, Hong W (1992) J Cell Biol 117:245-258

# ORGANIZATION OF THE GLYCOPROTEIN AND POLYSACCHARIDE SYNTHETIC PATHWAYS IN THE PLANT GOLGI APPARATUS

L.A. Staehelin
Department of Molecular, Cellular and Developmental Biology
University of Colorado
Boulder, CO 80309-0347
U.S.A.

## INTRODUCTION

The Golgi apparatus serves as a processing and sorting station for secretory proteins as they pass from their site of origin, the endoplasmic reticulum, to their final destination, the cell surface or the lysosomal/vacuolar systems. There is general consensus that in animal cells the Golgi apparatus consists of at least four distinct functional compartments known as *cis*, medial and *trans* Golgi cisternae, and the *trans* Golgi network (TGN) (Farquhar, 1985; Griffiths and Simons, 1986). Proteins enter the stack at its *cis* face and depart from the opposite *trans* face (Dunphy and Rothman, 1985), and transport between the cisternal compartments is mediated by transport vesicles (Duden et al., 1991). The definition of these compartments is based on biochemical fractionation studies, as well as on the histochemical and immuno-cytochemical localization of specific glycosyltransferases and their products (Kornfeld and Kornfeld, 1985; Roth, 1987). *Cis* Golgi cisternae contain the enzyme N-acetylglucosamine (GlcNAc)-1-phosphodiester α-N-acetylglucosaminidase, which is involved in the addition of mannose 6-phosphate residues to the oligosaccharide side chains of lysosomal enzymes. The enzyme GlcNAc transferase I is found in the medial cisternae, whereas most of the galactosyltransferase activity is located in the *trans* cisternae. The addition of terminal sialic acid residues, finally, is largely confined to the TGN. Thus the presence of a specific type of Golgi compartment in a biochemical fraction can be defined by the presence of specific marker enzyme activities, and the types of cisternae through which a given N-linked glycoprotein has passed can be determined by the types of modifications that are present on its oligosaccharide side chain(s). This knowledge of the functional organization of the Golgi apparatus has proven immensely valuable for designing experiments to explore mechanisms of trafficking through the Golgi apparatus of animal cells, and for

NATO ASI Series, Vol. H 74
Molecular Mechanisms of Membrane Traffic
Edited by D. J. Morré, K. E. Howell, and J. J. M. Bergeron
© Springer-Verlag Berlin Heidelberg 1993

deciphering the sites of action of drugs such as brefeldin A that interfere with the secretory pathway.

Despite years of efforts, defined marker enzyme activities for the different subcompartments of the plant Golgi apparatus are still lacking. Biochemical analyses of plant Golgi fractions have identified virtually all of the enzyme activities involved in the processing of N-linked oligosaccharide sidechains (Faye et al., 1989), but separation of the cisternae into well-defined subfractions          has not been achieved (see e.g. Sturm et al., 1987). The reason for this lack of success appears to rest with the fact that the primary function of the Golgi apparatus of most types of plant cells is not to process glycoproteins but to produce large quantities of complex polysaccharides for the plant cell wall. An expected consequence of the presence of such polysaccharides would be the masking of subtle differences in buoyant density of the different subclasses of cisternae on which the biochemical fractionation of Golgi cisternae in animal cells is based. To circumvent these problems and to obtain a better understanding of the possible diversity of the plant Golgi apparatus in different tissues, we have employed state-of-the-art electron microscopical and immuno-cytochemical techniques to define the functional organization of the plant Golgi apparatus *in situ*.

RESULTS AND DISCUSSION

*Ultrastructure of Golgi stacks in high pressure frozen/freeze-substituted cells*
The Golgi apparatus of plant cells differs in several important aspects from its animal counterpart. Thus, although the Golgi stacks in both systems are composed of flattened cisternae, the number of stacks per cell, their spatial organization and their behavior during mitosis differ substantially. A typical interphase onion root tip cell contains about 400 individual Golgi stacks (Garcia-Herdugo et al., 1988), versus about 10 for animal cells (Rambourg and Clermont, 1990), and these stacks are usually evenly dispersed throughout the cytoplasm and not clustered in a juxtanuclear position as in animal cells. Furthermore, the plant Golgi stacks persist during mitosis, since they are needed for the formation of the cell plate and the new wall that divides the two daughter cells. To reevaluate the macromolecular architecture of plant Golgi stacks we have examined cells preserved by high pressure freezing/freeze-substitution techniques.

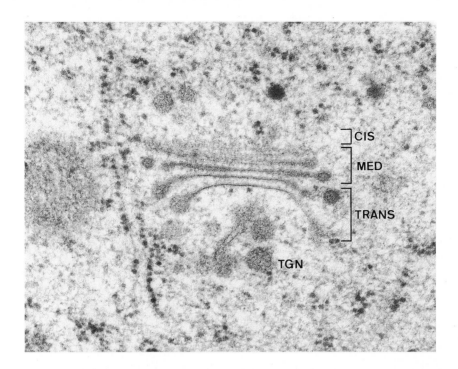

Fig. 1 High pressure frozen/freeze-substituted Golgi stack of a sycamore maple suspension culture cell. The different types of cisternae can be distinguished based on their position within the stack and the staining pattern of the cisternal lumen. (micrograph, G.F. Zhang, University of Colorado, Boulder)

Common to all plant Golgi stacks of such specimens (Fig. 1) are clear and discrete differences in staining patterns and width of cisternae that can be used to distinguish in spatial terms *cis*, medial, and *trans* types of Golgi cisternae (Staehelin et al., 1990; Zhang and Staehelin, 1992). Most Golgi stacks possess only one *cis* cisterna, that can be recognized not only based on its position within a stack, but also on its small size, its varied morphology, and its wide and lightly stained lumen. Medial cisternae are narrower and filled with more densely staining, mottled materials that appear to become concentrated in the bulbous margins. *Trans* cisternae are defined as having evenly stained, dense lumenal contents and a central domain where the lumen is completely collapsed, giving rise to a 4-6nm wide dark line in cross sectional views. In addition, all types of Golgi cisternae appear to bud vesicles around their margins that are covered by a non-clathrin type of coat. This non-clathrin type of coat is one of the main distinguishing features between Golgi cisternae and the more rounded and branched cisternae of the TGN, which gives rise to clathrin-coated vesicles. Not all Golgi stacks seem to have a direct association with a clearly defined TGN, and in some cases the trans-most of the trans cisternae exhibits branched bulbous margins that resemble TGN cisternae, hinting at the possibility that such cisternae may perform TGN functions. However, further studies are needed to confirm these ideas. A ribosome-excluding Golgi matrix zone seems to envelope each Golgi stack and its TGN, where present. Intercisternal fibrils (elements) are associated exclusively with *trans* Golgi cisternae of some cell types but not others.

*Tissue-specific morphology of Golgi stacks*

The root tip represents a useful model system for studying the relationship between plant cell differentiation and Golgi function. All root tip cells are produced by the apical meristem, which gives rise to the root proper on its distal side, and to the root cap on its proximal side. As the meristematic root cap cells become displaced towards the root tip surface, they first differentiate into gravity sensing columella cells, and then into "young" and "old", polysaccharide-secreting peripheral cells. Careful examination of high pressure frozen/freeze-substituted root tips of *Arabidopsis* and *Nicotiana* has revealed characteristic differences in the architecture of Golgi stacks of the different types of root cap cells, which appear to reflect a tissue-specific retailoring of the stacks as their functions change during development (Staehelin et al., 1990). Thus the Golgi stacks of meristematic cells are small and compact and differences in the morphology of *cis*, medial and *trans* cisternae are difficult to discern. However, as the meristematic cells are converted into columella cells, the diameter of the Golgi stacks increases, staining differences between *cis*, medial, and *trans* cisternae become pronounced, clusters of intercisternal filaments appear between *trans* cisternae, and the TGN enlarges. During the subsequent conversion of columella cells into the mucilage-secreting peripheral cells, the *trans* cisternae become filled with densely staining secretory products, while the *cis* and medial cisternal compartments decrease in size and lose their distinctive staining patterns. In older peripheral cells the appearance of the Golgi stacks becomes dominated by the mucilage-filled, and electron-dense *trans* cisternae and their large swollen margins. Most Golgi in these latter cells appear to lack a TGN.

*Immunocytochemical analysis of the assembly pathways of complex polysaccharides*

Plant cells synthesize two major classes of complex cell wall polysaccharides, the acidic pectic polysaccharides and the neutral hemicelluloses (McNeil et al., 1984).

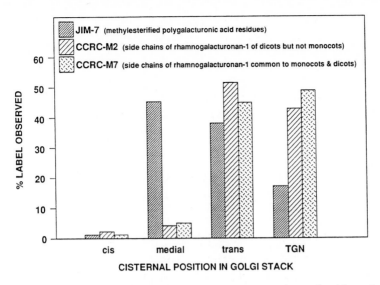

Fig. 2 Histogram depicting the distribution of pectic polysaccharide epitopes in Golgi cisternae and the TGN of immunolabeled sycamore maple suspension cultured cells. See text for details. (from Zhang and Staehelin, 1992)

Polygalacturonic acid/rhamnogalacturonan-I (PGA/RG-I) is the most abundant pectic polysaccharide of dicotyledonous plants. It consists of two covalently-linked domains, blocks of PGA interrupted with an occasional rhamnose residue, and blocks of RG-I. Most PGA blocks are methylesterified prior to secretion, and become deesterified during maturation of the cell walls. RG-I is composed of a backbone of alternating galactosyl and rhamnosyl residues, with about 50% of the rhamnosyl residues carrying oligosaccharide side chains. Pectic polysaccharides define the pore size and the degree of hydration of plant cell walls, and contribute to cell adhesion (Baron-Epel et al., 1988; Jarvis, 1984).

Xyloglucan (XG) is the most abundant hemicellulose of the cell walls of growing dicotyledonous plants. This large polysaccharide is composed of a $\beta$-1,4 linked glucosyl backbone that is decorated at regular intervals with xylosyl and xylosyl-galactosyl-fucosyl side chains to form typical hepta- and nona-saccharide repeats (Hayashi, 1989). XG has the ability to bind to the surface of cellulose fibrils, thereby forming crosslinks between the fibrils and contributing to the mechanical strength of the cell wall. Cell wall expansion involves enzymatic cleavage and possible rejoining of these crosslinks (Fry et al., 1992). Specific fragments derived

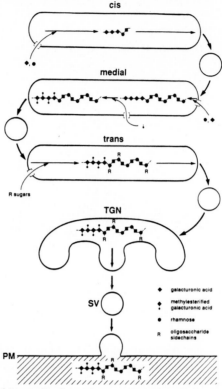

cis

medial

trans

R sugars

TGN

SV

◆ galacturonic acid

♦ methylesterified
: galacturonic acid

● rhamnose

R oligosaccharide
sidechains

PM

Fig. 3 Model of the biosynthesis pathway of PGA/RG-I in the Golgi apparatus of sycamore maple suspension cultured cells based on the quantitative immunolabeling experiments described in the text. (from Zhang and Staehelin, 1992).

from XG and PGA/RG-I possess hormone-like regulatory properties (Ryan and Farmer, 1991).

By using a battery of well characterized monoclonal and polyclonal antibodies that recognize defined sugar residues on XG and PGA/RG-I in conjunction with immunocytochemical labeling of high pressure frozen/freeze-substituted sycamore maple cells, we have been able to determine in which cisternae of Golgi stacks specific sugar groups are added to the growing XG and PGA/RG-I molecules (Zhang and Staehelin, 1992). The antibodies used in these experiments have been characterized as described in detail by Moore et al. (1986), Lynch and Staehelin (1992), and Zhang and Staehelin (1992), and have been shown to bind to the following epitopes: anti-XG, to the β-1,4 linked glucosyl backbone of XG; CCRC-

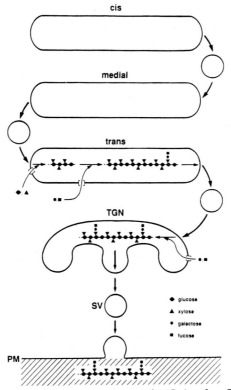

Fig. 4 Model of the biosynthesis pathway of XG in the Golgi apparatus of sycamore maple suspension culture cells as determined by the immunolabeling experiments described in the text. (from Zhang and Staehelin, 1992)

M1, to the terminal fucosyl residue of the trisaccharide side chain chain of XG; anti PGA/RG-I, to the deesterified PGA/RG-I transition region; JIM7, to methylesterified PGA; CCRC-M2 and M7, to side chains of RG-I.

The results of a subset of the quantitative anti PGA/RG-I labeling experiments are illustrated in the histogram Fig.2, and the findings of all of the pectic polysaccharide labeling studies are summarized in the interpretive diagram Fig. 3 (Zhang and Staehelin, 1992). Based on the observation that the anti PGA/RG-I antibodies label mostly *cis* and medial cisternae, but not *trans* cisternae and the TGN (data not shown; Moore et al., 1991; Zhang and Staehelin, 1992), we postulate that the synthesis of the unesterified PGA/RG-I backbone occurs in *cis* and medial cisternae. The sudden onset of labeling with the JIM7 antibodies in the medial cisternae (Fig. 2) suggests not only that methyl-esterification of PGA takes place in these cisternae, but that the lack of anti PGA/RG-I labeling in the *trans* cisternae could be caused by the methyl-esterification of PGA in the medial cisternae. This interpretation has been confirmed by means of labeling experiments of chemically deesterified samples (data not shown; Zhang and Staehelin, 1992). Finally, the confinement of the labeling of the PGA/RG-I sidechains with the CCRC-M2 and M7 antibodies to the *trans* cisternae and the TGN is consistent with the notion that these sidechains are added to the nascent pectic polysaccharides in the trans cisternae.

The interpretive diagram of the XG synthesis pathway shown in Fig. 4 is based on quantitative immunolabeling data obtained with the anti-XG and the CCRC-M1 antibodies (Zhang and Staehelin, 1992). Most intriguing is the finding that the assembly of XG seems to be limited exclusively to *trans* cisternae and the TGN. Thus the β-1,4 linked glucosyl backbone seems to be restricted to the *trans* cisternae, whereas the first fucose-containing trisaccharide side chains appear to be added in the *trans* cisternae and the remaining ones in the TGN. By being confined to *trans* Golgi cisternae and the TGN, this synthesis pathway appears to be unlike any Golgi associated assembly pathway described so far for animal cells. This has interesting experimental implications, since it provides opportunities for studying secretion from the *trans* Golgi compartment under conditions where protein secretion is inhibited by blocks prior to that compartment.

*Assembly of N- and O-linked glycans of glycoproteins*
Although we have employed only two anti glycoprotein-glycan antibodies in our immunolabeling studies to date, these antibodies have already provided important insights into the general organization of the synthesis pathways of N- and O-linked glycoproteins in the Golgi apparatus of plant cells. As mentioned in the Introduction, the processing of N-linked oligosaccharides in plant cells follows the pathways established for animal cells with several important modifications (Faye et al., 1989). For example they do not employ the mannose-6-phosphate system for targeting proteins to the vacuole (Vitale and Chrispeels, 1992), xylose is often linked to the β- linked core mannose, and sialic acid is not used as a terminal sugar. Employing anti-βXyl antibodies that specifically recognize the xylose residues linked β-1,2 to the β-linked mannose of the glycan core of N-linked oligosaccharides, we have demonstrated that these residues are added in the medial cisternae. Because the xylosyl transferase that carries out this reaction acts just after the N-acetylglucosamine transferase II (Faye et al., 1989), which in animal cells is one of the later acting medial cisterna enzymes (Kornfeld

and Kornfeld, 1985), our localization would be consistent with this latter enzyme also being present in medial cisternae of plant Golgi stacks. This being the case, one might expect the addition of the terminal fucosyl and galactosyl residues to occur in *trans* cisternae.

O-linked oligosaccharides are important components of two major types of cell wall glycoproteins, the hydroxyproline-rich glycoproteins (HRGPs), and the arabinogalactan proteins. We have produced anti-gE1 antibodies that recognize the terminal arabinose residues of the O-linked tetraarabinoside sidechains of the structural cell wall glycoprotein, extensin, an HRGP (Swords and Staehelin, submitted). Immunolabeling experiments with the anti-gE1 antibodies indicate that these O-linked glycans can be both initiated and completed in the *cis* cisternae (Moore et al., 1991). This places the addition of O-linked glycans into the same cisternae as in some types of animal cells, although the sugars types that are added are different.

This research was supported by NIH grant GM18639 and NSF grant DCB-8615763.

## REFERENCES

Baron-Epel O, Gharyal P , Schindler M (1988) Pectins as mediators of wall porosity in soybean cells. Planta 175:389-395

Duden R, Allan V, Kreis T (1991) Involvement of β-COP in membrane traffic through the Golgi complex. Trends Cell Biol. 1:14-19

Dunphy WG, Rothman JE (1985) Compartmental organization of the Golgi stack. Cell 42: 13-21

Farquhar MG (1985) Progress in unraveling pathways of Golgi traffic. Ann Rev Cell Biol 1:447-488

Faye L, Johnson KD, Stern, A, Chrispeels MJ (1989) Structure, biosynthesis, and function of asparagine-linked glycans of plant glycoproteins. Physiol Plant 75: 309-314

Fry SC, Smith RC, Renwick KF, Martin DJ, Hodge SK, Mathews KJ (1992) Xyloglucan endotransglycolase, a new wall-loosening enzyme activity from plants. Biochem J 282: 821-828

Garcia-Herdugo G, Gonzales-Reyes JA, Garcia-Navarro F, Navas P (1988) Growth kinetics of the Golgi apparatus during the cell cycle in onion root meristems. Planta 175: 305-312

Griffiths G, Simons K (1986) The trans Golgi network: Sorting at the exit site of the Golgi complex. Science 234: 438-443

Hayashi T (1989) Xyloglucans in the primary cell wall. Ann Rev Plant Physiol Mol Biol 40: 139-168

Jarvis MC (1984) Structure and properties of pectin gels in plant cell walls. Plant Cell and Envir 7:153-164

Kornfeld R, Kornfeld S (1985) Assembly of asparagine-linked oligosaccharides. Ann Rev Biochem 54: 631-664

Lynch MA, Staehelin LA (1992) Domain-specific and cell-type specific localization of two types of cell wall matrix polysaccharides in the clover root tip. J Cell Biol (in press)

McNeil M, Darvill AG, Fry SC, Albersheim P (1984) Structure and function of the primary cell walls of plants. Ann Rev Biochem 8: 625-663

Moore PJ, Darvill A, Albersheim P, Staehelin, LA (1986) Immunogold localization of xyloglucan and rhamnogalacturonan I in the cell walls of suspension-cultured sycamore cells. Plant Physiol 82: 787-794

Rambourg A, Clermont Y (1990) Three dimensional microscopy: structure of the Golgi apparatus. Eur J Cell Biol 51: 189

Roth J (1987) Subcellular organization of glycosylation in mammalian cells. Biochem Biophys Acta 906: 405-436

Ryan CA, Farmer EE (1991) Oligosaccharide signals in plants: a current assessment. Ann Rev Plant Physiol Mol Biol 42: 651-674

Staehelin LA, Giddings TH, Kiss JZ, Sack FD (1990) Macromolecular differentiation of Golgi stacks in root tips of *Arabidopsis* and *Nicotiana* seedlings as visualized in high pressure frozen and freeze-substituted samples. Protoplasma 157: 75-91

Sturm A, Johnson KD, Szumilo T, Elbein AD, Chrispeels MJ (1987) Subcellular localization of glycosidases and glycosyltransferases in the processing of N-linked oligosaccharides. Plant Physiol 85: 741-745

Vitale A, Chrispeels MJ (1992) Sorting of proteins to the vacuoles of plant cells. BioEssays 14: 151-160

Zhang GF, Staehelin LA (1992) Functional compartmentalization of the Golgi apparatus of plant cells. An immunocytochemical analysis of high pressure frozen and freeze-substituted sycamore maple suspension culture cells. Plant Physiol 99 (in press)

# CLIP-170, A CYTOPLASMIC LINKER PROTEIN MEDIATING INTERACTION OF ENDOSOMES WITH MICROTUBULES

J. Scheel, J. E. Rickard, P. Pierre, D. Hennig, P. I. Karecla, R. Pepperkok, B. Joggerst-Thomalla, A. Sawyer, R. G. Parton, T. E. Kreis

European Molecular Biology Laboratory
Meyerhofstrasse 1
D-6900 Heidelberg
Germany

Microtubules play a key role in the dynamic spatial organization of the cytoplasmic matrix. They are involved in regulating the structure and positioning of intracellular organelles like the Golgi apparatus and endosomes, and they provide the tracks for directed movement of such organelles [Kreis, 1990]. Membrane-bounded cytoplasmic organelles attach to microtubules, move along or remain stably associated with them, and eventually detach. These interactions of organelles with microtubules appear to be specific. For example, endosomes, but not Golgi elements, reverse their direction of translocation along microtubules upon acidification of the cytoplasm [Heuser, 1989; Parton et al., 1991]. These various interactions require different levels of

NATO ASI Series, Vol. H 74
Molecular Mechanisms of Membrane Traffic
Edited by D. J. Morré, K. E. Howell, and J. J. M. Bergeron
© Springer-Verlag Berlin Heidelberg 1993

regulation; specificity and timing of binding and release, positioning, and direction of movement. We postulate, therefore, that different proteins must be involved in the regulation of these interactions of cytoplasmic organelles with microtubules.

Various cell-free assays have been used to characterize factors mediating interactions of organelles with microtubules [see for example Mithieux et al., 1988; Pratt, 1986; Scheel and Kreis, 1991; Vale et al., 1985; van der Sluijs et al., 1990]. The microtubule-based motor proteins, kinesin and cytoplasmic dynein, which have been identified with such assays, are so far the best characterized of these factors [Schroer and Sheetz, 1991a; Vale, 1992; Vallee and Shpetner, 1990]. Non-motor proteins are, however, also required for organelle movement [Schnapp and Reese, 1989; Schroer and Sheetz, 1991b; Schroer et al., 1989], as well as for binding of organelles to microtubules [Karecla and Kreis, 1992; Mithieux and Rousset, 1989; Scheel and Kreis, 1991]. Several candidates for such proteins have been identified, but their roles in the interaction and movement of organelles along microtubules remain to be further elucidated: dynamin, a microtubule-activated GTPase [Collins, 1991], a 58 kDa microtubule binding protein localized to the Golgi apparatus [Bloom and Brashear, 1989], a 50 kDa lysosomal protein [Mithieux and Rousset, 1989], and dynactin, part of a cytosolic complex involved in cytoplasmic dynein dependent organelle movement [Gill et al., 1991; Schroer and Sheetz, 1991b]. We have recently identified a 170 kDa cytoplasmic linker protein (CLIP-170) which mediates binding of endosomes to microtubules. CLIP-170 belongs to a novel class of proteins mediating interactions of organelles with microtubules. Its structure, intracellular localization, and possible function are discussed here.

## BINDING OF CLIP-170 TO MICROTUBULES

CLIP-170 was originally identified in a fraction of ATP-eluted microtubule-binding proteins from HeLa cells (170K protein; Rickard and Kreis, 1990). The nucleotide sensitivity of the binding of CLIP-170 to microtubules is, however, in contrast to the microtubule-based motor proteins, not due to a direct interaction of the nucleotide-triphosphates with CLIP-170. Instead, its binding to microtubules is regulated by phosphorylation on serine residues which releases CLIP-170 from microtubules (pp170; Rickard and Kreis, 1991). Furthermore, the in vivo phosphorylation of CLIP-170 is affected by the state of the microtubule polymer, since treatment of cells with microtubule depolymerizing drugs leads to dephosphorylation of CLIP-170 [Rickard and Kreis, 1991]. The turnover of phosphate groups on CLIP-170 in vivo is rapid, suggesting that this modification may regulate dynamic changes in its activity.

## INTERACTION OF ENDOSOMAL MEMBRANES WITH MICROTUBULES

Cell-free assays measuring the binding of cytoplasmic organelles to microtubules [Scheel and Kreis, 1991; van der Sluijs et al., 1990] were used to test whether CLIP-170 is involved in these interactions. Depletion of CLIP-170 from cytosol abolishes its ability to promote binding of endocytic carrier vesicles to microtubules [Pierre et al., 1992], whereas binding of vesicles derived from the *trans*-Golgi network to microtubules remains unaffected (D. Hennig and T.E. Kreis, unpublished). The activity can be restored by adding back purified CLIP-170. Thus, CLIP-170 is required for the interaction of endocytic carrier vesicles with microtubules in this in vitro system. Since the purified protein alone is not sufficient to promote binding of the endosomal vesicles to microtubules, additional factors must be involved in this interaction. CLIP-170-dependent binding of endocytic carrier vesicles to microtubules depends on membrane-associated proteins, in addition to soluble factors, as shown by the abolition of binding upon trypsinization of the organelle fraction [Scheel and Kreis, 1991].

## STRUCTURE OF CLIP-170

Analysis of the primary structure of CLIP-170, its predicted secondary structure and comparison with other proteins indicate that CLIP-170 is a novel protein with an elongated structure consisting of three domains (Figure 1): an N-terminal head domain with a basic pI, a long central domain and a short C-terminal domain [Pierre et al., 1992]. The central domain, consisting of ~900 amino acids, is strongly predicted to form an $\alpha$-helix. Since this domain has a continuous pattern of heptad repeats and is homologous to the rod-like domains of kinesin and myosin, for example, it is very likely to form a dimeric coiled-coil rod [Lupas et al., 1991] between the two terminal domains. This structural prediction for CLIP-170 is supported by its hydrodynamic properties, which indicate that CLIP-170 is an elongated (110 x 2.5 nm) homodimer in solution [Pierre et al., 1992]. Thus, the typical head-and-tail organization

[Schliwa, 1989] of motor proteins like kinesin and myosin, which are involved in organelle-cytoskeleton interactions, is also found in CLIP-170, although no similarity (except in the coiled-coil domain) with these proteins is observed at the level of the primary structure.

The basic N-terminal head of CLIP-170 contains one of the functional domains of this protein, the microtubule-binding sites. Whereas neither the central rod nor the C-terminal tail of CLIP-170 bind to microtubules in vitro, the N-terminal head domain alone does so very efficiently [Pierre et al., 1992]. Thus, as in several microtubule-binding proteins, the microtubule-binding sites of CLIP-170 lie in a basic domain. They are, however, different from all of the so far known microtubule-binding motifs including tau [Lee et al., 1988], MAP2 [Lewis et al., 1988], and MAP1B [Noble et al., 1989].

Mutational analysis within the head domain suggests that a duplicated motif of ~40 amino acids is involved in binding to microtubules. The two copies of this motif are highly similar in sequence (54%) and are both followed by serine-rich regions. Both motifs are competent to mediate binding of CLIP-170 to microtubules, although their contributions might not be equivalent [Pierre et al., 1992]. This particular motif is homologous to a single motif (Figure 1) found in the BIK1 protein from yeast [Trueheart et al., 1987], and in 150 kDa proteins from rat (DP-150; Holzbauer et al., 1991) and D. melanogaster (Glued protein; Swaroop et al., 1987). Chicken dynactin, a protein involved in the regulation of dynein-based motility [Gill et al., 1991], is probably homologous to DP-150 and Glued protein.

FIGURE 1: PREDICTED DOMAIN STRUCTURE OF CLIP-170

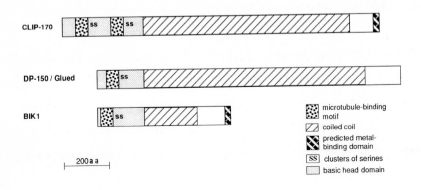

The BIK1 protein, originally identified as a bilateral karyogamy mutant [Trueheart et al., 1987], is a microtubule-binding protein from *S. cerevisiae* [Berlin et al., 1990]. In addition to the very similar microtubule-binding motif found in the head domains of both CLIP-170 and BIK1 protein, both proteins also contain a central coiled-coil rod and a highly similar C-terminal tail predicted to form a metal-binding motif (Figure 1) which could mediate protein-protein interactions [Pierre et al., 1992]. Thus, although the rod domain of BIK1 protein (200 aa) is much shorter than in CLIP-170 (900 aa) and the microtubule-binding motif occurs only once in BIK1 protein, CLIP-170 and BIK1 have a very similar domain organization and highly homologous motifs in their head and tail. Although it is not clear whether BIK1 protein is the yeast homologue of CLIP-170, their similarities suggest that they might have related functions.

## SUBCELLULAR LOCALIZATION OF CLIP-170

CLIP-170 can be isolated from high speed supernatant extracts of Hela cells, where it exists as a homodimer that appears not to be tightly associated with any other protein [Pierre et al., 1992; Rickard and Kreis, 1990; Rickard and Kreis, 1991]. Immunofluorescence microscopy of aldehyde-fixed cells shows a diffuse cytoplasmic staining indicative of a pool of soluble protein (J.E. Rickard and T.E. Kreis, unpublished observation). In addition, a patchy distribution throughout the cytoplasm which is particularly prominent after fixation of cells with methanol is observed. These patches are preferentially localized at the peripheral plus ends of a subset of microtubules [Rickard and Kreis, 1990]. They resist depolymerization of microtubules by nocodazole and brief extraction of cells with Triton X-100. The nature of these patches is not yet fully understood.

A subset of these patches colocalizes, however, with cellular membranes containing transferrin receptor (TFR) [Pierre et al., 1992]. This is consistent with preliminary localization of CLIP-170 on intracellular membranes by immunoelectron microscopy (R.G. Parton, J.E. Rickard and T.E. Kreis, unpublished). In addition, a more extensive colocalization of CLIP-170 and endosomes is observed in cells treated with brefeldin A, where TFR-positive endosomes form microtubule-dependent membrane tubules [Pierre et al., 1992]. Thus, we assume that CLIP-170 is localized to intracellular membranes, consistent with its involvement in the interaction of endosomes with microtubules in vitro. The extent of specificity of CLIP-170 for endosomes remains to be determined, possibly by immunoelectron microscopy and by biochemical studies of the binding of CLIP-170 to membranes and membrane associated receptors.

## FUNCTIONS OF CLIP-170

The properties of CLIP-170 suggest a role for this protein as a linker between endosomes and microtubules. Its structure - N-terminal domain which binds to microtubules, separated by a long rod domain from a C-terminal domain which might interact with other cellular structures - is consistent with this function.

The association of CLIP-170 with peripheral microtubule plus ends makes it a good candidate to interact with endocytic carrier vesicles in vivo. These vesicles, which derive from peripheral early endosomes, are thought to transport endocytosed material to more centrally located late endosomes [Gruenberg et al., 1989]. Whereas early events in endocytosis (budding of vesicles from the plasma membrane, fusion with early endosomes and recycling to the plasma membrane) occur normally in the absence of microtubules, delivery of endocytosed molecules to the pericentriolar late endosomes via endocytic carrier vesicles depends on microtubules [Gruenberg et al., 1989]. The accumulation of endocytosed markers in endocytic carrier vesicles in the absence of microtubules suggests that microtubules play an important role at this stage of the pathway and that the endocytic carrier vesicles interact with microtubules. A role for microtubules at this stage of the endocytic pathway is also suggested by the meeting of apical and basolateral endocytic carrier vesicles from MDCK cells, which in vitro depends on microtubules and microtubule-based motor proteins [Bomsel et al., 1990].

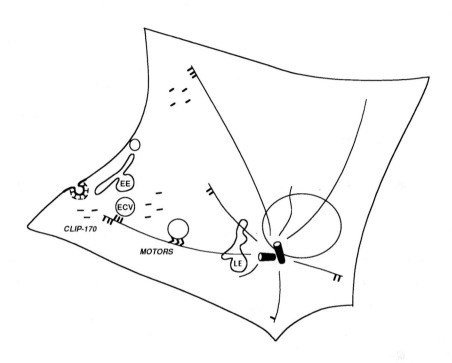

FIGURE 2: PUTATIVE FUNCTIONS OF CLIP-170

The schematic drawing illustrates the putative role of CLIP-170 in the interaction of endosomes with microtubules; details of the model are discussed in the text. EE: early endosome; ECV: endocytic carrier vesicle; LE: late endosome.

CLIP-170 is not a motor protein, since it is not directly sensitive to ATP or GTP, does not have any of the so far known consensus sequences for binding of nucleotide-triphosphates and has no similarity to the motor domains of known mechanochemical ATPases [Pierre et al., 1992]. It is also not tightly associated with motor or other proteins in HeLa high speed supernatants.

The existence of a significant soluble pool of CLIP-170 and the potential for dynamic regulation by phosphorylation suggest that it may interact transiently with cellular structures. Its localization at the peripheral ends of microtubules in interphase cells indicates that it may be involved in capturing specific membrane domains on early endosomes, to initiate contact with microtubules. This specificity would ensure efficient docking of endocytic carrier vesicles onto microtubules at their site of formation. After mediating binding, CLIP-170 must dissociate from the microtubules to allow the motor driven movement of the vesicles towards the microtubule minus ends in the cell center (Figure 2). This dynamic interaction of CLIP-170 with microtubules on the one hand and a membrane receptor on the other hand is probably controlled by phosphorylation. CLIP-170 is multiply phosphorylated in vivo, which provides a basis for differential regulation of its specific interactions [Rickard and Kreis, 1991]. Thus, "clipping" of (budding) endocytic carrier vesicles to microtubules and motor protein-mediated "clipping off" of these vesicles from early endosomes may be regulated by the combined action of a specific kinase and phosphatase.

The interaction of different cytoplasmic organelles with microtubules appears to depend on organelle specific cytosolic and membrane associated factors. CLIP-170 may be a paradigm for the characterization of these other linker proteins. Understanding the mechanisms and regulation of the association of CLIP-170 with subcellular structures may provide further insight into the role of microtubules in the spatial organization of the cytoplasmic matrix.

## REFERENCES

Berlin V, Styles CA, Fink GR, (1990) BIK1, a protein required for microtubule function during mating and mitosis in *Saccharomyces cerevisiae*, colocalizes with tubulin. J Cell Biol 111: 2573-2586

Bloom GS, Brashear TA (1989) A novel 58-kDa protein associates with the Golgi apparatus and microtubules. J Biol Chem 264: 16083-16092

Bomsel M, Parton R, Kuznetsov SA, Schroer TA, Gruenberg J (1990) Microtubule- and motor-dependent fusion in vitro between apical and basolateral endocytic vesicles from MDCK cells. Cell 62: 719-731

Collins CA (1991) Dynamin: a novel microtubule-associated GTPase. Trends Cell Biol 1: 57-60

Gill SR, Schroer TA, Szilak I, Steuer ER, Sheetz MP, Cleveland DW (1991) Dynactin, a conserved, ubiquitously expressed component of an activator of vesicle motility mediated by cytoplasmic dynein. J Cell Biol 115: 1639-1650

Gruenberg J, Griffiths G, Howell K (1989) Characterization of the early endosome and putative endocytic carrier vesicles in vivo and with an assay for vesicle fusion in vitro. J Cell Biol 108: 1301-1316

Heuser J (1989) Changes in lysosome shape and distribution correlated with changes in cytoplasmic pH. J Cell Biol 108: 855-864

Holzbauer ELF, Hammarback JA, Paschal BM, Kravit NG, Pfister KK, Vallee RB (1991) Homology of a 150K cytoplasmic dynein-associated polypeptide with the *Drosophila* gene *Glued*. Nature 351: 579-583

Karecla PI, Kreis TE (1992) Interaction of membranes of the Golgi complex with microtubules in vitro. Eur J Cell Biol 57: 139-146

Kreis TE (1990) Role of microtubules in the organisation of the Golgi apparatus. Cell Motil Cytoskel 15: 67-70

Lee G, Cowan N, Kirschner M (1988) The primary structure and heterogeneity of tau protein from mouse brain. Science 239: 285-288

Lewis SA, Wang DS, Cowan NJ (1988) Microtubule-associated protein MAP2 shares a microtubule binding motif with tau protein. Science 242: 936-939

Lupas A, Van Dyke M, Stock J (1991) Predicting coiled coils from protein sequences. Science 252: 1162-1164

Mithieux G, Auderbet C, Rousset B (1988) Association of purified thyroid lysosomes to reconstituted microtubules. Biochim Biophys Acta 969: 121-130

Mithieux G, Rousset B (1989) Identification of a lysosome membrane protein which could mediate ATP-dependent stable association of lysosomes to microtubules. J Biol Chem 264: 4664-4668

Noble M, Lewis SA, Cowan NJ (1989) The microtubule binding domain of microtubule-associated protein MAP1B contains a repeated sequence motif unrelated to that of MAP2 and tau. J Cell Biol 109: 3367-3376

Parton RG, Dotti CG, Bacallao R, Kurtz I, Simons K, Prydz K. (1991) pH-induced microtubule-dependent redistribution of late endosomes in neuronal and epithelial cells. J Cell Biol 113: 261-274

Pierre P, Scheel J, Rickard JE, Kreis TE (1992) CLIP-170 links endocytic vesicles to microtubules. Cell 70, in press

Pratt MM (1986) Stable complexes of axoplasmic vesicles and microtubules: protein composition and ATPase activity. J Cell Biol 103: 957-968

Rickard JE, Kreis TE (1990) Identification of a novel nucleotide-sensitive microtubule-binding protein in HeLa cells. J Cell Biol 110: 1623-1633

Rickard JE, Kreis TE (1991) Binding of pp170 to microtubules is regulated by phosphorylation. J Biol Chem 266: 17597-17605

Scheel J, Kreis TE (1991) Motor protein independent binding of endocytic carrier vesicles to microtubules in vitro. J Biol Chem 266: 18141-18148

Schliwa M (1989) Head and tail. Cell 56: 719-720

Schnapp BJ, Reese TS (1989) Dynein is the motor for retrograde axonal transport of organelles. Proc Natl Acad Sci USA 86: 1548-1552

Schroer TA, Sheetz MP (1991a) Functions of microtubule-based motors. Annu Rev Physiol 53: 629-652

Schroer TA, Sheetz MP (1991b) Two activators of microtubule-based vesicle transport. J Cell Biol 115: 1309-1318

Schroer TA, Steuer ER, Sheetz MP (1989) Cytoplasmic dynein is a minus end-directed motor for membranous organelles. Cell 56: 937-946

Swaroop A, Swaroop M, Garen A (1987) Sequence analysis of the complete cDNA and encoded polypeptide for the Glued gene of Drosophila melanogaster. Proc Natl Acad Sci USA 84: 6501-6505

Trueheart J, Boeke JD, Fink G (1987) Two genes required for cell fusion during cell conjugation: Evidence for a pheromone-induced surface protein. Mol Cell Biol 7: 2316-2328

Vale RD (1992) Motor proteins. In Guidebook to the Cytoskeletal and Motor Proteins. TE Kreis, RD Vale, eds. (Oxford University Press), in press

Vale RD, Reese TS, Sheetz MP (1985) Identification of a novel force-generating protein, kinesin, involved in microtubule-based motility. Cell 42: 39-50

Vallee RB, Shpetner HS (1990) Motor proteins of cytoplasmic microtubules. Annu Rev Biochem 59: 909-932

van der Sluijs P, Bennett MK, Anthony C, Simons K, Kreis TE (1990) Binding of exocytic vesicles from MDCK cells to microtubules in vitro. J Cell Sci 95: 545-553

# MATURATION OF SECRETORY GRANULES

Sharon A. Tooze
European Molecular Biology Laboratory
Meyerhofstrasse 1
Postfach 1022.09
6900 Heidelberg
Germany

Secretory granules are found in most endocrine, exocrine, and neuronal cells. Their function is to store secretory proteins, such as horomes, in a highly concentrated form within the cell. A hallmark of secretory granules (SGs) is the presence of an electron dense-cored interior which contains the stored secretory proteins. The release of these stored molecules can only be induced by an extracellular signal, which causes the SGs to fuse with the plasma membrane and thereby release the contents. This situation is in contrast to the secretion of proteins by the constitutive pathway: fusion of the constitutive vesicle with the plasma membrane and release of the content occurs very rapidly and is independent of extracellular signals (Burgess and Kelly, 1987). As has been well documented morphologically (for example see Tooze and Tooze, 1986), SGs form in the trans-Golgi network (TGN) when a condensed core of secretory proteins becomes enveloped by membrane, and buds into the cytoplasm. The formation of SGs from the TGN is also being dissected using biochemical approaches. The first event in the formation of SGs, the condensation, or aggregation, of the secretory proteins in the TGN, has been demonstrated in vitro to require conditions similiar to those thought to exist in the TGN, specifically 1-10mM Calcium and pH6.4 (Chanat and Huttner, 1991). The second event, the budding of the newly formed granules from the TGN has been reconstituted in a cell-free assay (Tooze and Huttner, 1990), and appears to be regulated by a heterotrimeric G-protein (Barr, Leyte et al., 1991; and see abstract from F. Barr).

The budding of SGs from the TGN is a very rapid event ($t_{1/2} \approx 5$ min), which occurs as rapidly as the budding of constitutive vesicles (Tooze and Huttner, 1990). The newly formed SG is referred to as an immature secretory granule (ISG), while the SGs stored for long periods are referred to as mature secretory granules (MSGs). The properites of the ISG have been studied in PC12 cells, a neuroendocrine cell line, using [35S]sulfate labelled Secretogranin II (SgII:Tooze et al., 1991) as a granule-specific marker (Huttner, Gerdes et al., 1991). To determine the half-life of the ISG,

NATO ASI Series, Vol. H 74
Molecular Mechanisms of Membrane Traffic
Edited by D. J. Morré, K. E. Howell, and J. J. M. Bergeron
© Springer-Verlag Berlin Heidelberg 1993

PC12 cells were pulse-labelled with [35S]sulfate for 5min and then chased for various times from 15min to 6hrs. A 15min chase is sufficient to allow transport of the labelled SgII into an ISG, while the MSG is labelled by a long (6hr) label and long (16hr) chase. A post-nuclear supernatant is prepared from the labelled cells and the ISG and MSG are resolved by velocity controlled sucrose gradients. This analysis has shown that the ISG undergoes a conversion to a MSG with a $t_{1/2}$ of ≈45 min (Tooze et al., 1991). To eliminate any artefacts due to dehydration of the SGs in high concentrations of sucrose, and to demonstrate that the ISG and the MSG were two distinct populations of particles, the ISG and MSG identified by the [35S]sulfate label were subjected to analytical differential centrifugation in isotonic sucrose (Fig. 1).

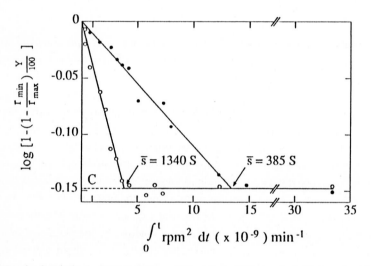

$$\int_0^t rpm^2 \, dt \, ( \times 10^{-9} ) \, min^{-1}$$

Figure 1. Analytical differential centrifugation (Slinde and Flatmark, 1973) of the ISGs (closed circles) and MSGs (open circles). A post-nuclear supernatant was prepared from cells after either a 5min pulse followed by a 15min chase (ISG) or a 6hr label followed by a 16hr chase (MSG) and subjected to analytical differential centrifugation. Results were quantitated and plotted as detailed in Tooze et al., 1991.

The results of the velocity centrifugation and the analytical differential centrifugation showed that during the chase the ISG was converted to a MSG. One component of this conversion appeared to result in change in the sedimentation value of the ISG from 385s to 1340s. To determine the changes that the ISG undergoes after budding from the TGN, preparations enriched in ISG and MSGs were prepared from PC12 cells employing sequential velocity and equilibrium sucrose gradients (Tooze et al., 1991). The preparations of the ISGs and MSGs were examined by thin section electron microscopy (Fig. 2), and morphometric analyses of the dense-cored structures was performed (Table 1).

The results of the morphological analysis show that the ISGs are smaller (average core size of 80 mn) than the MSGs (average core size of 120 mn). The increase in SG size during maturation may reflect a fusion event between ISG and ISG. This

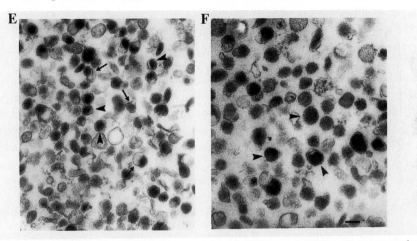

Figure 2. EM analysis of preparations enriched in ISGs and MSGs. Representative thin sections are shown of ISGs (panel E) and MSGs (panel F). Note the presence of dense-cored structures in both preparations, and that the dense cores of the ISGs (arrowhead in E) are smaller than those of the MSGs (arrowhead in F). The membrane of the ISG is often more irregularly shaped and loosely surrounding the core (arrow right panel).

hypothesis is supported by the pioneering studies of Farquhar in the pituitary mammotroph cell (Farquhar et al., 1978) which demonstrated that the nascent SGs

TABLE 1. Comparison of the diameter of the dense core of ISG and MSG

| | DENSE CORE DIAMETER | |
| SECRETORY GRANULE TYPE | measured mean (nm) | true mean (nm) |
| --- | --- | --- |
| Immature | 79.9 ± 12.8 (n=120) | 101.6 |
| Mature | 120.0 ± 11.7 (n=134) | 152.9 |

ISGs and MSGs were isolated from PC12 cells using sequential velocity and equilibrium gradient centrifugation. The core diameter was measured from micrographs of the granule preparations taken at a magnification of x17,000. Only those dense cores were measured where the granule membrane was visible surrounding the entire core. The calculations made to obtain the measured and true core diameter of the ISG and MSG were done according to Aherne and Dunnill, 1982.

undergo a morphological change in size which involves the fusion of individual granules. It is also supported by recent morphological data in AtT20 cells which suggests the ISG is a dynamic intermediate which may be extensively remodeled during maturation (Tooze and Tooze, 1986).

Several important questions are raised by the hypothesis that ISGs fuse during their maturation (Tooze, 1991). Of particular interest are the molecules involved in this fusion event, specifically the different classes of GTP-binding proteins (Pfeffer, 1992; Barr, Leyte et al., 1992). Since the ISGs and MSGs can fuse at the plasma membrane (Tooze et al., 1991) in a process mediated by heterotrimeric G-protein(s) (Tatham and Gomperts, 1991), it is of interest to determine if the same, or different GTP-binding proteins are involved in the ISG-ISG fusion event. To address this question a cell-free system is being developed to reconstitute the ISG-ISG fusion event and thereby allow the biochemical identification of the molecules involved.

Aherne, W. A. and M. S. Dunnill. (1982). Methods of estimating size distributions. Morphometry, Edward Arnold Ltd., London
Barr, F. A., A. Leyte and W. B. Huttner (1992). Trimeric G proteins and vesicle formation. Trends Cell Biol. 2:91-94
Barr, F. A., A. Leyte, S. Mollner, T. Pfeuffer, S. A. Tooze and W. B. Huttner (1991). Trimeric G-proteins of the trans-Golgi network are involved in the formation of constitutive secretory vesicles and immature secretory granules. FEBS 294:: 239-243
Burgess, T. L. and R. B. Kelly (1987). Constitutive and regulated secretion of proteins. Ann. Rev. Cell Biol. 3:243-293
Chanat, E. and W. B. Huttner (1991). Milieu-induced, selective aggregation of regulated secretory proteins in the trans-Golgi network. J. Cell Biol. 115:1505-1519
Farquhar, M. G., J. J. Reid and L. W. Daniell (1978). Intracellular transport and packaging of prolactin: a quantitative electron microscope autoradiographic study of mammotrophs dissociated from rat pituitaries. Endocrinology 102:296-311
Huttner, W. B., H.-H. Gerdes and P. Rosa (1991). The granin (chromogranin/ secretogranin) family. TIBS 16:27-30
Pfeffer, S. R. (1992). GTP-binding proteins in intracellular transport. Trends Cell Biol. 2:41-46
Slinde, E. and T. Flatmark (1973). Determination of sedimentation coefficients of subcellular particles of rat liver homogenates. Anal. Biochem. 56:324-340
Tatham, P. E. R. and B. D. Gomperts (1991). Late events in regulated exocytosis. Bioessays 13:397-401
Tooze, J. and S. A. Tooze (1986). Clathrin-coated vesicular transport of secretory proteins during the formation of ACTH-containing secretory granules in AtT20 cells. J. Cell Biol. 103:839-850
Tooze, S. A. (1991). Biogenesis of secretory granules. Implications arising from the immature secretory granule in the regulated pathway of secretion. FEBS 285: 220-224
Tooze, S. A., T. Flatmark, J. Tooze and W. B. Huttner (1991). Characterization of the immature secretory granule, an intermediate in granule biogenesis. J. Cell Biol. 115:1491-1503
Tooze, S. A. and W. B. Huttner (1990). Cell-free protein sorting to the regulated and constitutive secretory pathways. Cell 60:837-847

# FATTY ACYLATION IN MEMBRANE TRAFFICKING AND DURING MITOSIS

Dorothy I. Mundy
Department of Neurology
Baylor College of Medicine
One Baylor Plaza
Houston, Texas  77030

To date, there are four known classes of lipid modifications to proteins: palmitoylation, myristoylation, the addition of a GPI anchor and isoprenylation.  Of these modifications, it appears that only palmitoylation is readily reversible, perhaps providing an additional mechanism by which the activity of a protein can be modulated.  Palmitate is usually found attached via a thioester bond to cysteine residues, which can reside at any point in the primary structure of the protein, but is usually found at positions that are close to the membrane.  There are two classes of proteins that contain covalently bound palmitate.  A relatively minor subset are transmembrane glycoproteins and these are usually acylated shortly after synthesis, in a compartment between the ER and the Golgi.  The majority of the palmitoylated proteins, appear to be otherwise hydrophilic proteins, that are tightly associated with the cytoplasmic face of cellular membranes.  A number of palmitoylated proteins have been shown to undergo reversible fatty acylation, including several members of the ras super family, several erythrocyte cytoskeletal proteins, and in 1987 there was a report by Glick and Rothman (1987) that a cycle of reversible acylation might be involved in Golgi transport.  At that time I was interested in how transport to the Golgi apparatus was inhibited during mitosis and thought it might be possible to identify possible target proteins for acylation, by comparing the acylation patterns in mitotic and interphase cells.

Membrane traffic, in both the endocytic and the secretory pathways is inhibited during mitosis (Warren, 1985).  Therefore, studying trafficking in mitotic cells provides a unique system to identify components of the transport machinery.  In interphase cells, the Golgi apparatus exists as a single copy organelle, in a perinuclear position.  In mitotic cells, however, the Golgi is disassembled and spread throughout the cytoplasm.  Transport between the ER and the Golgi apparatus is inhibited perhaps because there is no functional acceptor compartment under these conditions and intermediates may accumulate.  When I labeled mitotic and interphase cells with $^3$H palmitate there was a dramatic increase in the labeling of a protein with an apparent molecular weight of 62 kD in the mitotic cells (Mundy and Warren, 1992).  Pretreatment of cells with cycloheximide had no affect on the acylation of p62, indicating that

NATO ASI Series, Vol. H 74
Molecular Mechanisms of Membrane Traffic
Edited by D. J. Morré, K. E. Howell, and J. J. M. Bergeron
© Springer-Verlag Berlin Heidelberg 1993

it was an existing protein, and not a newly synthesized protein that was being acylated. The acylation of p62 is tightly associated with the cell cycle. Mitotic cells are prepared using the microtubule inhibitor, nocodazole to arrest them at metaphase. If the nocodazole is removed, the cell completes mitosis, and enters G1. If cells are labeled for one hour immediately after the nocodazole is removed there is a dramatic decrease in the labeling, and by 30 minutes after wash out, when most of the cells have entered G1, there is no longer any labeling of p62. Acylated p62 is tightly associated with cellular membranes and is not found in the cytosol. p62 cannot be extracted from membranes by treatment with high salt, such as 1M KCl, which is typical for fatty acylated proteins. p62 can, however, be removed from membranes by urea and is partially solubilized at 4 M and completely solubilized at 5-6 M.

Another inhibitor of membrane transport that has dramatic affects on the morphology of the Golgi apparatus is the fungal metabolite brefeldin A (BFA), and it may be that some of the mechanisms that mediate this disassembly are similar to what occurs in mitosis. When I examined the effect of brefeldin on fatty acylation I found that the same 62 kD protein was acylated. When mitotic cells were treated with BFA there was no additional increase in the acylation of p62, which suggests that they are the same protein. Pretreatment of cells with cycloheximide did not abolish acylation, again indicating that it is an existing and not a newly synthesized protein that is being acylated, also demonstrating that p62 is not a mitosis-specific protein. To confirm that the proteins were identical, peptide digests were compared, using the Cleveland method. Two different enzymes generate the same $^3$H-palmitate labeled fragments, whether the protein was isolated from mitotic or brefeldin treated cells. Since one of the effects of BFA is to cause the redistribution of Golgi enzymes to the ER or rather, to a compartment just after the ER, I wanted to test the possibility that p62 was being acylated because it was redistributed to a site where the enzyme resides. In the presence of high concentrations of nocodazole, the redistribution of the Golgi enzymes is inhibited. However, p62 is acylated equally well in the presence or absence of this microtubule inhibitor. Nocodazole alone has no effect on acylation. The acylation of p62 can also be affected by other inhibitors of intracellular transport but with different efficiencies. CCCP interferes with transport out of the ER and is quite effective at increasing the acylation of p62. BFA remains the most effective, but both monensin and aluminum fluoride increase acylation to some extent.

These studies were done in CHO cells but a p62-like protein exists in many other cell-lines. Acylation is dramatically increased in NRK cells, A431 cells, Hela cells, and BHK cells in the present of a BFA. However, there is no effect of BFA on the acylation of p62 in PtK1 cells, a kangaroo kidney cell line. This is particularly interesting because this cell type is known to be resistant to the effects of BFA on the Golgi apparatus. This suggests that the effects of BFA on acylation of p62, is related specifically to the early part of the pathway and not to its effects on the TGN or the endosomal/lysosomal system. Where is p62? Subcellular localization was accomplished using sucrose gradients, where it was found to co-fractionate with a marker for the intermediate compartment or the cis-Golgi network. Sucrose gradient fractions of $^3$H-palmitate labeled proteins were precipitated and fluorographed or blotted with an antibody to p58, which is a marker protein for the cis-Golgi network, kindly provided by Jaakko Saraste (Saraste et al, 1987). p62 co-fractionates exactly, with p58 and with no other marker protein.

To further investigate the hypothesis that p62 is involved in membrane trafficking, I have reconstituted the acylation of p62 in permeabilized cells. When CHO cells are perforated by scraping them from tissue culture dishes and then labeled with $^3$H-palmitate in the presence of ATP and cytosol, there is a time-dependent increase in the labeling of p62, which plateaus between 1 and 2 hours. The labeling of p62 is specifically dependent on added cytosol and almost all acylation is dependent on the addition of ATP and an ATP-regenerating system. In addition, it appears that GTP-$\gamma$-S is able to block acylation. All of these results are consistent with acylation being dependent on transport but many processes are reconstituted in permeabilized cells and so this is just the beginning.

How might this fit in with models for budding and fusion? The current model suggests that budding of vesicles is dependent upon ATP, GTP and cytosolic proteins, such as COP's and ARF, and possibly others. One of the earliest effects of brefeldin is to interfere with the binding of at least two of these cytosolic proteins, the 110 kD $\beta$ COP protein and ARF. It has been shown that GTP-$\gamma$-S blocks the effects of brefeldin, possibly by locking ARF and $\beta$ COP onto membranes. One possibility is that the acyl group on p62 is constantly turning over in response to the binding of one or more of these proteins. In other words, binding results in the acylation of p62 and the dissociation of those proteins results in reacylation of p62. When BFA blocks the binding of cytosolic proteins, the result is that p62 is no longer deacylated and is trapped on the membrane in an acylated form. The effect of GTP-$\gamma$-S then, is to lock these cytosolic proteins on the membrane and thereby trap p62 in the deacylated form. This makes no assumptions about whether p62 is involved at the stage of budding, which has been suggested by the work with acyl CoA on isolated Golgi membranes, or if it is required at the site of docking or fusion, since GTP-$\gamma$-S is thought to accumulate vesicles at this stage in the transport pathway.

# REFERENCES

Glick, BS and Rothman, JE (1987) Possible role for fatty acyl-coenzyme A in intracellular protein transport. Nature 326:309-312

Mundy, DI and Warren, ·G (1992) Mitosis and inhibition of intracellular transport stimulate palmitoylation of a 62-kD protein. J Cell Biol 116:135-146

Saraste, J, Palade, GE and Farquhar, MG (1987) Antibodies to rat pancrease Golgi subfractions: Identification of a 58 kD cis-Golgi protein. J Cell Biol 105:2021-2029

Warren, G (1985) Membrane traffic and organelle division. TIBS 10:439-443

# RATES OF SYNTHESIS AND SELECTIVE LOSS INTO THE BILE OF FOUR RAT LIVER PROTEINS AND THE POLYMERIC IgA RECEPTOR

Laura J. Scott and Ann L. Hubbard,
Department of Cell Biology and Anatomy
Johns Hopkins University School of Medicine,
725 N. Wolfe St.
Baltimore, MD 21205

## INTRODUCTION

Hepatocytes have three distinct plasma membrane (PM) surfaces: apical (bile canaliculus), basal (blood) and lateral. Each domain contains distinct membrane proteins, such as the apical proteins, dipeptidyl peptidase IV (DPPIV) and HA4 (Ecto-ATPase) (Hubbard et al., 1985, Bartles et al., 1985), a basolateral protein, CE9, (Hubbard et al., 1985) and a basolateral protein which is concentrated on the lateral surface, HA321 (Braiterman and Hubbard, 1985). We have determined the half-lives and cellular amounts of these four PM molecules, their rates of synthesis and the amount of loss from the cell that can be accounted for by their release into bile. We have compared these values to those for the pIgA-receptor (pIgA-R), a sacrificial receptor, which travels first to the basolateral surface where it can bind pIgA, is then transported to the apical surface, and is subsequently cleaved externally and released into the bile as secretory component (SC) (Brown and Koppel, 1989).

## METHODS

Detailed methods are described in (Scott and Hubbard, 1992). Briefly, half-lives were determined by metabolically labeling rats with [35S]-methionine, [35S]-cysteine or Tran 35S-Label and then chasing with unlabeled methionine and/or cysteine for varying times. Following homogenization of the liver, specific antigens were immunoprecipitated

NATO ASI Series, Vol. H 74
Molecular Mechanisms of Membrane Traffic
Edited by D. J. Morré, K. E. Howell, and J. J. M. Bergeron
© Springer-Verlag Berlin Heidelberg 1993

homogenization of the liver, specific antigens were immunoprecipitated with monoclonal antibody coupled to Sepharose, run on SDS-PAGE, autoradiographed, and analyzed by densitometry. To determine the number of molecules per hepatocyte, PM proteins and (SC) were affinity-purified from extracts of PMs or bile. Immunoblot analysis was used to compare a standard curve of purified protein to homogenate. Rates of synthesis were determined by two separate methods: 1) calculation of the rate of synthesis from the number of molecules and the rate of degradation; or 2) direct comparison of the amount of radiolabel incorporated into each protein following a 30 or 60 min labeling. To determine the amount of each protein released into bile/day, bile was collected for 20 min and compared to homogenate by immunoblot analysis.

## RESULTS/DISCUSSION

The average half-life of pIgA-R is 1.1 hrs assuming a first order decay. In contrast, the half-lives of four resident PM proteins are much longer (4-9 days) (Table 1). All of the molecules are abundant components of the hepatocyte ($1-8 \times 10^6$ molecules/hepatocyte) (Table1).

### TABLE 1[a]

HALF-LIVES AND AMOUNTS OF pIgA-R AND PM PROTEINS

| Molecule | Half-Lives (Average) | Number or Molecules molecules/hep. x $10^{-6}$ | Rate of synthesis molecules/hep./hr x $10^{-4}$ |
|---|---|---|---|
| pIgA-R | 1.1 hours | $2.6 \pm 0.2$ | 160 |
| Apical | | | |
| HA4 | 5 days | $3.7 \pm 1.0$ | 2 |
| DPPIV | 9 days | $5.5 \pm 1.6$ | 2 |
| Basolateral | | | |
| CE9 | 5 days | $7.8 \pm 1.6$ | 5 |
| HA321 | 4 days | $1.3 \pm 0.4$ | 0.9 |

a.  Adapted from (Scott and Hubbard, 1992)

Using these two sets of values, we calculated that pIgA-R is synthesized at a rate of 1.6 x $10^6$ molecules/hepatocyte/hr, which is ~1/10th that of albumin, the most abundant secretory protein made by the hepatocyte (Lewandoski et al, 1988). The four PM proteins are synthesized at rates we calculated to be 30 to 150 fold lower than pIgA-R. However, direct metabolic labeling measurements suggest somewhat higher relative synthesis rates for the PM proteins.

To determine if release into bile could account for the degradation of apical PM molecules, we compared their levels in homogenate and bile (Figure 1). pIgA-R's rate of release into bile as SC, 670%/day, accounts for at least 50-75% of the total loss of the molecule from the cell. In contrast, the loss of HA4 and DPPIV via this route, although measurable, (2-3 %/day), is much less. (Figure 1). The loss of these two resident apical proteins into bile accounts for <20-30% of their loss from the cell, suggesting that the vast majority of these molecules must be endocytosed from the apical surface prior to their degradation internally. Apical endocytosis has been observed in a variety of polarized epithelial cells (Christensen, 1982, Bomsel et al., 1989) and retrograde perfusion of tracers into the rat bile duct has also suggested that membrane retrieval occurs at the apical surface of hepatocytes (Matter et al., 1969, Jones et al.,1984).

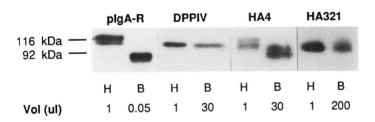

Figure 1. Release of pIgA-R, DPPIV, HA4 and HA321 into bile. Bile was collected for 20 min and the liver subsequently homogenized. 1µ of homogenate (H) and 0.05, 20 or 200 µl of bile (B) were analyzed by immunoblot detection. DPPIV was released into bile with an unchanged molecular weight and HA4 was found in two forms, one of which migrated more quickly than the homogenate HA4. Of the basolateral proteins only a very small amount of intact HA321 could be detected in bile. Reproduced from (Scott and Hubbard, 1992).

# REFERENCES

Bartles JR, Braiterman LT, Hubbard AL (1985) J Cell Biol 100: 1126- 1138

Bomsel M, Prydz K, Parton RG, Gruenburg J, Simons K (1989) J Cell Biol 109: 3243-3258

Braiterman LT, Hubbard AL (1985) J Cell Biol 101: 61

Brown WR, Kloppel TM (1989) Immunol Invest 18: 269-285.

Christensen EI (1982) Eur J Cell Biol. 29: 43-49

Hubbard AL, Bartles JR, Braiterman LT (1985) J Cell Biol 100:1115- 1125

Jones AL, Hradek GT, Schmucker DL, Underdown BJ (1984) Hepatology 4:1173-1183

Lewandoski AE, Liao WSL, Stinson-Fisher CA, Kent JD, Jefferson LS (1988) Am J Physiol 254: C634-C642

Matter A, Orci L, Rouiller C (1969) J Ultrastruct Res 11: (suppl.) 1-71

Scott LJ, Hubbard AL (1992) J Biol Chem 267: 6099-6106

# Effect of Brefeldin A Treatment on the Resident Golgi Protein, MG160

Patricia A. Johnston, and Nicholas K. Gonatas
Dept. of Pathology and Laboratory Medicine
U. of Pennsylvania School of Medicine
Philadelphia, Pa. 19104-6079

**INTRODUCTION:** The Golgi apparatus (GA) plays a crucial role in the post-translational processing and sorting of many cellular proteins. A small number of Golgi-specific proteins have been identified, including MG160 (Gonatas et al., 1989), an integral membrane sialoglycoprotein concentrated in the medial cisternae. Analysis of the biochemical properties of these Golgi-specific proteins should provide insight into the molecular mechanisms which underlie GA function. Here we examine the effects of BFA treatment on the lifecycle of MG160 in order to determine if it undergoes unique carbohydrate processing or targeting within the GA. Our data indicate that BFA promotes the accumulation of a 150 Kd precursor to MG160 which is Endo-H resistant, but not sialylated. These results raise interesting questions about the transit of MG160 through the GA.

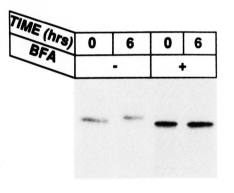

**Figure 1.** *Cells were treated with 5 µg/ml BFA for 30 minutes, pulse-labelled, and immunoprecipitated with an anti-MG160 monoclonal antibody, 10A8.*

**RESULTS:** MG160 was immunoprecipitated from PC12 cells pulse-labelled with $^{35}$S-methionine. During the 30 minute labelling period MG160 attained an apparent molecular weight of 150 Kd, and after 6 hours of chase the protein had attained its mature form of 160 Kd. In the presence of BFA, MG160 never matured to its 160 Kd form (figure 1). To determine at what step BFA inhibits the post-translational modification of MG160, immunoprecipitates were treated with Endo-H (figure 2). In control cells, MG160 becomes resistant to Endo-H by 2 hours of chase, and in the presence of BFA, at least partial Endo-H

resistance is achieved by 2 hours, but this process does not shift the molecular weight of the protein to 160 Kd. The processing step responsible for this final shift in molecular weight is apparently sialylation, as the 150 Kd precursor of MG160 immuno-purified from BFA-treated cells does not cross-react with a lectin (Limax Flavus lectin) specific for sialic acid residues (figure 3).

NATO ASI Series, Vol. H 74
Molecular Mechanisms of Membrane Traffic
Edited by D. J. Morré, K. E. Howell, and J. J. M. Bergeron
© Springer-Verlag Berlin Heidelberg 1993

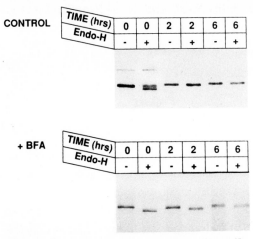

**Figure 2.** *MG160 was immunoprecipitated from $^{35}$S-labelled control and BFA-treated cells at the indicated times of chase and treated with Endo-H (0.1 U/ml).*

**CONCLUSION:** We have utilized BFA to characterize the biochemical maturation of a resident protein of the GA, MG160, in order to determine if this protein is uniquely processed. The observation that newly synthesized MG160 becomes Endo-H resistant in BFA treated cells suggests that it is exposed to enzymes of the cis and medial golgi cisternae. Numerous reports have proposed that the enzymes of these cisternae enter the endoplasmic reticulum (ER) during BFA exposure (Doms et al., 1989; Lippincott-Schwartz et al., 1989). The failure of MG160 to be sialylated in the presence of BFA suggests that it does not encounter a sialyltransferase, presumably because this enzyme is localized in the trans golgi network (TGN) of PC12 cells and does not relocate to the ER during BFA treatment. Sialyltransferases have been localized to the TGN of other cell types (Roth et al., 1985). The observation that BFA inhibits sialylation of MG160 is in agreement with previous results obtained for other proteins (Doms et al., 1989; Lippincott-Schwartz et al., 1989). MG160, however, is uniquely different from these proteins in that it is a permanent resident of the GA. If, in the intact GA, MG160 enters the TGN to be sialylated, then does it undergo retrograde transport to the medial cisternae? We are currently addressing this question by examining the distribution of MG160 during recovery from the effects of BFA.

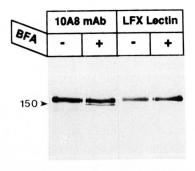

**Figure 3.** *MG160 immunopurified from control and BFA-treated cells was separated by PAGE, transferred, and probed with either 10A8 supernatent or Limax Flavus Lectin.*

**REFERENCES:**

Doms RW, Russ G and Yewdell JW (1989) Brefeldin A Redistributes Resident and Itinerent Golgi Proteins to the Endoplasmic Reticulum. J. Cell Biol 109:61-72.

Gonatas JO, Mezitis SGE, Stieber A, Fleischer and Gonatas NK (1989) MG160: A Novel Sialoglycoprotein of the Medial Cisternae of the Golgi Apparatus. JBC 264:646-653.

Lippincott-Schwartz J, Yuan LC, Bonificano, JS and Klausner RD (1989) Rapid Redistribution of Golgi Proteins into the ER in Cells Treated with Brefeldin A: Evidence for Membrane Cycling from Golgi to ER. Cell 56:801-813.

Roth J, Taatjes DJ, Lucocq JM, Weinstein J, Paulson JC (1985) Demonstration of an Extensive Trans-Tubular Network Continuous with the Golgi Apparatus Stack that may Function in Glycosylation. Cell 43:287-295.

# INHIBITION OF CHOLERA TOXIN BY BREFELDIN A

Palmer A. Orlandi, Patricia Currran, and Peter H. Fishman
Membrane Biochemistry Section
Laboratory of Molecular and Cellular Neurobiology, NINDS
The National Institutes of Health
Bethesda, MD  20892

Many aspects of the mechanism of action of cholera toxin (CT), the causative agent of cholera are well established (Fishman, 1990). CT consists of a pentameric B subunit which binds to ganglioside $G_{M1}$ on the cell surface and an A subunit which activates adenylyl cyclase. The latter process involves the reduction of A to $A_1$ peptide which ADP-ribosylates the stimulatory G protein, $G_s$ of adenylyl cyclase. There is a distinct lag phase of 15 to 20 minutes between toxin binding and its activation of adenylyl cyclase. Little is known, however, about events during this lag including where $A_1$ is generated and how it gains access to $G_s$ on the cytoplasmic side of the plasma membrane. In an attempt to identify those steps involved in the intracellular processing of CT and the site of CT-A reduction to the $A_1$ peptide, we have employed several known inhibitors of intracellular trafficking to study their effects on the response of human SK-N-MC neurotumor and CaCo-2 intestinal tumor cells to CT. Chloroquine and monensin affect intracellular trafficking by increasing endosomal and lysosomal pH. Monensin is also known to effect glycoprotein processing in the Golgi. Concentrations of chloroquine and monensin as high as 400 $\mu M$ and 10 $\mu M$ respectively, had little or no effect on CT stimulation of cyclic AMP in either cell line. Brefeldin A (BFA) however, totally inhibited the response to CT in a time, dose-dependent, and reversible manner with an $IC_{50}$ of 30 ng $mL^{-1}$. When added simultaneously with CT or shortly thereafter, the inhibitory effects of BFA were still evident. The effects of BFA however, diminished considerably when added >10 min following CT addition suggesting that BFA exerted its effects at an unknown step preceding toxin activation. BFA had no effect on $G_{M1}$, as cells treated for 30 min with BFA bound $^{125}I$-CT as well as control cells. ß-adrenergic agonist-stimulation of cyclic AMP was also unaffected by the presence of BFA as was the activation of adenylyl cyclase by CT-$A_1$ and NAD in cell membrane preparations indicating that it did not directly alter adenylyl cyclase. Furthermore, adenylyl cyclase was fully activated by $A_1$ and NAD in membranes from BFA-treated cells which indicated that BFA treatment did not deplete the plasma membrane of some component (such as ADP-ribosylation factor) required for modification of $G_s$ by CT. Whereas control cells generated small amounts of CT-$A_1$ in a time-dependent manner that paralleled the activation of adenylyl cyclase, no $A_1$ was detected in BFA-treated cells up to 60 min following toxin exposure. Toxin internalization as monitored by the reactivity of anti-CT-$A_1$ antiserum, however, indicated that control and BFA-treated cells internalized CT at

NATO ASI Series, Vol. H 74
Molecular Mechanisms of Membrane Traffic
Edited by D. J. Morré, K. E. Howell, and J. J. M. Bergeron
© Springer-Verlag Berlin Heidelberg 1993

identical rates. In contrast to the comparable rates of toxin up-take, intracellular CT was degraded much slower in BFA-treated cells. Under these conditions, CT may be accumulating intracellularly at an unknown site which was directly affected by the presence of BFA.

BFA has been shown to induce dramatic morphological and biochemical changes in organelles of the secretory pathway (Doms, et al., 1989, Lippincott-Schwartz, et al., 1989). Specifically, BFA causes a morphological disassembly of the Golgi apparatus with a redistribution of Golgi-resident proteins into the ER (Lippincott-Schwartz, et al., 1989) and functionally dissects the Golgi cisternae from the *trans* Golgi network (Chege and Pfeffer, 1990). Additionally, BFA prevents membrane assembly of cytosolic coat proteins which participate in the formation of non-clathrin-coated vesicles required for vesicular transport between Golgi cisternae (Orci, et al., 1991) as well coat proteins associated with the clathrin-coated vesicles that bud from the *trans*-Golgi network (Robinson and Kreis, 1992). These properties have allowed for extensive use of BFA in deducing the intricacies of intracellular membrane protein trafficking. Several possibilities exist to explain the dramatic inhibitory effects of BFA on the action of CT in contrast to chloroquine or monensin: 1. disruption of the intracellular transport mechanism responsible for the delivery of the internalized toxin to the site of reduction; 2. BFA-induced inhibition of the reductase necessary to generate $A_1$ peptide; or, 3. mislocation of the reductase caused by the BFA-induced uncoupling of the Golgi apparatus. Further studies on the activation of CT in the presence of BFA may allow us to understand more fully the steps and sites involved in the intoxication process.

Chege, N., and Pfeffer, S.R. (1990) Compartmentation of the Golgi complex: Brefeldin-A distinguishes *trans*-Golgi cisternae from the *trans*-Golgi network. J. Cell Biol., 111: 893-899.

Doms, R.W., Russ, G., and Yewdell, J.W. (1989) Brefeldin A redistributes resident and itinerant Golgi proteins to the endoplasmic reticulum. J. Cell Biol., 109, 61-72.

Fishman, P.H. (1990) in *ADP-Ribosylating Toxins and G Proteins: Insights into Signal Transduction* (Moss, J., & Vaughan, M., Eds.) pp 127-140, American Society of Microbiology, Washington.

Lippincott-Schwartz, J., Yuan, L.C., Bonifacino, J.S., and Klausner, R.D. (1989) Rapid redistribution of Golgi proteins into the ER in cells treated with Brefeldin A: evidence for membrane cycling from Golgi to ER. Cell, 56, 801-813.

Orci, L., Tagaya, M., Amherdt, M., Perrelet, A., Donaldson, J.G., Lippincott-Schwartz, J., Klausner, R.D., and Rothman, J.E. (1991) Brefeldin A. a drug that blocks secretion, prevents the assembly of non-clathrin -coated buds on Golgi cisternae. Cell, 64, 1183-1195.

Robinson, M.S., and Kreis, T.E. (1992) Recruitment of coat proteins onto Golgi membranes in intact and permeabilized cells: effects of Brefeldin A and G protein activators. Cell, 69, 129-138.

# SELECTIVE ASSOCIATION WITH GOLGI OF ADP-RIBOSYLATION FACTORS, 20-kDa GUANINE NUCLEOTIDE-BINDING PROTEIN ACTIVATORS OF CHOLERA TOXIN

S.-C. Tsai, R. Adamik, R. Haun, J. Moss, and M. Vaughan, Laboratory of Cellular Metabolism, National Heart, Lung, and Blood Institute, National Institutes of Health, Bethesda, MD 20892 USA.

ADP-ribosylation factors are 20-kDa GTP-binding proteins initially identified as factors required for cholera toxin-catalyzed ADP-ribosylation of $G_{s\alpha}$, the stimulatory GTP-binding protein of the adenylyl cyclase system (Kahn and Gilman, 1984; Bobak et al., 1990). The GTP-dependent ARFs enhance activity of the toxin with all of its substrates, and are believed physiologically to participate in protein transport by Golgi (Tsai et al., 1988; Donaldson et al., 1991).

ARF structures are highly conserved throughout eukaryotes from *Giardia* to primates (Murtagh et al., 1992). Six mammalian ARFs identified by cDNA cloning fall into three classes based on size, deduced amino acid sequences, and phylogenetic analysis (Tsuchiya et al., 1991): class I, ARFs 1, 2, and 3; class II, ARFs 4 and 5; class III, ARF 6. Recently, additional evidence consistent with this classification was obtained from the structure of ARF genes. Class I ARF genes contain four introns and five exons, with coding region (exons 2 to 5) intron/exon junctions in identical locations and the same distribution of consensus sequences for GTP-binding and hydrolysis. Class II ARF genes differ from the class I genes in the location of introns in relation to these consensus sequences.

Six mammalian ARFs were expressed as recombinant proteins in *E. coli*. All activate cholera toxin and are dependent on GTP for activity (Price et al., 1992). Amino acid sequences from tryptic peptides of purified sARF I and sARF II isolated from bovine brain cytosol were identical to deduced amino acid sequences of ARF 1 and ARF 3 cDNAs, respectively. Polyclonal antibodies against sARF II reacted with ARFs 1, 2, and 3 equally well, but reacted poorly with ARFs 4 and 5. Polyclonal antibodies against rARF 5 reacted with ARF 5, less well with ARF 4 and only slightly with class I ARFs; neither reacted with rARF 6.

Rat brain homogenate (850 xg supernatant) was incubated with ATP and a regenerating system at 37°C, and then centrifuged in sucrose density gradients to yield six fractions: soluble proteins (1 and 2); light microsomes; microsomes and Golgi; Golgi; Golgi and mitochondria. Golgi fractions contained galactosyltransferase activity. After incubation with GDPßS, ~90% of total ARF activity was present in the two soluble fractions. Whereas, after incubation with GTPγS, ~54% was in the soluble fractions, and 44% in the organelle fractions. Proteins (50 μg) from each fraction were separated by SDS-PAGE, transferred to nitrocellulose, and incubated with sARF II and rARF 5 antibodies. After incubation with GTPγS,

NATO ASI Series, Vol. H 74
Molecular Mechanisms of Membrane Traffic
Edited by D. J. Morré, K. E. Howell, and J. J. M. Bergeron
© Springer-Verlag Berlin Heidelberg 1993

a prominent ARF 5 band was associated with the Golgi fraction and considerable ARF 1 appeared in the microsomes plus Golgi and Golgi fractions. Immunoreactive ARF 3 was the major band in soluble fractions, but was present in relatively small amounts and equally distributed in organelle fractions.

When the Golgi fraction (50 $\mu$g protein) was incubated with purified bovine brain ARF 1 or 3 (10 or 20 $\mu$g), binding of ARF 1 increased with increasing concentration, but ARF 3 binding was unchanged. After incubation of Golgi fraction with ARF fractions from kidney or heart cytosol, ARF activity was associated with Golgi when GTP$\gamma$S (but not GDP$\beta$S) was present, as were immunoreactive ARFs 1 and 5. On incubation with rat brain ARFs and GTP$\gamma$S, and an ATP regenerating system, Golgi binding of ARF activity and immunoreactive ARFs 1 and 5 increased with increasing ATP concentration. ARFs 1, 3, and 5 have similar biochemical properties, yet behave differently at the cellular level. In this regard, the ARFs may resemble the rab proteins (Charrier et al., 1990), each with its specific and selective role in physiological regulation.

**References**

Bobak DA, Tsai S-C, Moss J, Vaughan M (1990) Enhancement of cholera toxin ADP-ribosyltransferase activity by guanine nucleotide-dependent ADP-ribosylation factors. In: Moss J, Vaughan M (eds) ADP-ribosylating toxins and G proteins: Insights into signal transduction. American Society for Microbiology, Washington, DC, pp 439-456

Donaldson JG, Kahn RA, Lippincott-Schwartz J, Klausner RD (1991) Binding of ARF and $\beta$-COP to Golgi membranes: possible regulation by a trimeric G protein. Science 254:1197-1199

Charrier P, Parton RG, Hauri HP, Simons K, Zerial M (1990) Localization of low molecular weight GTP-binding proteins to exocytic and endocytic compartments. Cell 62:317-329

Kahn RA, Gilman AG (1984) Purification of a protein cofactor required for ADP-ribosylation of the stimulatory regulatory component of adenylate cyclase by cholera toxin. J Biol Chem 259:6228-6234

Murtagh JJ Jr, Mowatt MR, Lee C-M, Lee F-JS, Mishima K, Nash TE, Moss J, Vaughan M (1992) Guanine nucleotide-binding proteins in the intestinal parasite Giardia lamblia, isolation of a gene encoding a ~20 kDa ADP-ribosylation factor. J Biol Chem 267:9654-9662

Price SR, Welsh CF, Haun RS, Stanley SJ, Moss J, Vaughan M (1992) Effects of phospholipid and GTP on recombinant ADP-ribosylation factors (ARFs). Molecular basis for differences in requirements for activity of mammalian ARFs. J Biol Chem, in press

Tsai S-C, Noda M, Adamik R, Chang P, Chen H-C, Moss J, Vaughan M (1988) Stimulation of choleragen enzymatic activities by GTP and two soluble proteins purified from bovine brain. J Biol Chem 263:1768-1772

Tsuchiya M, Price SR, Tsai S-C, Moss J, Vaughan M (1991) Molecular identification of ADP-ribosylation factor mRNAs and their expression in mammalian cells. J Biol Chem 266:2772-2777

# INTRACELLULAR TRANSPORT AND POST-TRANSLATIONAL MODIFICATIONS OF A SECRETORY HEAT SHOCK PROTEIN OF SACCHAROMYCES CEREVISIAE

M. Simonen, M. Wikström[1], B. Walse[1] and M. Makarow
Institute of Biotechnology
University of Helsinki
Valimotie 7
SF-00380 Helsinki
Finland

Several heat shock proteins serve as molecular chaperons, assisting in protein maturation by reducing malfoding, mostly in the cytoplasm, nucleus or mitochondria. Some of them, like the BiP protein, which belongs to the 70 kD heat shock proteins, are resident ER-proteins. We have cloned and characterized the first gene, the HSP150 gene of S. cerevisiae, which codes for a heat shock protein which is secreted to the exterior of the cell. The function of the protein is as yet unknown, however, it is conserved in divergent yeasts (Russo et al., 1992).

The HSP150 gene is preceded by a promoter region which contains several sequences resembling heat responsive DNA-elements (HSE). One of them, located between the TATA-box and transcription start sites, is sufficient to confer heat-regulation to transcription. Thus, the HSP150 gene is regulated by heat shock like e.g. the 70 kD heat shock genes.

Information on the deduced amino acid sequence, direct amino acid sequencing of N-termini of the purified protein and glycan analysis allow the following suggestions to be made on the maturation of the hsp150 protein. When the apoprotein of about 42 kD is translocated into the ER, the signal sequence is clea-

[1] Department of Physical Chemistry, University of Lund, Sweden

NATO ASI Series, Vol. H 74
Molecular Mechanisms of Membrane Traffic
Edited by D. J. Morré, K. E. Howell, and J. J. M. Bergeron
© Springer-Verlag Berlin Heidelberg 1993

ved off and many serine and threonine residues are glycosylated with single mannose residues. In the Golgi, the O-glycans are extended and the polypeptide chain is cleaved into two subunits. The N-terminal side of the cleavage site has a pair of basic amino acids, and thus the processing is probably perfomed by the late Golgi-located kex2 protease. The protein is secreted to the growth medium as a dimer of both subunits, which remain non-covalently attached to each other.

Unexpectedly, the hsp150 protein was secreted to the growth medium as efficiently in the sec7 mutant and the parental strain both at 25°C and 37°C. In the thermosensitive sec7 strain, other secretory proteins accumulate in the Golgi at 37°C. In a sec7 strain back-crossed with the parental strain (from Dr. R. Schekman, Berkeley), variable results were obtained. We crossed the original sec7 mutant with the back-crossed strain to examine the secretion characteristics of hsp150 in the progeny to see whether its secretion needs the SEC7 gene product.

In contrast to most yeast secretory proteins which remain intercalated in the cell wall, hsp150 diffuses efficiently to the growth medium. The larger subunit of hsp150 protein has a highly repetitive structure. Two thirds of it is composed of a 19 amino acid peptide, repeated 11 times in tandem. NMR analysis of the synthetic consensus peptide suggested that it has no conformation, and CD analysis of the purified subunit suggested that it occurs largely as a random coil. An extended structure could contribute to facile penetration of the cell wall by the hsp150 protein. According to our preliminary data, the repetitive region of the larger hsp150 subunit could serve as a leader, which does not interfere with the conformation of a heterologous protein attached to it, but is able to lead the fusion protein efficiently to the growth medium.

Russo P, Kalkkinen N, Sareneva H, Paakkola J and Makarow M (1992) A heat shock gene from Saccharomyces cerevisiae encoding a secretory glycoprotein. Proc. Natl. Acad. Sci USA 89, 3671-3675.

# TRANSPORT TO THE CELL SURFACE OF *CHLAMYDOMONAS*: MASTIGONEMES AS A MARKER FOR THE FLAGELLAR MEMBRANE

Mitchell Bernstein and Joel L. Rosenbaum
Department of Biology
Yale University
P.O. Box 6666
New Haven, CT   06511

We have begun experiments to determine how proteins are targeted to different cell surface domains of the unicellular green alga, *Chlamydomonas*. Though contiguous, the flagellar membrane and plasma membrane of *Chlamydomonas* define physically and biochemically distinct compartments.  The plasma membrane is surrounded by the cell wall, constructed from hydroxyproline rich glycoproteins, and a set of inducible enzymes, such as carbonic anhydrase and arylsulfatase,are specifically secreted into the periplasmic space.  In contrast,  the flagellar membrane is not surrounded by the cell wall and contains proteins not found on the plasma membrane, including agglutinins used for adhesion during mating, and a 350 kD protein required for gliding motility on solid substrates. (For reviews, see Harris, 1989, and Bloodgood, 1990.)

As a marker for the flagellar surface we have focused on mastigonemes, structures which extend from the flagellar membrane, but are absent from the plasma membrane (Figures 1A and 1B). *Chlamydomonas* mastigonemes  are simple structures which are 0.9 µm long and 16 nm wide (Witman et al.,1972). They are stable structures which can be isolated from the flagellar membrane after extraction with NP-40 and CsCl centrifugation. Electron microscopy of negatively stained mastigonemes reveals a repeat structure with a periodicity of 20 nm along the length of the mastigoneme (Figure 1C). Where mastigonemes are assembled and how they are transported to the flagellar membrane is unknown.

We isolated flagella from 1200 liters of cells and obtained a pure fraction of mastigonemes. The major protein of this fraction (Figure 2) was a protein of 220 kD (gp220). Previously, the major protein of mastigonemes has been described as a glycoprotein with a molecular weight between 170- 250 kD (Witman,1972; Monk et al.,1983).   Removal of carbohydrate from gp220 with trifluoromethanesulfonic acid yielded a product of 200 kD (data not shown). Thus, approximately 20 kD of gp220 is carbohydrate, which could correspond to approximately 10 N-linked oligosaccharide side chains, or an unspecified number of 0-linked sugars.

NATO ASI Series, Vol. H 74
Molecular Mechanisms of Membrane Traffic
Edited by D. J. Morré, K. E. Howell, and J. J. M. Bergeron
© Springer-Verlag Berlin Heidelberg 1993

We are currently isolating cDNAs for gp220 and generating immunological probes for studies on how this protein is assembled into mastigonemes and transported to the cell surface. The ability to combine biochemical, genetic, and molecular studies in *Chlamydomonas* should allow use of this simple eukaryote to describe how polarized, non-epithelial cell types can establish and maintain stable cell surface domains.

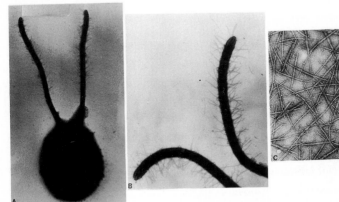
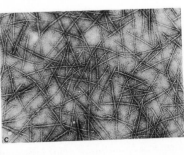

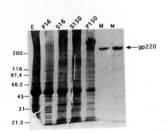

Figure1. (above) Mastigonemes. A) *Chlamydomonas* cell, x4000 B) Flagella with mastigonemes, x6000 C) Purified mastigonemes, x13,800. Samples were stained with 1% phosphotungstic acid. Panel C reproduced from Witman et al. 1972.

Figure 2. (left) Purification of mastigonemes. E, 1% NP-40 extract of flagella; M, mastigoneme fraction isolated after P150 was loaded on an equilibrium CsCl gradient. SDS-PAGE was on 5%-15% acrylamide gradient gel.

## References

Bloodgood,R A (1990) Gliding motility and flagellar glycoprotein dynamics in *Chlamydomonas*. In Ciliary and Flagellar Membranes. R A Bloodgood (ed.). Plenum Press, New York. 91-128

Harris,E H (1989) The *Chlamydomonas* Sourcebook: A Comprehensive Guide to Biology and Laboratory Use. Academic Press, Inc., New York. 780 pp.

Monk,B C , W S Adair, R A Cohen, and U W Goodenough (1983) Topography of *Chlamydomonas*: Fine structure and polypeptide components of the gametic flagellar membrane surface and the cell wall. Planta 158:517-533

Witman,G , K Carlson, J Berliner, and J L Rosenbaum (1972) *Chlamydomonas* flagella: I. Isolation and electrophoretic analysis of microtubules, matrix, membranes, and mastigonemes. J Cell Biol 54:507-539

# LIPID TRANSPORT FROM THE GOLGI TO THE PLASMA MEMBRANE OF EPITHELIAL CELLS

G. van Meer, I.L. van Genderen, W. van 't Hof, K.N.J. Burger and P. van der Bijl
Department of Cell Biology, Utrecht University Medical School
AZU H02.314
Heidelberglaan 100, 3584 CX Utrecht
The Netherlands

The two plasma membrane domains of epithelial cells display unique protein and lipid compositions. The general feature of the lipid polarity appears to be an enrichment of glycosphingolipids on the apical, and of the phospholipid phosphatidylcholine, PC, on the basolateral surface. The difference is maintained by the tight junctions, the zone of cell-cell contacts that encircles the apex of each epithelial cell. This structure acts as a barrier to lipid diffusion in the outer leaflet of the plasma membrane (van Meer, 1989b). The differences in lipid composition must therefore reside in the outer leaflets of the apical and basolateral plasma membrane domain. While the compositional differences appear to be generated by sorting of newly synthesized lipids before they reach the cell surface, lipid sorting along the transcytotic route would also seem necessary to prevent lipid intermixing by that pathway.

Lipid sorting in the exocytic pathway of epithelial cells

After synthesis a variety of glucosylceramide analogs was preferentially delivered to the apical surface of MDCK and Caco-2 cells, with an apical/basolateral polarity of delivery of ≥ 2. In contrast, sphingomyelins arrived at the surface with polarities ≤ 1 (van Meer et al., 1987; van 't Hof and van Meer, 1990; van 't Hof et al., 1992). Recent work has demonstrated that in MDCK cells also the C6-NBD-analog of the simple glycosphingolipid galactosylceramide, which differs from glucosylceramide only in the orientation of one single hydroxyl group, is preferentially transported to the basolateral domain (van der Bijl et al., in prep). The simplest mechanism to generate the different polarities of delivery would be a sorting event driven by glycosphingolipid microdomain formation in the lumenal bilayer leaflet of the TGN, followed by vesicular traffic to either plasma membrane domain (van Meer and Simons, 1988). Such a mechanism is supported by the available evidence, especially the synthesis of the sphingolipids in the Golgi (Pagano, 1990), the tendency of sphingolipids and not glycerolipids to form

NATO ASI Series, Vol. H 74
Molecular Mechanisms of Membrane Traffic
Edited by D. J. Morré, K. E. Howell, and J. J. M. Bergeron
© Springer-Verlag Berlin Heidelberg 1993

hydrogen bonds (Pascher, 1976), and a vesicular mode of transport to the cell surface (Kobayashi and Pagano, 1989; van Meer, 1989a; Schwarzmann and Sandhoff, 1990; Kobayashi et al., 1992). The strongest evidence for the vesicular nature of the transport is the observation by Kobayashi and Pagano (1989), that both glucosylceramide and sphingomyelin were unable to reach the cell surface in mitotic cells. In line with this, we have observed that transport of both lipids to the apical cell surface, but not to the basolateral surface, was reduced by microtubule depolymerization. This made us conclude that both are present in the secretory pathway before it bifurcates to the two cell surface domains at the TGN (van Meer and van 't Hof, 1992). Concerning the nature of the transport of the higher glycosphingolipids, we have established by immuno-electronmicroscopy that the complex glycosphingolipid Forssman antigen exclusively occurred in organelles that are connected by vesicular transport. It was absent from mitochondria and peroxisomes (van Genderen et al., 1991).

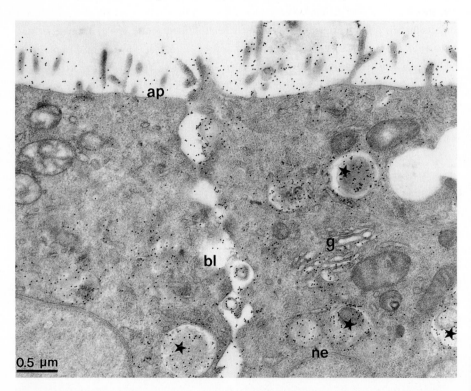

Figure 1. Distribution of the Forssman glycolipid labeled with the monoclonal antibody 33B12 (Sonnenberg et al., 1986) and 15 nm protein A-gold in freeze-substituted and Lowicryl HM 20 embedded MDCK strain II cells grown on filters. ap: apical; bl: basolateral; g: Golgi; ne: nuclear envelope; *: multivesicular structures (van Genderen et al. [1991] J Cell Biol 115: 1009; by copyright permission of the Rockefeller University Press).

Sidedness of GlcCer synthesis in the Golgi

An unexpected complexity has been encountered in the finding by various research groups that the synthesis of glucosylceramide occurs on the cytoplasmic surface of the Golgi. On the one hand it was found that in isolated intact Golgi membranes the glucosyl transferase activity was sensitive to protease treatment whereas the sphingomyelin synthase was not (Coste et al., 1986; Futerman and Pagano, 1991; Trinchera et al., 1991; Jeckel et al., 1992). In addition, newly synthesized glucosylceramide was quantitatively accessible from the cytosolic face of isolated intact Golgi membranes and in permeabilized cells, whereas sphingomyelin was completely protected (Jeckel et al., 1992). Glucosylceramide must therefore be translocated to the Golgi lumen where conversion to the more complex glycosphingolipids occurs. Evidence has been presented suggesting that also the second step in the synthesis of complex glycosphingolipids, the transfer of galactose onto the glucosylceramide to yield lactosylceramide, would be a cytosolic event (Trinchera et al., 1991). In that case both glucosylceramide and lactosylceramide would have to cross a cellular membrane to reach the surface of the plasma membrane. In order for the sorting model to be true, the translocation event ought to occur in the trans Golgi network, or before. Our observation that microtubule depolymerization affected apical but not basolateral glucosylceramide transport and that the side of precursor ceramide addition had a striking effect on the resulting polarity of glucosylceramide delivery, made us conclude that, at no stage of transport, glucosylceramide seems to be free to exchange through the cytoplasm (van Meer and van 't Hof, 1992). Our preliminary evidence suggests that glucosylceramide indeed becomes trapped in a vesicular compartment before it reaches the plasma membrane. Glucosylceramide has been found on the luminal surface of the Golgi (Karrenbauer et al., 1990) and of transport vesicles originating from the trans-Golgi network (Kobayashi et al., 1992). Purification and immunolocalization of the responsible enzymes may be necessary for a proper understanding of the early events in glycosphingolipid processing.

Lipid domains and protein sorting

We have hypothesized that microdomain formation of glycosphingolipids may be an indispensable part of the machinery that generates the lateral segregation of membrane glycoproteins in the trans-Golgi network of epithelial cells (Simons and van Meer, 1988). The same hypothesis has been proposed for the sorting of proteins that are anchored in the membrane by a glycosylphosphatidylinositol (GPI) tail. Since in MDCK cells all GPI-proteins were found to be apical, it was proposed that they partition into the putative glucosylceramide domain (Lisanti and Rodriguez-Boulan, 1990). Indeed, in FRT cells both the polarity of delivery of GPI-proteins and that of glucosylceramide were reversed (Zurzolo et al., in prep). Consistent with this notion, both GPI-proteins and sphingolipids displayed a selective resistance against extraction by Triton-X-100 in the cold (Brown and Rose, 1992). Analogous to the proposed self-aggregation of glycosphingolipids in the trans-Golgi network of epithelial cells, we have suggested that microdomain formation of the sphingolipid sphingomyelin with cholesterol plays a part in the sorting event on the *cis*-side of the Golgi (van Meer, 1989a). Lipid domains provide an appealing scenario for lateral sorting. However, the idea is based on mere correlations. First, one will have to come up with direct proof for the occurrence of sphingolipid microdomains in cellular membranes. After that, the major challenge will be to elucidate their role in intracellular lipid and protein sorting.

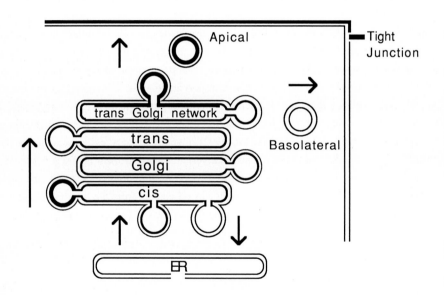

Figure 2. Potential sorting events in the secretory pathway that involve the lateral segregation of sphingolipids. Sphingomyelin aggregation may be required for outward transport from the *cis*-Golgi. Sphingomyelin would be excluded from return traffic to the ER. Glycosphingolipids would assemble into apical precursor domains in the TGN of epithelial cells.

References

Brown DA, Rose JK (1992) Sorting of GPI-anchored proteins to glycolipid-enriched membrane subdomains during transport to the apical cell surface. Cell 68: 533-544

Coste H, Martel MB, Got R (1986) Topology of glucosylceramide synthesis in Golgi membranes from porcine submaxillary glands. Biochim Biophys Acta 858: 6-12

Futerman AH, Pagano RE (1991) Determination of the intracellular sites and topology of glucosylceramide synthesis in rat liver. Biochem J 280: 295-302

Jeckel D, Karrenbauer A, Burger KNJ, van Meer G, Wieland F (1992) Glucosylceramide is synthesized at the cytosolic surface of various Golgi subfractions. J Cell Biol 117: 259-267

Karrenbauer A, Jeckel D, Just W, Birk R, Schmidt RR, Rothman JE, Wieland FT (1990) The rate of bulk flow from the Golgi to the plasma membrane. Cell 63: 259-267

Kobayashi T, Pagano RE (1989) Lipid transport during mitosis. Alternative pathways for delivery of newly synthesized lipids to the cell surface. J Biol Chem 264: 5966-5973

Kobayashi T, Pimplikar, SW, Parton, RG, Bhakdi, S, Simons, K (1992) Sphingolipid transport from the trans-Golgi network to the apical surface in permeabilized MDCK cells. FEBS Lett 300: 227-231

Lisanti MP, Rodriguez-Boulan E (1990) Glycophospholipid membrane anchoring provides clues to the mechanism of protein sorting in polarized epithelial cells. TIBS 15: 113-118

Pagano RE (1990) Lipid traffic in eukaryotic cells: mechanisms for intracellular transport and organelle-specific enrichment of lipids. Curr Opin Cell Biol 2: 652-663

Pascher I (1976) Molecular arrangements in sphingolipids. Conformation and hydrogen bonding of ceramide and their implication on membrane stability and permeability. Biochim Biophys Acta 455: 433-451

Schwarzmann G, Sandhoff K (1990) Metabolism and intracellular transport of glycosphingolipids. Biochemistry 29: 10865-10871

Simons K, van Meer G (1988) Lipid sorting in epithelial cells. Biochemistry 27: 6197-6202

Sonnenberg A, van Balen P, Hengeveld T, Kolvenbag GJC, van Hoeven RP, Hilgers J (1986) Monoclonal antibodies detecting different epitopes on the Forssman glycolipid hapten. J Immunol 137: 1264-1269

Trinchera M, Fabbri M, Ghidoni R (1991) Topography of glycosyltransferases involved in the initial glycosylations of gangliosides. J Biol Chem 266: 20907-20912

van Genderen IL, van Meer G, Slot JW, Geuze HJ, Voorhout WF (1991) Subcellular localization of Forssman glycolipid in epithelial MDCK cells by immuno-electronmicroscopy after freeze-substitution. J Cell Biol 115: 1009-1019

van Meer G (1989a) Lipid traffic in animal cells. Annu Rev Cell Biol 5: 247-275

van Meer G (1989b) Polarity and polarized transport of membrane lipids in a cultured epithelium. Modern Cell Biology (Satir BH, ed), Functional Epithelial Cells in Culture. (Matlin KS, Valentich JD, eds), Alan R Liss Inc, New York 8: 43-69

van Meer G, Simons K (1988) Lipid polarity and sorting in epithelial cells. J Cell Biochem 36: 51-58

van Meer G, Stelzer EHK, Wijnaendts-van-Resandt RW, Simons K (1987) Sorting of sphingolipids in epithelial (Madin-Darby canine kidney) cells. J Cell Biol 105: 1623-1635

van Meer G, van 't Hof (1992) Epithelial sphingolipid sorting is insensitive to reorganization of the Golgi by nocodazole, but is abolished by monensin and by brefeldin A. Submitted

van 't Hof W, van Meer G (1990) Generation of lipid polarity in intestinal epithelial (Caco-2) cells: Sphingolipid synthesis in the Golgi complex and sorting before vesicular traffic to the plasma membrane. J Cell Biol 111: 977-986

Epithelial sphingolipid sorting allows for extensive variation of the fatty acyl chain and the sphingosine backbone. Biochem J 283: 913-917

# LIPID TRAFFIC TO THE PLASMA MEMBRANE OF LEEK CELLS. SORTING BASED ON FATTY ACYL CHAIN LENGTH.

P. Moreau, B. Sturbois, L. Maneta-Peyret, D.J. Morré[*] and C. Cassagne

Institut de Biochimie Cellulaire du CNRS
Université de Bordeaux II
1 rue Camille Saint-Saëns
33077 Bordeaux-cedex
France

INTRODUCTION

The mechanisms by which lipids are sorted and transferred to the plasma membrane are poorly understood as compared to the situation of glyco- proteins but are now under intense investigations particularly in animal cells (Van Meer 1989, Voelker 1991, Koval and Pagano 1991).

The lipids can be transported from one membrane to another one by at least four different mechanisms (Sleight 1987):

1. A non-protein mediated cytosolic transfer. This possibility is almost ruled out on the basis of kinetic data.

2. A lipid transfer through transient interconnections between membranes. Such a mechanism should play a minor role in lipid traffic to the plasma membrane.

3. A cytosolic protein mediated transfer of monomers between two membranes. This hypothesis well studied in vitro has not been clearly confirmed in vivo. Moreover, recent evidences indicate that lipid transfer proteins could also be located outside of some plant cells or in the lumen of the ER ( Bernhard et al 1991, Sterk et al 1991, Madrid 1991, Madrid and Von Wettstein 1991), or be involved in specialized functions associated with the regulation of vesicular transfer from the Golgi apparatus (Bankaitis et al 1990).

4. The vesicular lipid transfer(s). It is clear that this pathway is at least involved in the transport of glycoproteins , more and more informations are available in animal cells ( Van Meer 1989, Voelker 1991, Koval and Pagano 1991).

[*] Dept. Med. Chem., Purdue University, West-Lafayette, Indiana, USA.

NATO ASI Series, Vol. H 74
Molecular Mechanisms of Membrane Traffic
Edited by D. J. Morré, K. E. Howell, and J. J. M. Bergeron
© Springer-Verlag Berlin Heidelberg 1993

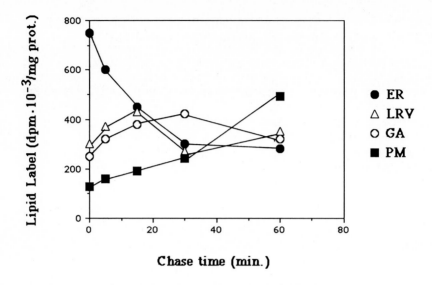

**Fig. 1.** Lipid labeling of the various membrane fractions as a function of chase time following a 120 min labeling period. ( Data replotted from Bertho et al 1991).

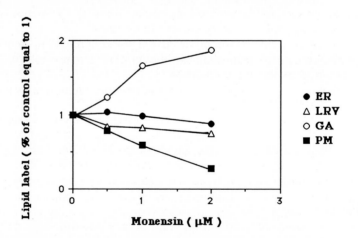

**Fig. 3.** Variation of lipid radioactivity in the different membrane fractions upon monensin addition. ( Data replotted from Bertho et al 1991).

On the contrary, little is known on the intracellular pathways of lipid traffic in plant cells.

The leek seedling model is the first system that allowed to analyse the vesicular transfer of lipids and particularly of very long chain fatty acids (VLCFA)-containing lipids to the plasma membrane of plant cells. VLCFA have more than 18 carbon atoms and are wax precursors synthesized by the elongases in the ER and the Golgi apparatus (Moreau et al 1988a). The plasma membrane which is the main cellular site of VLCFA accumulation does not synthesize these molecules (Moreau et al 1988a, Moreau et al 1988b). Different studies will be summarized in this review which demonstrate the vesicular nature of the transfer of VLCFA to the plasma membrane. These studies lead to the concept that lipids could be sorted, at least partly, according to their fatty acyl chain length.

## Kinetics of lipid transfer through the ER-GA-PM pathway of leek cells

Microsomes, prepared from 7 day-old leek seedlings pulse-chased with acetate, were further fractionated into ER (endoplasmic reticulum), GA (Golgi apparatus), LRV (lipid rich vesicles) and PM (plasma membrane) and the label of the lipids of these purified membrane fractions was determined for the various chase times (5 min, 15 min, 30 min and 60 min).

Fig.1 shows the radioactivity of the total lipids as a function of the chase time following a 120 min labeling period. At the end of the labeling period the ER is the most radioactive fraction. Following a 15 min chase, a high decrease of the radioactivity corresponding to an increase of labeling in the LRV and the Golgi is observed in the ER . Then, after a 30 min chase, the radioactivity still decreases in the ER, a high decrease is found for the LRV, and the Golgi apparatus shows the highest radioactivity. Finally, after a 60 min chase, a high decrease of radioactivity is observed for the Golgi apparatus and the plasma membrane fraction is the most radioactive one.

The kinetics are in agreement with a vesicular transport of lipids to the plasma membrane and not with the involvement of lipid transfer proteins. Moreover, the label variation of the LRV fraction is also in favour of the involvement of lipid rich vesicles as intermediates in the transport pathway.

Numerous observations have shown that ER, GA and PM fractions are reasonably clean from a biochemical point of view and homogeneous from a functional point of view ( Moreau et al 1988a, Moreau et al 1988b, Bertho et al 1989, Bertho et al 1991).

This is not the case of the LRV fraction which could be a mixture of at least three types of vesicles. Fig. 2 shows a tentative model to explain

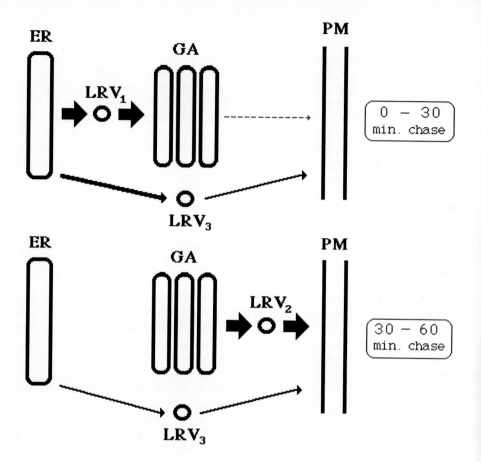

**Fig. 2.** Tentative model to explain the different origins of the vesicles present in the LRV fraction according to the chase time period. ( For details see the text ).

the origin of these different vesicles. The kinetics of the chase experiments between 0 and 30 min of chase suggest that the LRV fraction contains vesicles originating from the ER and destinated to fuse with the Golgi apparatus. Then, between 30 and 60 min of chase, the kinetic data of the LRV fraction compared with that of the Golgi and the plasma membrane are best explained by the presence among LRV of vesicles related to the GA-PM step. Consequently, we probably evidence the existence of two different populations of lipid rich vesicles named LRV1 and LRV2 in Fig.2. Moreover, there is also another type of vesicles named LRV3 transferring the lipids directly from the ER to the plasma membrane that must be included in this discussion. Effectively, the kinetic data are not unconsistent with a direct transfer from the ER to the plasma membrane for at least the short times of chase ( Fig. 1, see the small and linear increase of radioactivity in the plasma membrane observed between 0 and 30 min of chase).

Further experiments using monensin and low temperatures will also argue in favor of the existence of such vesicles involved in a direct ER-PM pathway (see below).


## Monensin block of the lipid transfer to the plasma membrane

Monensin applied externally to leek seedlings enters the cells rather efficiently (ca 10% of administered, Bertho et al 1989).

Up to 2μM, monensin did not affect greatly the lipid synthesis (Bertho et al 1989) and consequently the microsome labeling, whereas important changes occured in the distribution of the label between the various membrane fractions (Fig. 3). Increasing the external concentration of monensin led to a higher accumulation of labeled lipids in the Golgi (+87% at 2μM). At the same time, an important decrease of the radioactivity of the lipids was observed in the plasma membrane (-73% at 2μM).

These results strongly suggest that at least some lipids are transported to the plasma membrane via the Golgi apparatus. Moreover, the small decrease of radioactivity observed for the LRV fraction (-26% at 2μM) would be in agreement with the presence of vesicles named LRV2 in Fig.2. That a small decrease is only found for the total LRV fraction could be explained by an eventual increase of radioactivity associated with the LRV1 vesicles after monensin treatment.

Finally, we have analysed the fatty acid labeling of the plasma membrane and microsome fractions of the leek seedlings incubated with or without 1 μM monensin ( Fig.4, data from Bertho et al 1991).

We have observed an important decrease of the radioactivity of the VLCFA found in the plasma membrane (-70%) whereas only a small decrease was obtained for the C16 fatty acids (-25%) and no significant variation was found for the C18 fatty acids (-5%).

The lipid analyses showed no significant variation of lipid radioactivity in the crude microsomes whereas we observed a 45% decrease of labeled phospholipids and only a 25% decrease of neutral lipids in the plasma membrane (Bertho et al 1991).

These results suggested that the deficit of labeled VLCFA was rather correlated to that of labeled phospholipids. Consequently, it is likely that the monensin block concerned chiefly the VLCFA-containing phospholipids.

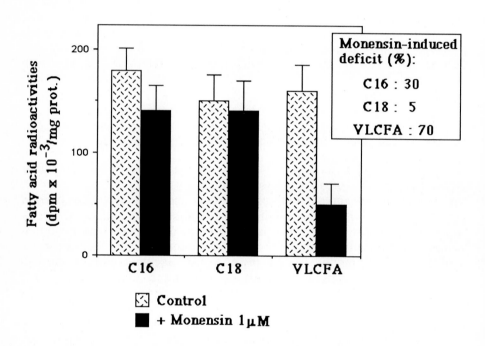

**Fig.4.** Effect of monensin on the fatty acid transfer to the plasma membrane. Selective block concerning the VLCFA.

## Low temperature effects on vesicular lipid transfer to the plasma membrane

Lowering the temperature from 24°C to 12°C leads to a small decrease of lipid synthesis but to a 60% decrease of radioactivity associated with the plasma membrane. Since the decrease observed for the plasma membrane is higher than that of the total lipid synthesis, lowering the temperature must have affected lipid transfer to this membrane.

The analysis of the radioactivity of the lipids of the different membrane fractions revealed an accumulation of phospholipids and sterols in the ER (+110% and +60% respectively) and the Golgi apparatus (+50% and +110% respectively).

The accumulation of the phospholipids was higher in the ER than in the Golgi but the contrary was observed for the sterols. These results indicate that, even if the main block at 12°C is localized between the Golgi and the plasma membrane, there is also a disturbance of the ER-GA step at this temperature.

By analysing the ratio of plasma membrane to the crude microsomes for the radioactivity of the various fatty acids, we found that this ratio was not or only slightly affected by lowering the temperature for C16 and C18 fatty acids, whereas this ratio was dramatically decreased for the VLCFA.

Consequently, it appears that the delivery of the VLCFA to the plasma membrane is really affected by lowering the temperature.

Concerning the various membrane fractions, we found that whatever the fatty acyl-chain length, an accumulation of the total fatty acids was observed for the ER and the Golgi.

However, the accumulation of VLCFA was 2 or 3 times more important than that of C16 and C18 fatty acids.

For the LRV fraction, only a small decrease was observed for the C16 and C18 fatty acids but a higher decrease (twice) was obtained for the VLCFA. The situation is rather similar for the plasma membrane with a more pronounced decrease for the VLCFA reaching -75%.

The VLCFA are primarily inserted into PC, PE and the neutral lipids; we have analysed the fatty acids of the lipids of the plasma membrane and the crude microsomes after in vivo incubations at 12°C and 24°C. At 12°C, only C16 and C18 fatty acids were detected in the lipids of the plasma membrane whereas VLCFA were undetectable. Since VLCFA were clearly present in the lipids of the microsomes at 12°C, we concluded that the decrease of VLCFA in the plasma membrane was the consequence of a non delivery of VLCFA-containing lipids to the plasma membrane at 12°C.

## CONCLUSIONS AND PERSPECTIVES

That lipids and particularly VLCFA are transferred to the plasma membrane via a vesicular ER-GA-PM pathway in leek cells is supported by many lines of evidences ( Moreau et al 1988a, Moreau et al 1988b, Bertho et al 1989, Bertho et al 1991):

1. The elongases ( synthesizing the VLCFA ) are localized in the ER and the Golgi.
2. The VLCFA accumulate in the plasma membrane which is devoid of any elongating activity.
3. A sequence of transfer of the VLCFA was demonstrated by pulse chase experiments between the ER, the LRV fraction, the Golgi and the plasma membrane.
4. Monensin led to an accumulation of the phospholipids in the Golgi and to a non delivery of these lipids to the plasma membrane.
5. Low temperatures (as 12°C)  block the transport of VLCFA-containing lipids to the plasma membrane and these lipids accumulate in the ER and the Golgi fractions.

The effect of low temperature on the transfer of VLCFA to the plasma membrane is in agreement with the results obtained by pulse-chase experiments or with monensin. These preliminary results on the effect of low temperature are similar to what is found in animal cells for the transfer of glycoproteins at low temperatures (Tartakoff 1986) and consequently, are in good agreement with the vesicular nature of the transport pathway.

The vesicular transfer of lipids appears to be an important pathway in plant cells as in animal cells. New concepts emerge and particularly that of a lipid sorting involving the fatty acyl chain length of the transfered lipids and that of an eventual direct vesicular transfer of lipids from the ER to the plasma membrane.

We must reach now the molecular level to better understand the mechanisms involved and to determine the molecules required, as carried out for glycoshingolipids (Simons and Van Meer  1988) and glycoproteins (Rothman and Orci  1992) in animal cells.

The future investigations will have to deal with the cell-free reconstitution of lipid transfers between various plant cell membrane compartments.

## REFERENCES

Bankaitis VA, Aitken JR, Cleves AE, Dowhan W (1990) Nature 347: 561-562
Bernhard WR, Thoma s, Botella J, Somerville CR (1991) Plant Physiol.  95: 164-170
Bertho P, Moreau P, Juguelin H, Gautier M, Cassagne C (1989) Biochim. Biophys. Acta  978: 91-96
Bertho P, Moreau P, Morré DJ, Cassagne C (1991) Biochim. Biophys. Acta  1070: 127-134

Ferrel JE, Lee KJ, Huestis WH (1985) Biochemistry 24: 2857-2864

Koval M, Pagano RE (1991) Biochim. Biophys. Acta 1082: 113-125

Madrid S (1991) Plant Physiol. Biochem. 29: 695-703

Madrid S, Von Wettstein (1991) Plant Physiol. Biochem. 29: 705-711

Moreau P, Bertho P, Juguelin H, Lessire R (1988a) Plant Physiol. Biochem. 26: 173-178

Moreau P, Juguelin H, Lessire R, Cassagne C (1988b) Phytochem. 27: 1631-1638

Pownall HJ, Bick DLM, Massey JB (1991) Biochemistry 30: 5696-5700

Rothman JE, Orci L (1992) Nature 355: 409-415

Simons K, Van Meer G (1988) Biochemistry 27: 6197-6202

Sleight RG (1987) Annu. Rev. Physiol. 49: 193-208

Sterk P, Booij H, Schellekens GA, Van Kammen A, De Vries SC (1991) The plant cell 3: 907-921

Tartakoff AM (1986) EMBO J. 5: 1477-1482

Van Meer G (1989) Annu. Rev. Cell Biol. 5: 247-275

Voelker DR (1991) Microbiol. Rev. 55: 543-560

# HEXADECYLPHOSPHOCHOLINE AS AN USEFUL TOOL FOR INVESTIGATING PHOSPHATIDYLCHOLINE BIOSYNTHESIS AND SORTING

C. C. Geilen, Th. Wieder and W. Reutter
Institut fur Molekularbiologie und Biochemie
Freie Universitat Berlin
Arnimallee 22
D-1000 Berlin 33 (Dahlem)
Germany

Phosphatidylcholine is the most abundant phospholipid in mammalian tissues. The main route of its biosynthesis is the CDP-choline pathway with CTP : choline phosphate cytidylyltransferase (CT) (EC 2.7.7.15) as the rate-limiting enzyme ( for review see Vance, 1990). Recently, we report about a new inhibitor of this pathway hexadecylphosphocholine (HePC), which directly inhibits the CT (Haase et al., 1990; Geilen et al., 1991). HePC belongs to the group of alkylphosphocholines, which are structurally different from the well known alkyllysophosphoglycerides and acyllysophosphoglycerides. Figure 1 shows the chemical structure of HePC. Treatment of MDCK cells with micromolar concentrations of HePC leads to an in-hibition of the phos-phatidylcholine bio-synthesis without an effect on the other phospholipid classes. This inhibition is paralleled by the disturbance of the translocation of CT. This enzyme is inactive in the lipid-free, cyto-solic form. Therefore, HePC may serve as an useful tool for investigating processes in

Figure 1: Chemical structure of HePC

which phosphatidylcholine is involved. Figure 2 shows the inhibitory effect of HePC on phosphatidylcholine biosynthesis (2a) and two examples of the use of HePC as a tool in biochemistry studies. First (2b), the regulation between both pathways of phospha-tidylcholine biosynthesis, the CDP-choline pathway and via methylation of phospha-tidylethanolamine. Inhibition of the CDP-choline pathway by HePC leads to an increased biosynthesis by methylation of phosphatidylethanolamine, which support the studies on a balanced synthesis of glycerolipids (for review see Tijburg et al., 1989). Second (2c), HePC antagonizes the phorbol ester-induced stimulation of phosphatidylcholine biosynthesis in Hela cells by antogonizing the increased membrane translocation of CT, which confirmed the mechanism recently suggested by Utal et al. (1991). Further targets of the use of HePC may be the sorting process of phosphoglycerides and sphingolipids, the secretion of lipoproteins or the phosphatidylcholine biosynthesis of the regenerating liver.

NATO ASI Series, Vol. H 74
Molecular Mechanisms of Membrane Traffic
Edited by D. J. Morré, K. E. Howell, and J. J. M. Bergeron
© Springer-Verlag Berlin Heidelberg 1993

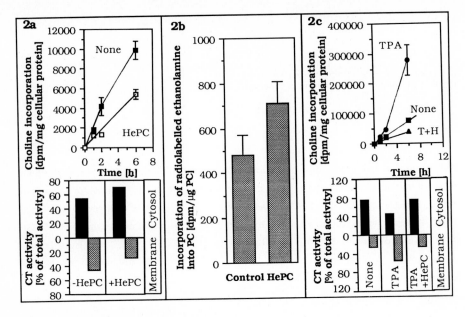

Figure 2: Different applications of HePC

a) Inhibition of [3H]-choline incorporation into phosphatidylcholine of MDCK cells by treatment with 50 μM HePC, and the subcellular distribution of CT after 4 h of preincubation with 50 μM HePC measured by digitonin release (1 min).

b) Effect of HePC on the regulation of both biosynthetic pathways of phosphatidylcholine (PC) by measuring [3H]-ethanolamine incorporation into phosphatidylcholine after treatment with 50 μM HePC for 4h.

c) Antagonization of phorbol ester-induced stimulation of phosphatidylcholine biosynthesis in Hela cells by HePC, and its effect on the subcellular distribution of CT after 4 h of preincubation with 50 μM HePC measured by digitonin release (1 min). [T+H = 100 nM TPA plus 50 μM HePC]

References

Geilen CC, Wieder Th, Reutter W (1992) Hexadecylphosphocholine inhibits translocation of CTP : choline phosphate cytidylyltransferase in Madin-Darby canine kidney cells. J Biol Chem 267: 6719 - 6724

Haase R, Wieder Th, Geilen CC, Reutter W. (1990) The phospholipid analogue hexadecylphosphocholine inhibits phosphatidylcholine biosynthesis in Madin-Darby canine kidney cells. FEBS Lett 288: 129 - 132

Tijburg LBM, Geelen MJH, van Golde LMG (1989) Regulation of the biosynthesis of triacylglycerol, phosphatidylcholine and phosphatidylethanolamine in the liver. Biochim Biophys Acta 1004: 1 - 19

Utal AK, Jamil H, Vance DE (1991) Diacylglycerol signals the translocation of CTP : choline phosphate cytidylyltransferase in Hela cells treated with 12 - O - tetradecanoyl-phorbol - 13 - acetate. J Biol Chem 266: 24084 - 24091

Vance DE (1990) Phosphatidylcholine metabolism: masochistic enzymology, metabolic regulation, and lipoprotein assembly. Biochem Cell Biol 68: 1151 - 1165

# BIOSYNTHESIS OF THE SCRAPIE PRION PROTEIN IN SCRAPIE-INFECTED CELLS.

Taraboulos A, Borchelt DR, Raeber A, Avrahami D and § Prusiner SB
Departments of Neurology HSE 781 and § Biochemistry and Biophysics
University of California, San Francisco, CA 94143-0518, USA

Compelling evidence argues that the major component of the scrapie prion is a host- encoded protein, the scrapie prion protein (PrPSc) (reviewed in Prusiner, 1991). The events of PrPSc biosynthesis are thus of central importance to understanding the biology of scrapie. PrPSc is the abnormal isoform of a normal phosphoinositol glycolipid (GPI)-anchored plasma membrane sialoglycoprotein PrPC. While the two PrP isoforms differ strikingly in many of their properties, their structural differences have not yet been elucidated. PrPSc is insoluble in detergents and possesses a protease-resistant core, while PrPC is readily soluble in most detergents and is completely degraded by proteases. In infected cells in culture, PrPSc becomes protease-resistant as a result of an as yet unidentified post-translational event (Borchelt et al., 1990) unrelated to PrP N-linked glycosylation (Taraboulos et al., 1990a). In scrapie infected cells in culture, PrPSc accumulates primarily within secondary lysosomes (Taraboulos et al., 1990b; McKinley et al., 1991). All PrP antibodies described to date react equally well with both PrPC and denatured PrPSc, but are much less reactive with native PrPSc. To study further the biosynthesis of protease-resistant PrPSc, we have used scrapie-infected ScN2a and ScHaB cells in pulse-chase radiolabeling experiments.

1) Brefeldin A (BFA) caused the redistribution of the Golgi marker MG160, but not of PrPSc, into the endoplasmic reticulum (ER); it also blocked the export of PrPC to the cell surface. Presence of BFA during both the pulse and the chase periods completely but reversibly inhibited PrPSc synthesis. Transport of PrP along the secretory pathway is thus required for PrPSc synthesis, and the ER-Golgi is not competent for the formation of protease-resistant PrPSc (Taraboulos et al., 1992). A possible reason for the inability of the ER-Golgi to synthesize PrPSc may be the absence, in this compartment, of existing PrPSc that may be required as a 'template' for the formation of new PrPSc.

2) Incubation of ScN2a cells with the protease dispase or with phosphatidylinositol- specific phospholipase C (PIPLC) inhibited the synthesis of PrPSc, suggesting that PrPSc is derived from a GPI-anchored, protease-sensitive plasma membrane (PM) precursor, perhaps PrPC itself (Borchelt et al., 1992).

3) Using the ionophore monensin as an inhibitor of *mid* Golgi glycosylation we determined that PrP traverse the *mid* Golgi stack **before** becoming PrPSc.

4) PrPSc synthesis was not inhibited by lysosomotropic amines or monensin, and thus does not require acidic pH. In addition, PrPSc was *N*-terminally trimmed 1-2 hours **after** acquiring its protease-resistant core. In contrast to PrPSc synthesis, this degradation could be inhibited by lysosomotropic amines as well as by monensin, indicating that it occurs in an acidic degradative compartment such as endosomes or lysosomes (Taraboulos et al., 1992). In addition, we detected a 18°C-sensitive step that has to be crossed prior to the formation of protease-resistant PrPSc (Borchelt

NATO ASI Series, Vol. H 74
Molecular Mechanisms of Membrane Traffic
Edited by D. J. Morré, K. E. Howell, and J. J. M. Bergeron
© Springer-Verlag Berlin Heidelberg 1993

et al., 1992), pointing to the possible involvement of the endocytic system in PrP$^{Sc}$ synthesis.

These results as well as data by others (Caughey and Raymond, 1991; Caughey et al, 1991) suggest that PrP$^{Sc}$ synthesis occurs in a compartment that PrP, a GPI-anchored protein, can efficiently access from the PM. The plasma membrane, the endocytic system and the caveolar system are possible candidates for the site of PrP conversion (Figure 1).

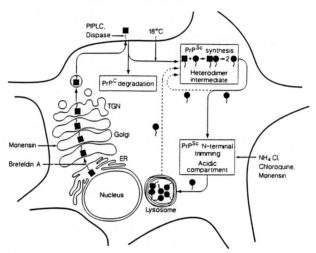

**Figure 1:** Pathways of prion protein synthesis and degradation in cultured cells. PrP$^{Sc}$ is denoted by circles; squares designate PrP$^C$ and the PrP$^{Sc}$ precursor, which may be indistinguishable. Rectangular boxes denote as yet unidentified subcellular compartments. PrP$^{Sc}$ synthesis probably occurs through the interaction of PrP$^{Sc}$ precursor with existing PrP$^{Sc}$ (cf Prusiner 1991). Dotted lines denote possible PrP$^{Sc}$ feedback pathways. (Reproduced from Taraboulos et al., 1992)

**References**

Borchelt DR et al. (1990). Scrapie and cellular prion proteins differ in their kinetics of synthesis and topology in cultured cells. J Cell Biol 110: 743-752.

Borchelt DR et al. (1992). Evidence fo synthesis of scrapie prion proteins in the endocytic pathway. J Biol Chem, in press.

Caughey B et al. (1991). N-terminal truncation of the scrapie-associated form of PrP by lysosomal protease(s): implications regarding the site of conversion of PrP. J Virol 65: 6597-6603.

Caughey B and Raymond G. (1991). The scrapie-associated form of PrP is made from a cell surface precursor that is both protease- and phospholipase-sensitive. J Biol Chem 256: 18217

McKinley MP et al. (1991). Ultrastructural localization of scrapie prion proteins in cytoplasmic vesicles of infected cultured cells. Lab Invest 65: 622-630.

Prusiner SB (1991). Molecular biology of prion diseases. Science 252: 1515-1522.

Taraboulos A et al. (1990a). Acquisition of protease-resistance by prion proteins in scrapie-infected cells does not require asparagine-linked glycosylation. Proc Nat Acad Sci (USA) 87: 8262.

Taraboulos A et al. (1990b). Scrapie prion proteins accumulate in the cytoplasm of persistently-infected cultured cells . J Cell Biol 110: 2117-2132.

Taraboulos A et al. (1992). Synthesis and trafficking of prion proteins in cultured cells. Mol Biol Cell, in press.

# PROTEIN PHOSPHORYLATION REGULATES THE CELLULAR TRAFFICKING AND PROCESSING OF THE ALZHEIMER BETA/A4 AMYLOID PRECURSOR PROTEIN

G. L. Caporaso, S. E. Gandy, J. D. Buxbaum, T. Suzuki, C. Nordstedt, K. Iverfeldt, T. V. Ramabhadran, A. J. Czernik, A. C. Nairn, and P. Greengard
Laboratory of Molecular and Cellular Neuroscience
The Rockefeller University
1230 York Avenue, New York, NY 10021

Cerebral deposition of the beta/A4 amyloid peptide is an important feature of Alzheimer disease. The beta/A4 peptide is derived from the amyloid precursor protein (APP), a transmembrane glycoprotein whose function is unknown. Abnormal processing of APP has been causally linked to Alzheimer disease since mutations in the coding sequence of APP are associated with an inherited early-onset form of the disease. We have employed the rat neuroendocrine PC12 cell line to examine the normal cellular trafficking and proteolytic processing of APP and to investigate the role of protein phosphorylation in the regulation of normal and aberrant APP metabolism.

Treatment of PC12 cells with either monensin (which inhibits trafficking of proteins from the proximal to distal Golgi) or brefeldin A (which causes both resorption of the Golgi into the endoplasmic reticulum and reticulation of the trans-Golgi network and endosomal system) prevented the normal post-translational modification of APP and completely abolished APP proteolytic degradation, secretory cleavage, and secretion. These data suggest that APP proteolysis and secretory cleavage occur in the distal Golgi or a subsequent compartment, and argue against the occurrence of APP proteolysis in the endoplasmic reticulum, as has been shown for other proteins.

The lysosomotropic drug chloroquine had no effect on the post-translational modification or secretion of APP. However, the drug exerted marked inhibitory effects on the degradation of APP holomolecules and the cell-associated carboxyl-terminal APP fragment resulting from secretory cleavage. Therefore, acidic organelles such as endosomes and lysosomes may be the site of degradation of these species. This implies that there are at least two cellular pathways for

NATO ASI Series, Vol. H 74
Molecular Mechanisms of Membrane Traffic
Edited by D. J. Morré, K. E. Howell, and J. J. M. Bergeron
© Springer-Verlag Berlin Heidelberg 1993

APP proteolytic processing: a chloroquine-insensitive compartment where secretory cleavage occurs and a chloroquine-sensitive compartment where terminal degradation occurs.

In PC12 cells, activation of protein kinase C (PKC) with phorbol ester or inhibition of protein phosphatases 1 and 2A with okadaic acid resulted in an increased rate of proteolytic processing of APP that could almost entirely be accounted for by an increase in APP secretion. These findings are especially relevant because they suggest that stimulation of protein phosphorylation might divert APP molecules away from the amyloidogenic pathway(s), since the site of proteolytic cleavage that results in APP secretion occurs within the beta/A4 domain of APP. However, we cannot exclude the possibility that protein phosphorylation may also regulate an alternative, amyloidogenic pathway, since simultaneous treatment with phorbol ester and okadaic acid results in production of additional, larger APP carboxyl-terminal fragments that might contain an intact beta/A4 domain.

The phorbol ester- and okadaic acid-induced effects on APP secretion may be due to phosphorylation of APP by PKC (substrate activation), activation of an APP "secretase" (enzyme activation), or redirected cellular trafficking that increases APP contact with its protease(s). Support for the "substrate activation" model comes from studies in our laboratory demonstrating that PKC can phosphorylate synthetic APP carboxyl-terminal peptides and that PKC can phosphorylate APP at a specific site in semi-intact PC12 cells. Additional studies employing APP phosphorylation-site mutants will hopefully clarify the role of protein phosphorylation in the cellular trafficking and proteolysis of APP.

Buxbaum JD, Gandy SE, Cicchetti P, Ehrlich ME, Czernik AJ, Fracasso RP, Ramabhadran TV, Unterbeck AJ, Greengard P (1990) Proc Natl Acad Sci USA 87:6003-6006
Caporaso GL, Gandy SE, Buxbaum JD, Greengard P (1992) Proc Natl Acad Sci USA 89:2252-2256
Caporaso GL, Gandy SE, Buxbaum JD, Ramabhadran TV, Greengard P (1992) Proc Natl Acad Sci USA 89:3055-3059
Gandy SE, Czernik AJ, Greengard P (1988) Proc Natl Acad Sci USA 85:6218-6221
Suzuki T, Nairn AC, Gandy SE, Greengard P (1992) Neuroscience 48:755-761

# TARGETING OF THE POLYMERIC IMMUNOGLOBULIN RECEPTOR IN TRANSFECTED PC12 CELLS

F. Bonzelius[*][¶], G. A. Herman[*][¶], M. H. Cardone[§], K. E. Mostov[§] and R. B. Kelly[*]
Department of Biochemistry and Biophysics[*] and Department of Anatomy[§]
University of California
San Francisco, CA 94143
USA

## INTRODUCTION

Synaptic vesicles are among the best characterized membranous organelles. A rat pheochromo-cytoma cell line, PC12, makes vesicles of the same size, density and membrane composition as authentic synaptic vesicles, isolated from rat brain (Clift-O'Grady et al., 1990). These organelles have been termed "synaptic vesicle-like vesicles" (SVLVs). In recent years PC12 cells have been used by several investigators as a model system for studying synaptic vesicle biogenesis and the targeting of synaptic vesicle-specific proteins. Evidence has accumulated from these studies that synaptic vesicles can be derived by endocytosis and that the precursor organelle in the nerve terminal is either the plasma membrane or an early endosomal compart-ment (Johnston et al., 1989; Clift-O'Grady et al., 1990; Regnier-Vigouroux et al., 1991).

The same biogenetic origin has been postulated for another class of membrane-bounded vesicles, the transcytotic vesicles of polarized epithelial cells. One of the best-studied examples of transcytosis is the basolateral to apical transport of IgA and IgM across various mucosa by binding to the polymeric immunoglobulin receptor (pIgR). Targeting of pIgR ($M_r$ 100 - 105 kD) to the basolateral plasma membrane, endocytotic uptake and subsequent transcytosis have been described in detail in transfected MDCK cells (Apodaca et al., 1991).

To test whether the pathways for the biogenesis of synaptic vesicles and transcytotic vesicles are related, we asked whether pIgR would be recognized by synaptic vesicle sorting machinery and sorted to SVLVs in PC12 cells. For that purpose, the cells were transfected with wild-type or mutant c-DNA encoding pIgR. Targeting to synaptic vesicles was investigated by assaying for co-sedimentation with synaptophysin, a synaptic vesicle-specific protein of $M_r$ 38 kD.

## RESULTS AND CONCLUSIONS

Polymeric immunoglobulin receptor c-DNA was subcloned into the polylinker site of PCB6 (D. Russell, M. Roth, C. Brewer), a mammalian expression vector containing a CMV promoter driving the pIgR gene. PC12 cells were stably transfected with wild-type c-DNA by lipofection using Lipofectin (BRL) or DOTAP (Boehringer). Prior to subcellular fractionation, stable

---

¶ The first two authors made equivalent contributions.

clones were grown in 6 mM sodium butyrate for 24 hours to increase the expression level of the polymeric immunoglobulin receptor. Control experiments have shown that this treatment does not change the subcellular distribution of pIgR as detected by our approach. To isolate SVLVs, the cells were homogenized in an EMBL cell cracker and subjected to a subcellular fractionation protocol, including differential centrifugation and velocity gradient centrifugation (5 - 25% glycerol on top of a 50% sucrose pad) as a final purification step. The gradient fractions were analyzed by SDS-PAGE and western blotting, using monoclonal antibodies against pIgR and synaptophysin. Most of the SVLVs, as identified by their synaptophysin-immunoreactivity, sedimented as a defined peak in the upper half of the gradient, whereas the main peak of pIgR-immunoreactivity was detected on the sucrose pad.

Phosphorylation of a serine residue at position 664 of pIgR has been demonstrated to be required for efficient transcytosis of the receptor (Casanova et al., 1990). Substitution of $Ser^{664}$ by aspartic acid (which is believed to mimic the negative charge of the phosphate group) results in a higher transcytosis efficiency as compared to the wild-type pIgR. To test, whether in PC12 cells the $Asp^{664}$ receptor is targeted differently than the wild-type pIgR, we transiently transfected these cells - by electroporation - with c-DNA encoding for the receptor mutant. 24 hours after transfection the cells were incubated in 6mM sodium butyrate for 14 hours. The subsequent cell fractionation and the analysis of the velocity gradient were carried out as described above. Again, the main peak of pIgR-immunoreactivity did not co-sediment with SVLVs, but was detected on the sucrose pad at the bottom of the gradient.

These data show that the transcytotic vesicle-specific protein pIgR, as well as the $Asp^{664}$ mutant, is segregated away from SVLVs in transfected PC12 cells. It most likely is not recognized and sorted by synaptic vesicle sorting machinery.

## ACKNOWLEDGEMENTS

Dr. Frank Bonzelius was supported by a NATO/DAAD fellowship.

## REFERENCES

Apodaca G, Bomsel M, Arden J, Breitfeld PP, Tang K, Mostov KE (1991) The polymeric immunoglobulin receptor. J Clin Invest 87: 1877 - 1882

Casanova JE, Breitfeld PP, Ross SA, Mostov KE (1990) Phosphorylation of the polymeric immunoglobulin receptor required for its efficient transcytosis. Science 248: 742 - 745

Clift-O'Grady L, Linstedt AD, Lowe AW, Grote E, Kelly RB (1990) Biogenesis of synaptic vesicle-like structures in a pheochromocytoma cell line PC-12. J Cell Biol 110: 1693 - 1703

Johnston PA, Cameron PL, Stukenbrok H, Jahn R, De Camilli P, Suedhof TC (1989) Synaptophysin is targeted to similar microvesicles in CHO and PC12 cells. EMBO J 8: 2863 - 2872

Regnier-Vigouroux A, Tooze SA, Huttner WB (1991) Newly synthesized synaptophysin is transported to synaptic-like microvesicles via constitutive secretory vesicles and the plasmamembrane. EMBO J 10: 3589 - 3601

# Signals for Receptor-Mediated Endocytosis

I. S. Trowbridge, J. Collawn, S. White, and A. Lai
Department of Cancer Biology
The Salk Institute
P.O. Box 85800
San Diego, California   92186-5800

Receptor-mediated endocytosis is the mechanism by which a variety of nutrients, hormones and growth factors are selectively removed from the circulation and rapidly taken up by cells (reviewed in Goldstein et al., 1985). Studies of the low density lipoprotein receptor (LDLR) provided the first evidence that the initial step in this process is the concentration of specific cell surface receptors in clathrin-coated pits (Anderson et al., 1977). Analysis of naturally-occurring internalization-defective mutant LDLRs from patients with familial hypercholesterolemia indicated that the cytoplasmic domain of the receptor was essential for high efficiency endocytosis. Mutant receptors from the patient J.D. differed from wild-type LDLR only by the substitution of a cysteine residue for tyrosine at position 807 within the cytoplasmic domain implying that this tyrosine was essential for rapid internalization (Davis et al., 1986). Subsequently, the cytoplasmic domains of other receptors including the transferrin receptor (TR) (Rothenberger et al., 1987; Jing et al., 1990), cation-independent mannose-6-phosphate receptor (Man-6-PR) (Lobel et al., 1989), polymeric immunoglobulin receptor (poly-IgR) (Mostov et al., 1986) and asialoglycoprotein receptor (ASGPR) (Fuhrer et al., 1991) were shown to be important for clustering in coated pits and specific tyrosine residues within their cytoplasmic domains were also identified as critical for high efficiency endocytosis (Jing et al., 1990; Lobel et al., 1989; Breitfeld et al., 1990; Fuhrer et al., 1991). Together, these data led to the concept that constitutively recycling receptors have a tyrosine-containing internalization signal located within their cytoplasmic domains. However, the primary structures of receptor cytoplasmic tails were sufficiently dissimilar that inspection of their sequences did not reveal a shared feature that might provide a clue as to the nature of internalization signals and early attempts to identify

NATO ASI Series, Vol. H 74
Molecular Mechanisms of Membrane Traffic
Edited by D. J. Morré, K. E. Howell, and J. J. M. Bergeron
© Springer-Verlag Berlin Heidelberg 1993

internalization signals based on sequence patterns met with limited success (Vega and Strominger, 1989; Ktistakis et al., 1990). It was necessary, therefore, to determine the sequences of internalization signals of individual receptors experimentally by functional analysis of mutant receptors derived by *in vitro* mutagenesis. The internalization signals of several receptors have now been identified and, as described below, their properties can be accounted for by a model that invokes a common structural chemistry and implicates a tight turn, as defined in Collawn et al. (1990), as the conformational motif for high efficiency endocytosis. Predictions based on this model, which is now supported by direct structural evidence, have been substantiated and have led to further insight into the properties of internalization signals that may have potential practical implications.

## Sequences of Internalization Signals

The sequence patterns of internalization signals of the human LDLR (Chen et al., 1990), TR (Collawn et al., 1990) and cation-independent Man-6-PR (Canfield et al., 1991) have been determined. The LDLR and Man-6-PR are both type I membrane proteins with carboxy-terminal cytoplasmic tails. Mutant receptors with truncated cytoplasmic domains were used to define the minimum number of residues required for rapid internalization and then the contribution of individual residues to the internalization signal was assessed by alanine scanning mutations. Only the first 22 amino acids (residues 790-811) of the 50-residue LDLR cytoplasmic tail were required for rapid endocytosis (Davis et al., 1987). Within this region, a six-residue sequence [802]FXNPXY[807] (where X stands for any amino acid) located 12 residues from the transmembrane region was shown by alanine scanning to be required for high efficiency endocytosis (Chen et al., 1990). Initially, the importance of the NPXY sequence was emphasized, because it was conserved in all members of the LDLR family and also found in the cytoplasmic tails of other cell surface proteins including the insulin receptor and epidermal growth factor receptor (EGFR). However, the tetrapeptide sequence, NPXY, is not sufficient to promote rapid internalization of either the LDLR or the TR (see below) strongly suggesting that the LDLR internalization signal is the six residue sequence FXNPXY. For the Man-6-PR, it was shown that only the membrane-proximal 29 amino acids of the 163 residue cytoplasmic tail were required for rapid endocytosis and that the internalization signal was the six residue sequence, YXYXKV, located 23 residues from the transmembrane region (Canfield et al., 1991). In this case, the first tyrosine residue in the signal appears less critical for activity, as altering it to alanine only reduced the internalization rate by about 25%.

We took a similar approach to identify the internalization signal in the 61 residue amino-terminal tail of the human TR, a type II membrane protein, seeking first to localize the signal to a specific region of the receptor by analysis of a series of deletion mutants and then to identify the specific residues involved by alanine scanning. A key finding was that mutant receptors with either residues 3 - 18 or residues 29 - 59 deleted were internalized as efficiently as wild-type receptors, indicating that amino acids within these two regions are not required for rapid internalization and that the internalization signal is localized to the region spanning residues 19 - 28 (Jing et al., 1990). Insertion of this 10-residue region into a tailless receptor lacking all but four amino acids of the cytoplasmic domain restored wild-type activity, demonstrating that this region was not only required but sufficient for high efficiency endocytosis. The only tyrosine residue in the TR cytoplasmic tail is located within this region at position 20 and it was shown that changing this residue to glycine severely impaired endocytosis, implying that tyrosine was an important element of the TR internalization signal (Jing et al., 1990). To define the minimum length of TR cytoplasmic tail required for rapid internalization, incremental deletion of residues from the carboxy-terminal end of the 10-residue region in the tailless receptor was performed (Collawn et al., 1990). Deletion of one residue led to partial loss of activity, and deletion of three or more residues generated mutant receptors that were internalized as slowly as tailless receptors. However, the carboxy-terminal four residues were not specifically required for high efficiency endocytosis, as they could be deleted from the wild-type receptor without loss of activity (Collawn et al., 1990). These data indicate that the internalization signal has to be separated by a minimum number of residues from the transmembrane region for activity. We presume this spacer region is necessary to allow the internalization signal to interact effectively with a recognition structure in coated pits. The internalization signal of the TR was then identified as the tetrapeptide [20]YXRF[23] by alanine scanning mutations of the remaining six amino-terminal amino acids in the 10 residue region (Collawn et al., 1990). Although other studies have been interpreted to indicate that additional residues in the amino-terminal region of the receptor cytoplasmic tail directly contribute to the TR internalization signal (McGraw et al., 1991; Girones et al., 1991), our deletion of residues 3 - 18 without loss of activity excludes this possibility. Alteration of residues within this region could reduce TR internalization efficiency by indirectly affecting either the conformation or accessibility of the YXRF internalization signal. The YTRF internalization sequence is conserved in human, mouse and chicken TRs (see Gerhardt et al., 1991), but not apparently in Chinese hamster TR in which a single base change is reported to alter the tyrosine residue to cysteine (Alvarez et al., 1990).


A Tight Turn is the Conformational Motif for High Efficiency Endocytosis


Mutational analysis indicated that the activity of the TR internalization signal was relatively independent of position within the cytoplasmic domain provided it was separated by at least 7 residues from the transmembrane region. This suggested that the conformation of the internalization signal must be determined by the YXRF tetrapeptide sequence independent of the structure of adjacent residues. Moreover, the localization of the internalization signal to this tetrapeptide suggested YXRF may have a preferred local structure that could be determined independently of the rest of the TR protein structure. This reasoning led us to try to obtain evidence that YXRF has a preferred conformation by examining structures of tetrapeptide analogues of the YXRF in proteins of known three-dimensional crystallographic structure in the Brookhaven Protein Data Bank (Collawn et al., 1990). The sequence patterns we chose were either known or predicted from functional analysis of mutant TRs to have at least 50% of the internalization activity of the YTRF sequence. Strikingly, we found that of the 28 structures matching the TR internalization sequence pattern, 23 had backbone conformations very similar to a type I β-turn. Further, the tight turn propensity of eight tetrapeptides most closely matching YTRF with the sequence pattern (Y,F) hydrophilic, (R,K), (F,Y) did not appear to be influenced by the conformation of adjacent residues and were all surface-exposed. As the probability of finding this conformational preference by chance was very low, we concluded that it was likely that the TR internalization signal was a tight turn. We then asked whether the tight turn structural motif might be a common feature of internalization signals despite their disparate sequences. As NPXY had been shown to be required for rapid internalization of the LDLR, we searched the protein structure data base for analogues of this sequence and found that 4 of 5 related sequences identified were also in tight turns. This finding led us to propose that an exposed tight turn is the recognition motif for high efficiency endocytosis (Collawn et al., 1990). Subsequently, experimental support for the tight turn model has been obtained from nuclear magnetic resonance studies of synthetic peptides corresponding to the proposed internalization signals of the LDLR (Bansal and Gierasch, 1991) and lysosomal acid phosphatase (LAP) (Eberle et al., 1992). Most significantly, the NPVY sequence of the LDLR internalization signal was shown to adopt a type I β-turn conformation in a synthetic nonapeptide, whereas inactive variants of this sequence did not. However, further structural studies are required to substantiate the tight turn model, as a limitation of the LDLR study is that the NMR data was collected from peptides at pH3 and the LAP tetrapeptide sequence shown to be a β-turn may not be active as an internalization signal (see below).

## Internalization signals are Interchangeable Self-Determined Structural Motifs

One prediction of the tight turn model is that, as different internalization signals share a common self-determined conformational motif, it might be possible to transplant the internalization signal of one receptor into the cytoplasmic domain of another and retain activity. To directly determine whether internalization signals were interchangeable, self-determined structural motifs, we replaced the TR internalization signal with four- or six-residue signals from the LDLR and Man-6-PR (Collawn et al., 1991). Six residue signals from both receptors promoted rapid internalization of the TR, as did the four-residue signal from the Man-6-PR, establishing that all three signals are functionally equivalent and are active in both type I and type II membrane proteins. This observation is consistent with the tight turn model, as the orientation of a turn in three-dimensional space is independent of the polypeptide's orientation with respect to the cell membrane. The four-residue LDLR sequence NPVY was not active when transplanted into the TR cytoplasmic tail implying that the complete LDLR internalization signal is the six-residue sequence, FDNPVY.

## Internalization Activity of Predicted Internalization Signals in Other Receptors and Lysosomal Membrane Proteins

A comparison of the internalization signals of LDLR, TR, and Man-6-PR and active variants of these signals suggests that despite diverse sequences, internalization signals have a common structural chemistry. Thus, for both four- and six-residue signals, an amino-terminal aromatic residue and either an aromatic or large hydrophobic carboxy-terminal residue are required for activity (Canfield et al., 1991; Trowbridge, 1991). Intervening residues at positions 2 and 3 of the proposed tight turn are usually hydrophilic with a preference for a positively charged residue at one of these positions. Although specific tyrosine residues in the cytoplasmic tails of a variety of receptors and lysosomal membrane glycoproteins have been identified as important for rapid internalization, their internalization signals have not yet been fully defined. However, based on the sequence patterns of the internalization signals for LDLR, TR and Man-6-PR, we identified putative internalization signals that included the critical tyrosine residues in these molecules (Trowbridge, 1991). We then assayed the internalization activity of these putative signals by transplantation into the cytoplasmic tail of the TR (Collawn et al., 1991 and unpublished results). We found that predicted internalization signals from the poly-Ig receptor, YSAF, and asialoglycoprotein receptor subunit H1, YQDL, efficiently promoted internalization of TR. Interestingly, the predicted internalization signal YRVH, from the lysosomal membrane glycoprotein LAP was also highly active, whereas the sequence PPGY which shares the same tyrosine and had previously been identified as being in a tight turn in the LAP cytoplasmic tail was completely inactive. The putative internalization signal, YQTI, from the lysosomal membrane glycoprotein, lamp-1, was also active when transplanted into the TR cytoplasmic domain.

In addition to transplanting heterologous internalization signals into the TR, we have also examined whether insertion of multiple signals in the cytoplasmic domain increases internalization efficiency. Secondary structure prediction using the Chou-Fasman algorithm suggested that the TR cytoplasmic domain consists of five helices separated by four turns with YTRF at turn 2 (numbered from the amino-terminus). Additional YTRF sequences were substituted independently at turns 1, 3 and 4 and the internalization rate of each of the mutant TRs containing two signals were assayed. The most striking effect was seen with a second signal inserted in turn 3 (residues 31-34) which led to a two-fold increase in the rate of internalization of TR. Further, mutant receptors with only one active internalization signal inserted in turn 3 were internalized as wild-type TRs. These results suggest multiple signals can increase internalization efficiency and may provide an explanation for the puzzling observation that substitution of Tyr for Ser at position 34 can restore to wild-type levels the internalization rate of a mutant TR with Tyr-20 altered to Cys (McGraw et al., 1991).

## Concluding Remarks

The fact that internalization signals are self-determined structural motifs that can be transplanted from one receptor to another has potentially important implications for modulating the rate of uptake of ligands, including drugs, into the cell. It should be possible to enhance the endocytosis of cell surface molecules lacking a signal by transplanting an internalization motif into their cytoplasmic tails. Further, the structural information now available may allow the design of small molecules that mimic the structure of internalization signals and inhibit endocytosis.

## Acknowledgements

This work was supported by National Cancer Institute Grant CA34787 and by NATO Grant 880393 (IST), and by the Arthritis Foundation (JC).

## References

Alvarez, E, Girones, N, Davis, RJ (1990) A point mutation in the cytoplasmic domain of the transferrin receptor inhibits endocytosis. Biochem. J. 267:31-35.

Anderson, RGW, Goldstein, JL, Brown, MS (1977) A mutation that impairs the ability of lipoprotein receptors to localise in coated pits on the cell surface of human fibroblasts. Nature 270:695-699.

Bansal, A, Gierasch, LM (1991) The NPXY internalization signal of the LDL receptor adopts a reverse-turn conformation. Cell 67:1195-1201.

Breitfeld, PP, Casanova, JE, McKinnon, WC, Mostov, KE (1990) Deletions in the cytoplasmic domain of the polymeric immunoglobulin receptor differentially affect endocytotic rate and postendocytotic traffic. J. Biol. Chem. 265:13750-13757.

Canfield, WM, Johnson, KF, Ye, RD, Gregory, W, Kornfeld, S (1991) Localization of the signal for rapid internalization of the bovine cation-independent mannose-6-phosphate/insulin-like growth factor-II receptor to amino acids 24-29 of the cytoplasmic tail. J. Biol. Chem. 266:5682-5688.

Chen, W-J, Goldstein, JL, Brown, MS (1990) NPXY, a sequence often found in cytoplasmic tails, is required for coated pit-mediated internalization of the low density lipoprotein receptor. J. Biol. Chem. 265:3116-3123.

Collawn, JF, Kuhn, LA, Liu, L-FS, Tainer, JA, and Trowbridge, IS (1991) Transplanted LDL and mannose-6-phosphate receptor internalization signals promote high-efficiency endocytosis of the transferrin receptor. EMBO J. 10:3247-3253.

Collawn, JF, Stangel, M, Kuhn, LA, Esekogwu, V, Jing, S, Trowbridge, IS, Tainer, JA (1990) Transferrin receptor internalization sequence YXRF implicates a tight turn as the structural recognition motif for endocytosis. Cell 63:1061-1072.

Davis, CG, Lehrman, MA, Russell, DW, Anderson, RGW, Brown, MS, Goldstein, JL (1986) The J.D. mutation in familial hypercholesterolemia: amino acid substitution in cytoplasmic domain impedes internalization of LDL receptors. Cell 45:15-24.

Davis, CG, van Driel, IR, Russell, DW, Brown, MS, Goldstein, JL (1987) The low density lipoprotein receptor: identification of amino acids in cytoplasmic domain required for rapid endocytosis. J. Biol. Chem. 262:4075-4082.

Eberle, W, Sander, C, Klaus, W, Schmidt, B, von Figura, K, Peters, C (1991) The essential tyrosine of the internalization signal in lysosomal acid phosphatase is part of a $\beta$ turn. Cell 67:1203-1209.

Fuhrer, C, Geffen, I, Spiess, M (1991) Endocytosis of the ASGP receptor H1 is reduced by mutation of tyrosine-5 but still occurs via coated pits. J. Cell Biol. 114:423-431.

Gerhardt, EM, Chan, L-N L, Jing, S, Qi, M, Trowbridge, IS (1991) The cDNA sequence and primary structure of the chicken transferrin receptor. Gene 102:249-254.

Girones, N, Alvarez, E, Seth, A, Lin, I-M, Latour, DA, Davis, RJ (1991) Mutational analysis of the cytoplasmic tail of the human transferrin receptor. J. Biol. Chem. 266:19006-19012.

Goldstein, JL, Brown, MS, Anderson, RGW, Russell, DW, Schneider, WJ (1985) Receptor-mediated endocytosis: concepts emerging from the LDL receptor system. Annu. Rev. Cell Biol. 1:1-39.

Jing, S, Spencer, T, Miller, K, Hopkins, C, Trowbridge, IS (1990) Role of the human transferrin receptor cytoplasmic domain in endocytosis: Localization of a specific signal sequence for internalization. J. Cell Biol. 110:283-294.

Ktistakis, NT, Thomas, D, Roth, MG (1990) Characteristics of the tyrosine recognition signal for internalization of transmembrane surface glycoproteins. J. Cell. Biol. 111:1393-1407.

Lobel, P, Fujimoto, K, Ye, RD, Griffiths, G, Kornfeld, S (1989) Mutations in the cytoplasmic domain of the 275 kd mannose-6-phosphate receptor differentially alter lysosomal enzyme sorting and endocytosis. Cell 57:787-796.

McGraw, TE, Pytowski, B, Arzt, J, Ferrone, C (1991) Mutagenesis of the human transferrin receptor: Two cytoplasmic phenylalanines are required for efficient internalization and a second-site mutation is capable of reverting an internalization-defective phenotype. J. Cell Biol. 112:853-861.

Mostov, KE, de Bruyn Kops, A, Deitcher, DL (1986) Deletion of the cytoplasmic domain of the polymeric immunoglobulin receptor prevents basolateral localization and endocytosis. Cell 47:359-364.

Rothenberger, S, Iacopetta, BJ, Kuhn, LC (1987) Endocytosis of the transferrin receptor requires the cytoplasmic domain but not its phosphorylation site. Cell 49:423-431.

Trowbridge, IS (1991) Endocytosis and signals for
    internalization. Curr. Opin. Cell Biol. 3:634-641.
    Erratum 3:1062.
Vega, MA, Strominger, JL (1989) Constitutive endocytosis of HLA
    class I antigens requires a specific protein portion of the
    intracytoplasmic tail that shares structural features with
    other endocytosed molecules. Proc. Natl. Acad. Sci. USA
    86:2688-2692.

# REGULATION OF EARLY ENDOSOME FUSION IN VITRO

Olivia Steele-Mortimer, Michael J. Clague,

Leo Thomas, Jean-Pierre Gorvel and Jean Gruenberg

European Molecular Biology Laboratory

Postfach 10.2209, D-6900 Heidelberg, Germany

Tel: +49-6221-387.288;  Fax: +49-6221-387.306

## MEMBRANE TRANSPORT IN THE ENDOCYTIC PATHWAY

It is well known that molecules entering the cell do so via the early endosome. Within five minutes of internalization into animal cells, lipids, solutes and trans-membrane proteins (including receptors and their bound ligands), appear in early endosomes located predominately at the cell periphery (Griffiths and Gruenberg, 1991; Murphy, 1991). Molecules which return to the plasma membrane, in particular cell surface receptors, are rapidly recycled from the early endosome. In

NATO ASI Series, Vol. H 74
Molecular Mechanisms of Membrane Traffic
Edited by D. J. Morré, K. E. Howell, and J. J. M. Bergeron
© Springer-Verlag Berlin Heidelberg 1993

contrast, molecules destined for degradation are transported to the perinuclear late endosomes, and eventually to the lysosomes. The endocytic pathway is also connected, via vesicular traffic, to the biosynthetic pathway (Kornfeld and Mellman, 1989).

The mechanism of transport from early to late endosomes, and then to the lysosomes, is still unclear. Two kinds of models have been proposed for this process, based on whether the endocytic pathway is made up of: 1) stable compartments connected by vesicular transport (Griffiths and Gruenberg, 1991) or 2) a series of transient compartments undergoing maturation (Murphy, 1991). Although the debate over the mechanism continues, components that mediate endocytic membrane transport are being identified, primarily by the use of cell-free assays that reconstitute single membrane transport events. These include: fusion of endocytic clathrin-coated vesicles with early endosomes (Woodman and Warren, 1991), lateral fusion between early endosomes (Gruenberg and Howell, 1989), meeting between apical and basolateral endosomes in polarized MDCK cells (Bomsel et al., 1990) and transfer from endosomes to either lysosomes (Mullock et al., 1989) or to the trans-Golgi network (Goda and Pfeffer, 1988). These assays are now being used to dissect the mechanisms that mediate endocytic membrane transport.

# EARLY ENDOSOME FUSION IN VITRO

Several groups have shown that early endosomes will fuse with one another in vitro (Gruenberg and Howell, 1989). This lateral fusion event occurs with a high efficiency (Gruenberg et al., 1989) and is specific, since early endosomes do not fuse with late endosomes (Gorvel et al., 1991) or other distal stages of the endocytic pathway (Gruenberg et al., 1989). This specificity has been clearly demonstrated in vitro with endosomes prepared from MDCK cells. These cells are polarized and have two populations of early endosomes, apical and basolateral, which are topologically and functionally distinct in vivo (Bomsel et al., 1989; Parton et al., 1989). In vitro, fusion only occurs between endosomes of the same population and apical and basolateral endosomes do not fuse with each other (Bomsel et al., 1990). We have previously proposed that the propensity of early endosomes to undergo lateral fusion in vitro suggests that, in vivo, the early endosome is a highly dynamic organelle, consisting of a network of elements connected by fusion and fission events (Gruenberg and Howell, 1989). However, the organization of the early endosome may vary in different cell types, as indicated by recent in vivo studies which have revealed that, morphologically, the structure of this organelle shows considerable variation (Marsh et al., 1986; Hopkins et al., 1990; Tooze and Hollinshead, 1991).

## REGULATION OF EARLY ENDOSOME FUSION

The use of in vitro assays has allowed the identification of some of the components involved in lateral fusion between early endosomes, including the N-ethylmaleimide-sensitive fusion protein or NSF (Diaz et al., 1989), originally identified as a factor necessary for intra-Golgi transport (Rothman and Orci, 1992). The regulation of early endosome fusion by phosphorylation/dephosphorylation events, both in mitosis (Tuomikoski et al., 1989) and interphase (Woodman et al., 1992), has also been demonstrated using cell-free assays. We have shown that, in vitro, fusion is inhibited by mitotic cytosol. This inhibition appears to be due to the activity of cdc2 kinase, a protein involved in cell cycle control, as interphase cytosol loses its ability to support fusion when it is supplemented with cdc2 (Tuomikoski et al., 1989). In addition, it appears that the inhibition is mediated by the association of cyclinB with the kinase (Thomas et al., 1992). We propose that the arrest of membrane traffic in mitotic animal cells (Warren, 1989) could be due, at least partially, to the activity of the cyclinB-cdc2 complex. Finally, in common with other steps of membrane transport, GTP-binding proteins are involved in the regulation of endosome fusion (Goud and McCaffrey, 1991).

# SMALL GTP-BINDING PROTEINS

The first GTP-binding proteins to be implicated in membrane transport were the ras-related yeast proteins YPT1 and SEC4, which are involved in transport from the ER to Golgi (Segev et al., 1988; Baker et al., 1990) and in the fusion of exocytic transport vesicles with the plasma membrane (Goud et al., 1988; Salminen and Novick, 1987), respectively. A second line of evidence for the involvement of GTP-binding proteins in membrane transport is the sensitivity of cell-free assays, reconstituting both exocytic and endocytic traffic, to the nonhydrolysable GTP analogue, GTPγS (Goud and Mc Caffrey, 1991). It is now presumed that at least two subfamilies of small GTP-binding proteins, SEC4/YPT1/rab and ARF/SAR, are involved in membrane transport (Pfeffer, 1992).

In mammalian cells a large number of YPT1/SEC4 related proteins, comprising the rab family, have been identified (Valencia et al., 1991). Localization studies have shown that each rab protein is restricted to the cytosolic face of one or two compartments within the cell (Chavrier et al., 1990a; Fischer von Mollard et al., 1990; Goud et al., 1990; van der Sluijs et al., 1991). This localization supports the theory, proposed by Bourne (1988), that these proteins are targeting molecules. According to this model the presence of a rab protein on each transport vesicle would ensure targeting of the vesicle to the correct acceptor membrane, thus confering specificity to the transport process. Integral to this model is the assumption that these small GTP-binding proteins exist in two conformational states, which are determined by the binding of nucleotide. Thus, the proteins are predicted to act as molecular "switches", which can be turned on or off by the binding of GTP or GDP respectively. By analogy with p21ras, it is likely that binding and hydrolysis of nucleotide by rab proteins is controlled by their interaction with a number of regulatory proteins. For example, the intrinsic GTPase activity of rab proteins is low (Kabçenell et al., 1990; Wagner et al., 1987) indicating a role for a GAP (GTPase activating protein). A GAP activity has been reported for rab3A (Burstein et al., 1991). Two other proteins predicted to interact with rabs are GDI (GDP-dissociation inhibitor) and GDS (GDP-dissociation stimulator).

We have shown that the rab5 protein, which localizes to the plasma membrane and to early endosomes (Chavrier et al., 1990), is involved in the regulation of early endosome fusion in vitro (Gorvel et al., 1991). Cell-free fusion of early endosomes is inhibited by antibodies against rab5 and this inhibition can be reversed by the addition of cytosol containing overexpressed wild type rab5. In contrast, two rab5 mutants, one with a point mutation in the GTP-binding domain and another lacking the C-terminal motif required for membrane attachment, were unable to reverse inhibition. More recently, rab5 has been shown to be involved at an early stage of the endocytic pathway in vivo (Bucci et al., in press).

The precise function of the rab5 protein in endosome-endosome fusion is still unknown and proteins with which rab5 interacts have not yet been identified. However, the presence of a signal in the C-terminal regions of the rab5 and rab7 proteins (Chavrier et al., 1991), which is both necessary and sufficient for localization to the correct membranes, suggests the presence of membrane proteins able to recognize, and presumably interact with, specific rab proteins. In addition, like other members of the YPT1/SEC4/rab family, which are synthesized as cytosolic proteins and require an isoprenoid modification of a C-terminal cysteine residue for membrane association, rab5 must interact with a prenyltransferase (Kinsella and Maltese, 1991).

One observation that may lead to a better understanding of the function of rab5 is that the rab5 protein is extremely sensitive to trypsin. Preliminary studies we have done indicate that it is this sensitivity which is responsible for loss of fusion activity after gentle trypsinization of early endosome membranes (Gorvel, J. P., Clague, M. J.,

Huber, L., Chavrier, P., Steele-Mortimer, O., and Gruenberg, J., submitted). We have treated early endosomes with trypsin, under the very mild conditions required to inhibit fusion, and then used two-dimensional gel electrophoreisis to compare the polypeptide composition of the membranes with that of untreated membranes. Under these conditions the only protein which showed detectable cleavage was rab5. Significantly, trypsin treatment of either "acceptor" or "donor" early endosomes was as effective in inhibiting fusion as treatment of both sets of membranes. These observations indicate that,for fusion to occur, functional rab5 is required on both membranes.

The elucidation of the mechanism of endosome fusion is made more difficult by the involvement of other GTP-binding proteins. Recently it has been shown that a second rab protein, rab4, is present on early endosome membranes (van der Sluijs et al., 1991). It is not known whether rab4 is involved in endosome fusion, or in another membrane transport event involving the early endosomes, however, the fact that two rab proteins share the same localization pattern is intriguing. It is also interesting that the rab4 protein can be phosphorylated by the cdc2 kinase in vitro, and in mitotic cells phosphorylation may be associated with a decrease in the ratio of membrane bound to cytosolic rab4 (Bailly et al., 1991). In contrast, the rab5 protein does not have a consensus site for cdc2 kinase and we have been unable to detect any evidence of phosphorylation.

The second familly of small GTP-binding proteins implicated in membrane transport is the ARF/SAR familly. In yeast, SAR1 is required for the formation of transport vesicles from the ER (D'Enfert et al., 1991) and in mammalian cells, ARF has been shown to be involved in intra Golgi vesicular transport (Serafini et al., 1991) and also in transport between the ER and the cis-Golgi (Balch et al., 1992). More recently, ARF has been implicated in the regulation of early endosome fusion in vitro (Lenhard et al., 1992). It is possible that the different proteins from the SEC4/YPT1/rab and ARF/SAR families regulate different sequential steps in early endosome fusion, as in other steps of membrane transport (Gruenberg and Clague, 1992).

## HETEROTRIMERIC G-PROTEINS

A role in endocytosis has also been postulated for the heterotrimeric G-proteins..This was first suggested by the effect of aluminum fluoride (Mayorga et al., 1989; Wessling-Resnick and Braell, 1990), which activates these proteins but not the monomeric GTP-binding proteins (Kahn, 1991). More recently, Colombo et al. (1992) have shown that the heterotrimeric G-proteins play a role in the regulation of early endosome fusion, in vitro. It is not yet known whether these G-proteins are regulated in the same way as the G-proteins involved in signal transduction, which are activated by the binding of ligands to specific receptors.

## CONCLUSIONS

Recent studies have yielded tantalizing glimpses into the complex mechanisms controlling the interactions between early endosomal membranes. NSF, phosphorylation/dephosphorylation events and GTP-binding proteins have all been showed to play a role. However, little is known about the functions of these factors in the recognition/fusion process or even at what steps in the process they are involved. Other proteins involved in endosome fusion, for example proteins interacting with GTP-binding proteins or those regulated by phosphorylation, remain to be identified. The cell-free assays now available are being used, in combination with techniques such as fractionation and protein purification, to identify and characterize these components. The development of more refined in vitro assays, in which the different steps of the recognition/fusion process can be dissected, will ultimately lead to a far better understanding of this complex process.

# References

Bailly, E., McCaffrey, M., Touchot, N., Zahraoui, A., Goud, B., and Bornens, M. (1991). Phosphorylation of two small GTP-binding proteins of the rab family by p34 cdc2. Nature *350*, 715-718.

Baker, D., Wuestehube, L., Scheckman, R. and Segev, N. (1990). GTP-binding YPT1 protein and Ca2+ function independently in a cell-free protein transport reaction. PNAS USA *87*,355-359.

Balch, W. E. (1992). From G minor to G major. Current Biology *2*, 157-160.

Burstein, E. S., Linko-Stentz, K., Lu, Z., and Macara, I. G. (1991). Regulation of the GTPase activity of the ras like protein p25-rab3a: evidence for a rab3a specific GAP. J. Biol. Chem. *266*, 2689-2692.

Bomsel, M., Parton, R., Kuznetsov, S. A., Schroer, T. A., and Gruenberg, J. (1990). Microtubule and motor dependent fusion in vitro between apical and basolateral endocytic vesicles from MDCK cells. Cell *62*, 719-731.

Bomsel, M., Prydz, K., Parton, R. G., Gruenberg, J., and Simons, K. (1989). Endocytosis in filter-grown Madin-Darby canine kidney cells. j. Cell Biol. *109*, 3243-3258.

Bourne, H. (1988). Do GTPases direct membrane traffic in secretion? Cell *53*, 669-671.

Bourne, H. R., Sanders, D. A., and McCormick, F. (1990). The GTPase superfamily: a conserved switch for diverse cell functions. Nature *348*, 125-132.

Bucci, C., Parton, R. G., Mather, I. H., Stunnenberg, H., Simons, K., Hoflack, B., and Zerial, M. (in press). The small GTP-ase rab5 functions as a regulatory factor in the early endocytic pathway. Cell

Chavrier, P., Gorvel, J. P., Stelzer, E., Simons, K., Gruenberg, J., and Zerial, M. (1991). Hypervariable C-terminal domain of rab proteins acts as a targetting signal. Nature *353*, 769-772.

Chavrier, P., Parton, R. G., Hauri, H. P., Simons, K., and Zerial, M. (1990). Localisation of low molecular weight GTP binding proteins to exocytic and endocytic compartments. Cell *62*, 317-329.

Colombo, M. I., Mayorga, L. S., Casey, P. J., and Stahl, P. D. (1992). Evidence of a role for heterotrimeric GTP-binding proteins in endosome fusion. Science *255*, 1695-1697.

D'Enfert, C., Wuestehube, L. J., Lila, T., and Schekman, R. (1991). Sec12p-dependent membrane binding of the small GTP-binding protein Sar1p promotes formation of transport vesicles from the ER. J. Cell Biol. *114*, 663-670.

Diaz, R., Mayorga, L. S., Weidman, P. J., Rothman, J. E., and Stahl, P. (1989). Vesicle fusion following receptor-mediated endocytosis requires a protein active in Golgi transport. Nature *339*, 398-400.

Fisher von Mollard, G., Mignerg, G., Baumert, M., Perin, M., Hanson, T., Burger, P., Jahn, R. and Sudhof, T. (1990). Rab3 is a small GTP-binding protein exclusively localized to synaptic vesicles. PNAS USA *87*, 1988-1992.

Goda, Y., and Pfeffer, S. R. (1988). Selective recycling of the mannose 6-phosphate/IGF-II receptor to the TGN in vitro. Cell *55*, 309-320.

Gorvel, J. P., Chavrier, P., Zerial, M., and Gruenberg, J. (1991). Rab 5 controls early endosome fusion in vitro. Cell *64*, 915-925.

Goud, B., Salminen, A., Walworth, N. C., and Novick, P. J. (1988). A GTP-binding protein required for secretion rapidly associates with secretory vesicles and the plasma membrane in yeast. Cell *53*, 753-768.

Goud, B., Zahraoui, A., Tavitian, A. and Saraste, J. (1990) Small GTP-binding protein associated with Golgi cisternae. Nature *345*, 553.

Goud, B., and McCaffrey, M. (1991). Small GTP-binding proteins and their role in transport. Curr. Opin. Cell Biol. *3*, 626-633.

Griffiths, G., and Gruenberg, J. (1991). The arguments for pre-existing early and late endosomes. Trends Cell Biol. *1*, 5-9.

Gruenberg, J., and Clague, M. (1992). Regulation of intracellular membrane transport. Curr. Op. Cell Biol. *in press*,

Gruenberg, J., Griffiths, G., and Howell, K. E. (1989). Characterisation of the early endosome and putative endocytic carrier vesicles in vivo and with an ssay of vesicle fusion in vitro. J. Cell Biol. *108*, 1301-1316.

Gruenberg, J., and Howell, K. E. (1989). Membrane traffic in endocytosis: insights from cell-free assays. Annu. Rev. Cell Biol. *5*, 453-481.

Hopkins, C. R., Gibson, A., Shipman, M., and Miller, K. (1990). Movement of internalized ligand-receptor complexes along a continuous endosomal reticulum. Nature *346*, 335-339.

Kabcenell, A. K., Goud, B., Northup, J. K., and Novick, P. J. (1990). Binding and hydrolysis of guanine nucleotides by sec4p, a yeast protein involved in the regulation of vesicular transport. J. Biol. Chem. *265*, 9366-9372.

Kahn, R. A. (1991). Fluoride is not an activator of the smaller (20-25 kDa) GTP-binding proteins. J. Biol. Chem. *266*, 15595-15597.

Kinsella, B. T., and Maltese, W. A. (1991). rab GTP-binding protein implicated in vesicular transport are isoprenylated in vitro at cysteines within a novel carboxy-terminal motif. J. Biol. Chem. *266*, 8540-8544.

Kornfeld, S., and Mellman, I. (1989). The biogenesis of lysosomes. Annu. Rev. Cell Biol. *5*, 483-526.

Lenhard, J. M., Kahn, R. A., and Stahl, P. D. (1992). Evidence for ADP-ribosylation factor (ARF) as a regulator of in vitro endosome-endosome fusion. *267*, 13047-13052.

Marsh, H. M., Griffiths, G., Dean, G. E., Mellman, I. and Helenius, A. (1986). Three-dimensional structure of endosomes in BHK-21 cells. PNAS USA *83*, 2899-2903.

Mayorga, L. S., Diaz, R., Colombo, M. I., and Stahl, P. D. (1989). GTPγS stimulation of endosome fusion suggests a role for a GTP-binding protein in the priming of vesicles before fusion. Cell Regulation *1*, 113-124.

Mullock, B. M., Branch, W. J., vanSchaik, M., Gilbert, L. K., and Luzio, J. P. (1989). Reconstitution of an endosome-lysosome interaction in a cell-free system. J. Cell Biol. *108*, 2093-2099.

Murphy, R. F. (1991). Maturation models for endosome and lysosome biogenesis. Trends Cell Biol. *1*, 77-82.

Parton, R. G., Prydz, K., Bomsel, M., Simons, K., and Griffiths, G. (1989). Meeting of the apical and basolateral endocytic pathways of the Madin-Darby canine kidney cell in late endosomes. J. Cell Biol. *109*, 3259-3272.

Pfeffer, S. (1992). GTP-binding proteins in intracellular transport. Trends Cell Biol. *2*, 41-46.

Rexach, M. F., and Schekman, R. W. (1991). Distinct biochemical requirements for the budding, targeting and fusion of ER-derived transport vesicles. J. Cell Biol. *114*, 219-229.

Rothman, J. E., and Orci, L. (1992). Molecular dissection of the secretory pathway. Nature *355*, 409-415.

Salminen, A., and Novick, P. J. (1987). A ras-like protein is required for a post-Golgi event in yeast secretion. Cell *49*, 527-538.

Segev, N., Mulholland, J., and Botstein, D. (1988). The yeast GTP-binding YPT1 protein and a mammalian counterpart are associated with the secretion machinery. Cell *52*, 915-924.

Serafini, T., Orci, L., Amherdt, M., Brunner, M., Kahn, R. A., and Rothman, J. E. (1991). ADP-ribosylation factor is a  subunit of the coat of Golgi-derived COP coated vesicles: a novel role for a GTP-binding protein. Cell *67*, 239-253.

Thomas, L., Clarke, P., Pagano, M., and Gruenberg, J. (1992). Inhibition of membrane fusion in vitro via cyclin B but not cyclin A. J. Biol. Chem. *267*, 6183-6187.

Tooze, J., and Hollinshead, M. (1991). Tubular early endosomal network in AtT20 and other cells. J. Cell Biol. *115*, 635-654.

Tuomikoski, T., Felix, M., Doree, M., and Gruenberg, J. (1989). Inhibition of endocytic vesicle fusion in vitro by the cell-cycle control protein kinase cdc2. Nature *342*, 942-945.

Valencia, A., Chardin, P., Wittinghofer, A., Sander, C. (1991). The ras protein family: evolutionary tree and role of conserved amino acids. Biochemistry *30*, 4637-4648.

van der Sluijs, P., Hull, M., Zahraoui, A., Tavitian, A., Goud, B., and Mellman, I. (1991). The small GTP-binding protein rab4 is associated with early endosomes. Proc. Natl. Acad. Sci. *88*, 6313-6317.

Wagner, P., Molenaar, C. M., Rauh, A. J., Brokel, R., Schmitt, H. D. and Gallwitz, D. (1997) Biochemical properties of the ras-related YPT1 protein in yeast: a mutational analysis. EMBO J. *6*, 2373-2379.

Warren, G. (1989). Mitosis and membranes. Nature *342*, 857-858.

Wessling-Resnick, M., and Braell, W. A. (1990). Characterization of the mechanism of endocytic vesicle fusion in vitro. J. Biol. Chem. *265*, 16751-16759.

Woodman, P., Mundy, D. I., Cohen, P., and Warren, G. (1992). Cell-free fusion of endocytic vesicles is regulated by phosphorylation. J. Cell Biol. *116*, 331-338.

Woodman, P., and Warren, G. (1991). Isolation of functional, coated endocytic vesicles. J. Cell Biol. *112*, 1133-1141.

# Adaptins and their Role in Clathrin-Mediated Vesicle Sorting

Margaret S. Robinson, Catriona L. Ball, and Matthew N.J. Seaman
Department of Clinical Biochemistry, University of Cambridge
Cambridge CB1 4UR, England

## Introduction

Adaptins are subunits of protein complexes known as adaptors, which are components of the coats on clathrin-coated vesicles. Several lines of evidence support the model shown in Figure 1, in which the adaptors are depicted as the components of the coats that give coated pits and vesicles their specificity, enabling them to concentrate selected membrane proteins which act as receptors for ligands on the other side of the membrane.

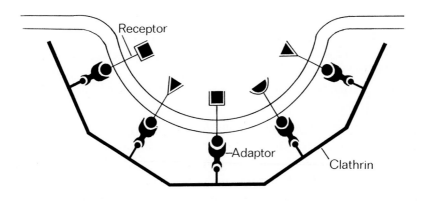

**Figure 1.** Model of a coated pit.

First, both biochemical and structural studies indicate that adaptors are located between the clathrin outer shell and the vesicle membrane, and thus are in the right position to interact with membrane proteins in the coated pit (Unanue et al., 1981; Vigers et al., 1986). Second, direct binding has been demonstrated in vitro between adaptors and some of the proteins known to be concentrated in coated pits, including the low density lipoprotein (LDL) receptor and the mannose-6-phosphate (M6P) receptor (Pearse, 1988; Glickman et al., 1989). Third, there are two populations of clathrin-coated vesicles in the cell, one associated with the plasma membrane and one associated with the *trans*-Golgi

NATO ASI Series, Vol. H 74
Molecular Mechanisms of Membrane Traffic
Edited by D. J. Morré, K. E. Howell, and J. J. M. Bergeron
© Springer-Verlag Berlin Heidelberg 1993

network (TGN), which are thought to concentrate different membrane proteins; and although the clathrin is identical for the two types of coated vesicles, the adaptors are different (Robinson, 1987; Ahle et al., 1988). Finally, the ability of the two different types of adaptors to bind to different membrane proteins is consistent with their localisation to the two membrane compartments: e.g., the plasma membrane adaptor interacts with the LDL receptor while the Golgi adaptor does not (Glickman et al., 1989).

## Adaptin diversity

Both the plasma membrane-associated and the TGN-associated adaptors are heterotetramers. Their subunits consist of two adaptins of ~100kD: a β-adaptin (β for the plasma membrane, β' for the TGN) and an adaptor-specific adaptin (α for the plasma membrane, γ for the TGN), plus one copy each of two smaller proteins of ~50kD and ~20kD. We have been investigating the possibility that there might be multiple isoforms of the α- and γ-adaptins, which could potentially have different functions and be associated with different subsets of plasma membrane and TGN coated vesicles. In the case of γ-adaptin, one-dimensional gel electrophoresis only reveals a single band. However, Southern blotting suggests that there may be a second γ-adaptin gene, which might have escaped detection either because the two γ-adaptin gene products have very similar electrophoretic mobilities, or because expression of the second γ-adpatin gene is low or non-existent in the tissues that have so far been investigated. We are now attempting to clone the putative second γ-adaptin gene from genomic libraries (Ball and Robinson, unpublished results).

In the case of α-adaptin, two bands can clearly be resolved on gels of coated vesicles purified from brain or on Western blots of brain homogenates probed with α-specific antibodies. These have been designated αA and αC for historical reasons: originally the three bands that could be seen in the 100K region were referred to as A, B, and C; but B is now called β-adaptin. In tissues other than brain, the slower migrating form of α-adaptin, αA, is not detectable (Robinson, 1987). When a brain cDNA library was originally screened for α-adaptin sequences, two distinct types of clones were obtained, clearly the products of different genes. The one encoding a protein of 108kD was assumed to be αA, while the one encoding a protein of 104kD was assumed to be αC. However, Northeren blotting revealed that there were significant amounts of the message encoding the 108kD protein in liver as well as brain (Robinson, 1989). We have recently found that the αA message in liver is spliced differently from that in brain, so that the protein is 22 amino acids shorter (Ball and Robinson, unpublished results). This difference presumably accounts for the observation that in tissues such as liver, only one α-adaptin band can be resolved: both proteins are expressed, but αA co-migrates with αC. The expression vector pGEX was used to construct fusion proteins containing α-adaptin sequences unique to αC, to αA, or to brain αA, and monospecific antibodies have now been raised. Preliminary immunofluorescence results indicate that the various isoforms of α-adaptin can co-exist in the same coated pit or vesicle; but so far only non-polarised tissue culture cells have been examined (Ball and Robinson, unpublished results). It will be of interest to compare the distribution of the different α-adaptins in more polarised cells such as epithelial cells or neurons.

**Targeting of engineered adaptins in vivo**
We have also been addressing the question of how the two adaptor complexes are targeted to the correct subcellular compartment: the plasma membrane or the TGN. It seems likely that adaptor targeting is independent of the ability of adaptors to bind to the cytoplasmic domains of proteins like the LDL receptor and the M6P receptor, since the distribution of adaptors is quite different from the distribution of the membrane proteins known to be concentrated in coated pits. For instance, the majority of the M6P receptors in most cells are found in late endosomes and pre-lysosomes; but adaptors do not appear to bind to these compartments. Thus, a more plausible scenario is one in which adaptors are pre-targeted to the plasma membrane or the TGN, after which they can bind to the cytoplasmic domains of selected membrane proteins in the same compartment.

To look for potential targeting signals on the two adaptor complexes, we have begun to swap domains between the plasma membrane and Golgi adaptors, beginning with the C termini of αA and αC. The C termini were chosen for two reasons. First, the model for adaptor structure (Figure 2) predicts that the C termini of the adaptins, which form the adaptor "ears", fold independently and do not interact with the other subunits of the adaptor complex (Heuser and Keen, 1988; Kirchhausen et al., 1989). Second, a comparison of the sequences of α- and γ-adaptins reveals that they are 25% identical, but the homology stops just before the putative ear domain (Robinson, 1990). This lack of similarity between the two ears suggests that they have adaptor-specific functions.

To switch the ears between the two adaptins, new restriction sites were engineered on either side of the α- and γ-adaptin hinges to facilitate the construction of chimeras. Five new constructs were created: γγα, γγ–, αγα, α γ γ, and α γ –. The nomenclature refers to the three adaptin domains: N terminus, hinge, and ear; thus, all the constructs contain hinges derived from bovine γ-adaptin. To localise the constructs by immunofluorescence, they were transfected into Rat 1 fibroblasts and visualised with mAb 100/3, a species-specific antibody that reacts with the γ-adaptin hinge but does not recognise the rodent protein. Labelling of both transiently and stably expressing cells indicates that the α- and γ-adaptin ears are not required for adaptor recruitment. Thus, targeting information is likely to be contained within the adaptor head (or possibly the β-adaptin ear) (Robinson, unpublished results).

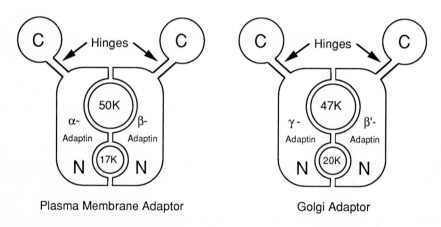

Plasma Membrane Adaptor                 Golgi Adaptor

**Figure 2.** Models of the two adaptor complexes.

Future experiments should allow the targeting signal to be identified. In addition, it should be possible to use the existing constructs to investigate the role of the α- and γ-adaptin ears by overproducing the mutant proteins to compete out most of the endogenous wild-type α- or γ-adaptin. One possibility is that the ears might be involved in the binding of adaptors to proteins like the LDL receptor and the M6P receptor. If so, cells expressing chimeras might be expected to mis-sort their membrane proteins.

**Effects of brefeldin A on adaptors**

The drug brefeldin A (BFA) causes the Golgi stack to disappear and resident Golgi membrane proteins to mix with the ER; in addition, changes can also be detected in the TGN, endosomes, and lysosomes (reviewed by Klausner et al., 1992). Although these effects occur within minutes, the drug has an even more rapid effect on a peripheral membrane protein of the Golgi stack, β-COP. β-COP is associated with the non-clathrin-coated (or COP-coated) vesicles that mediate traffic from one Golgi cisterna to the next, and is also present in the cytosol as a component of a protein complex known as the coatomer (Serafini et al., 1991). Coatomers are believed to cycle on and off the membrane during the COP-coated vesicle cycle, in much the same way that adaptors and clathrin cycle on and off the membrane during the clathrin-coated vesicle cycle. In BFA-treated cells, however, within 30 seconds essentially all the β-COP appears to be cytoplasmic rather than membrane-bound (Donaldson et al., 1991). Thus, BFA is believed to interfere with the COP-coated vesicle cycle, preventing the binding of coatomers to the membrane.

β-COP was recently cloned and sequenced and was shown to be homologous to β-adaptin (Duden et al., 1991). This finding suggested that there would be underlying similarities between clathrin-coated and COP-coated vesicles, one of which could be in the regulation of the association of their coat proteins with the appropriate membrane. To test this possibility, cells were treated with BFA and the distribution of α- and γ-adaptin was assayed by immunofluorescence. The drug was found to have no discernable effect on α-adaptin, even after one hour. However, the drug caused a very rapid change in the distribution of γ-adaptin: within less than a minute, the protein appeared to be cytoplasmic rather than associated with the TGN (Robinson and Kreis, 1992; Wong and Brodsky, 1992).

The similarity in effect of BFA on γ-adaptin and β-COP can be extended further. $AlF_4-$ prevents β-COP redistribution in cells that are subsequently treated with the drug, while in permeabilized cells, effects of BFA can be blocked by pretreating with GTPγS (Donaldson et al., 1991). These results indicate that G proteins (presumably heterotrimeric G proteins) participate in the association of coatomers with the Golgi membrane. Similarly, we found that pretreatment of intact cells with $AlF_4-$ and of permeabilized cells with GTPγS also has a protective effect on γ-adaptin when the drug is added subsequently (Robinson and Kreis, 1992).

The involvement of G proteins in the two coated vesicle cycles suggests that the recruitment of coat proteins onto their target membranes is likely to be a regulated rather than a constitutive process. It is possible that signals are transmitted across the membranes of the Golgi stack or the TGN, and that compartment-specific G proteins then trigger the binding of coat proteins as a first step in the process of vesicle budding. To try to identify the proteins that participate in coat protein recruitment, we have developed systems in which membrane binding of both TGN and plasma membrane adaptors can be reconstituted in vitro.

## Adaptor recruitment in vitro

The basic design of our in vitro system is shown in Figure 3. NRK cells are permeabilized by freezing and thawing, then incubated with pig brain cytosol. Newly recruited Golgi and plasma membrane adaptors can be visualised by immunofluorescence, using species-specific or tissue-specific antibodies against γ- or α-adaptin.

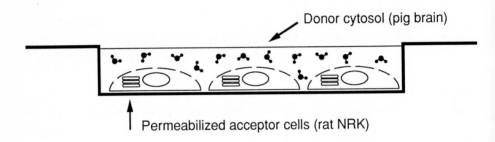

**Figure 3.** Design of the in vitro system.

Recruitment of Golgi adaptors (labelled with mAb 100/3) is enhanced by the addition of ATP and a regenerating system, and is increased still further by the addition of GTPγS (Robinson and Kreis, 1992). We have recently found that highly purified pig brain Golgi adaptors are also capable of binding to the TGN membrane in permeabilized cells, but increased labelling is seen when mouse or rat cytosol is added. A more quantitative assay has also been developed, using radioiodinated Golgi adaptors. GTPγS was found to cause a ~4-fold increase in the amount of label associated with the cell pellet, while cytosol caused a ~3-fold increase. We now plan to use this system as an assay for identifying both cytoplasmic and membrane-bound factors (Seaman and Robinson, unpublished results).

Plasma membrane adaptor recruitment can also be reconstituted in vitro using this system. Either a species-specific mAb against α-adaptin (AC2-M15) or the brain-specific antibody raised against the extra 22 amino acids in αA can be used to label newly recruited α-adaptin without any background from endogenous protein. Like Golgi adaptor recruitment, plasma membrane adaptor recruitment is enhanced by the addition of energy in the form of ATP and a regenerating system. However, the effect of GTPγS on plasma membrane adaptor recruitment is strikingly different. Addition of GTPγS appears to block the binding of plasma membrane adaptors to the plasma membrane, causing them to bind instead to a perinuclear compartment. Preliminary double labelling experiments suggest that this compartment is a subcompartment of the endosomal system (Seaman and Robinson, unpublished results). Our current working hypothesis is that plasma membrane adaptor receptors cycle between the plasma membrane of the other compartment, which may be a storage compartment. Normally the receptors are switched on when they are in the plasma membrane and off when they are in the other compartment, but GTPγS is somehow reversing the switch. By purifying the putative receptors for both plasma membrane and Golgi adaptors, using the in vitro system as an assay, we hope to be able to elucidate the role of G proteins in both recruitment processes.

## References

Ahle, S., Mann, A., Eichelsbacher, U., and Ungewickell, E. (1988) Structural relationships between clathrin assembly proteins from the Golgi and the plasma membrane. EMBO J. 4: 919-929

Donaldson, J.G., Lippincott-Schwartz, J., and Klausner, R.D. (1991) Guanine nucleotides modulate the effects of brefeldin A in semipermeable cells: regulation of the association of a 110kD peripheral membrane protein with the Golgi apparatus. J. Cell Biol. 112: 579-558

Duden, R., Griffiths, G., Frank, R., Argos, P., and Kreis, T.E. (1991) β-COP, a 110 kd protein associated with non-clathrin-coated vesicles and the Golgi complex, shows homology to β-adaptin. Cell 64: 649-665

Glickman, J.N., Conibear, E., and Pearse, B.M.F. (1989) Specificity of binding of clathrin adaptors to signals on the mannose-6-phosphate/insulin-like growth factor II receptor. EMBO J. 8: 1041-1047

Heuser, J.E., and Keen, J. (1988) Deep-etch visualization of proteins involved in clathrin assembly. J. Cell Biol. 107: 877-886

Kirchhausen, T., Nathanson, K.L., Matsui, W., Vaisberg, A., Chow, E.P., Burne, C., Keen, J.H., and Davis, A.E. (1989) Structural and functional division into two domains of the large (100- to 115-kDa) chains of the clathrin-associated complex AP-2. Proc. Natl. Acad. Sci. USA 86: 2612-2616

Klausner, R.D., Donaldson, J.G., and Lippincott-Schwartz, J. (1992) Brefeldin A: insights into the control of membrane traffic and organelle structure. J. Cell Biol. 116: 1071-1080

Pearse, B.M.F. (1988) Receptors compete for aaptors found in plasma membrane coated pits. EMBO J. 7: 3331-3336

Pearse, B.M.F., and Robinson, M.S. (1990) Clathrin, adaptors, and sorting. Ann. Rev. Cell Biol. 6: 151-171

Robinson, M.S. (1987) 100-kD coated vesicle proteins: molecular heterogeneity and intracellular distribution studied with monoclonal antibodies. J. Cell Biol. 104: 887-895

Robinson, M.S. (1989) Cloning of cDNAs encoding two related 100-kD coated vesicle proteins (α-adaptins). J. Cell Biol. 108: 833-842

Robinson, M.S. (1990) Cloning and expression of γ-adaptin, a component of clathrin-coated vesicles associated with the Golgi apparatus. J. Cell Biol. 111: 2319-2326

Robinson, M.S., and Kreis, T.E. Recruitment of coat proteins onto Golgi membranes in intact and permeabilized cells: effects of brefeldin A and G protein activators. Cell 69: 129-138

Serafini, T., Stenbeck, G., Brecht, A., Lottspeich, F., Orci, L., Rothman, J.E., and Wieland, F.T. (1991) A coat subunit of Golgi-derived non-clathrin-coated vesicles with homology to the clathrin-coated vesicle coat protein β-adaptin. Nature 349: 215-220

Unanue, E.R., Ungewickell, E., and Branton, D. (1981) The binding of clathrin triskelions to membranes from coated vesicles. Cell 26: 439-446

Vigers, G.P.A., Crowther, R.A., and Pearse, B.M.F. (1986) Location of the 100kd-50kd accessory proteins in clathrin coats. EMBO J. 5: 2079-2085

Wong, D.H., and Brodsky, F.M. (1992) 100-kD proteins of Golgi- and trans-Golgi network-associated coated vesicles have related but distinct membrane binding properties. J. Cell Biol. 117: 1171-1179.

# ENDOSOMES AND CELL SIGNALLING

Barry I. Posner[1] and John J.M. Bergeron[2]
Depts of Medicine[1] and Anatomy[2]
McGill University and the Royal Victoria Hospital
687 Pine Avenue, West
Montreal, Quebec, Canada    H3A 1A1

The biological responses to many peptide hormones and growth factors are currently under intense investigation. The main area of uncertainty encompasses the transducing events which link initial events at the cell surface to those processes influencing the proteins (egs. enzymes, transporters) effecting the biological response. We have pursued the study of insulin and EGF action by examining the events following ligand-receptor complex formation at the cell surface. Along with others we have observed that these complexes undergo rapid internalization (Posner et al.,1982; Bergeron et al., 1985). In the case of the EGF and insulin receptor this process is accompanied by activation of their respective receptor tyrosine kinases (Cohen et al., 1985; Kay et al., 1986; Khan et al., 1986; Posner et al., 1987).

## ENDOSOMES AND CELL SIGNALLING

The process of receptor-mediated endocytosis leads to the concentration of ligand-receptor complexes in a heterogeneous population of tubulovesicular structures now referred to as endosomes (ENs), a term which covers all the intracellular nonlysosomal components involved in the uptake of exogenous substances into cells (Bergeron et al., 1985). The analysis of isolated ENs has confirmed that they possess distinct biochemical characteristics. Thus the electrophoretic pattern of endosomal proteins differed greatly from that of plasmalemma (Khan et al., 1986; Marsh et al., 1987; Schmid et al., 1988)

NATO ASI Series, Vol. H 74
Molecular Mechanisms of Membrane Traffic
Edited by D. J. Morré, K. E. Howell, and J. J. M. Bergeron
© Springer-Verlag Berlin Heidelberg 1993

and lysosomes (Marsh et al., 1987). Furthermore several investigators have documented the capacity of ENs to lower their internal pH via an NEM-sensitive ATPase (Mellman et al., 1986; Al-Awqati et al., 1986). The initial rate and extent of proton transfer appears lowest in 'early' ENs and gradually increases in 'late' ENs and lysosomes (Schmid et al., 1988).

Three functions have, to date, been suggested for ENs, namely: the processing of internalized ligands; the sorting and recycling of receptors; and the modulation of transmembrane signalling.

(a)   Processing of Internalized Ligands.

Prelysosomal degradation of internalized ligand was demonstrated in macrophage ENs (Diment et al., 1985) by a protease recently identified as cathepsin D (Diment et al., 1988). Other proteolytic enzymes have been reported in ENs (Pease et al., 1985; Ajioka et al., 1987; Schaudies et al., 1987) suggesting that proteolytic processing within this system may be quite complex. Evidence for the prelysosomal processing of internalized EGF (Schaudies et al., 1987) and parathyroid hormone (Diment et al., 1989) has been adduced. We and our collaborators have shown that insulin is degraded in both 'early' and 'late' ENs (Hamel et al., 1988); and have characterized this degradation process in isolated ENs (Doherty et al., 1990).

(b) Sorting and Recycling of Receptors.

The precise fate of ligand-receptor complexes within ENs varies with the particular ligand and receptor (Klausner et al.,

1983, Mostov et al., 1986). With respect to receptor tyrosine kinase (egs. insulin, EGF, PDGF) it appears that their intrinsic tyrosine kinase activity is necessary for their targeting to the degradative pathway and hence downregulation (Russell et al., 1987; Honegger et al., 1987; Felder et al., 1990).

## (c) Transmembrane Signalling.

The insulin receptor kinase (IRK): Following the demonstration that insulin binding to its receptor augmented the intrinsic kinase activity of the ß-subunit (Rosen et al., 1983) it was shown that this augmented activity persisted after the removal of insulin (Khan et al., 1986; Klein et al., 1986). This activated state of the receptor was shown to be dependent on ß-subunit autophosphorylation (Khan et al., 1986). A variety of studies have shown that an active IRK is necessary to effect the insulin response.

Based on our studies of insulin internalization we suggested that an activated form of the insulin receptor may extend the signalling process initiated at the cell surface to a substantially larger cytosolic volume and/or carry out specific functions intracellularly (Posner et al., 1980). This initial proposal has been supported by our observation that insulin administration augmented IRK activity in rat liver ENs (Khan et al., 1986; Khan et al., 1989), an observation confirmed in isolated rat adipocytes (Klein et al., 1987). Furthermore IRK mutations which impaired tyrosine kinase activity dramatically reduced ligand-induced internalization (Chou et al., 1987; Hari et al., 1987) indicating a relationship between IRK activation and internalization into ENs.

We recently demonstrated that endosomal IRK was activated by as little insulin as 15ng/100 g.b.wt., attesting to the physiologic relevance of these observations. The activated state of endosomal IRK was due to autophosphorylation since alkaline phosphatase treatment of partially purified receptors in vitro abolished all IRK activity. Following insulin administration ß-subunit phosphotyrosine (P-Y) content measured by either <u>in vivo</u> $^{32}$P-labelling or immunoblotting with anti P-Y antibodies ($\alpha$P-Y) at the time of maximal kinase activity (PM, 30 sec; EN, 2 min post-injection), was significantly less for the endosomal compared to PM IRK (Table 1). Nevertheless maximum

Table 1. Phosphotyrosine content of endosomal IRK as a percent of that in PM.

| Procedure | MnATP | P-Y Content in ENs (%PM) |
|-----------|-------|--------------------------|
| $^{32}$P-labelling | − | 10 ± 5* |
|  | + | 430 ± 130 |
| P-Y content | − | 29 ± 5 |
|  | + | 256 ± 18 |

PM and ENs were prepared at 30 sec and 2 min after i.v. insulin (1.5 or 15 µg/100 g.b.wt.). Basal IRK $^{32}$P-labelling was determined by measuring the specific activity of IRs prepared from rats preinjected 60 min prior to insulin with $^{32}$P-orthophosphate (5mCi/rat) (Burgess <u>et al,</u> 1992). $^{32}$P-labelling consequent to autophosphorylation was measured with ATP-$\tau$-$^{32}$P as substrate as described elsewhere (Khan <u>et al,</u> 1989). P-Y content was measured by immunoblotting with antibodies to P-Y and the IR ß-subunit both before (−MnATP) and after autophosphorylation (+MnATP) (Burgess <u>et al,</u> 1992). All values are the mean ± SE (N=3 or 4) except (*) which is the mean ± ½ range of 2 experiments.

autophosphorylating activity (+MnATP) of endosomal IRK was much greater than that of the PM IRK measured by either $^{32}$P incorporation from ATP-$\tau$-$^{32}$P (4.3-fold, Khan et al, 1989) or by immunoblotting with $\alpha$P-Y (2.6-fold, Burgess et al, 1992). (Table 1). Furthermore, endosomal IRK activity against an exogenous substrate (expressed as Vmax/Km) was two-fold greater than that of PM (Khan <u>et al.</u>, 1989). These data raise the possibility that limited dephosphorylation may deactivate inhibitory P-Y residues selectively leading to full activation of the IRK; whereas further dephosphorylation would lead to inactivation of the IRK prior to its recycling to the cell surface. Our recent demonstration of IRK-associated phosphotyrosine phosphatase (PTPase) activity in ENs is consistent with this proposal (Faure <u>et al.</u>, 1992).

The EGF receptor kinase (ERK):    Studies of the EGF receptor
have shown that following EGF binding to its cell surface
receptor the ligand-receptor complex is internalized into ENs
(Lai et al., 1989). Cohen and Fava (Cohen et al., 1985)
demonstrated an active ERK in the endocytic compartment of A431
cells and we demonstrated ligand-dependent internalization of
active ERK in rat liver ENs in vivo (Kay et al., 1986; Lai et
al., 1989).    We observed that the endosomal ERK was
phosphorylated to a level, which greatly exceeded that attained
by the cell surface ERK (Wada et al., 1992) (Table 2).

Table 2. $^{32}$P-labelling of the EGF receptor following EGF
administration as a percent of that in PM receptors from
uninjected rats.

| Time after EGF (mins) | Receptor Specific Activity (% PM at t=0) | |
|---|---|---|
| | PM | EN |
| 0 | 100 ± 40 | 165 ± 125 |
| 0.5 | 510 ± 335 | 230 ± 100 |
| 5.0 | 630 ± 245 | 1840 ± 535 |
| 15 | 835 ± 210 | 1880 ± 200 |

Rats were injected with $^{32}$P-orthophosphate (5 mCi/100 g.b.wt.)
followed by EGF (10 μg/100 g.b.wt.) before sacrificing at 60 min
after $^{32}$P administration (Wada et al, 1992).    Subcellular
fractions were prepared and receptor content, radiolabelling,
and specific activity were determined as described in detail
elsewhere (Wada et al, 1992).    Each value is the mean ± S.D. of
3 separate studies.

This continuation, in vivo, of autophosphorylation of the
internalized ERK argues against internalization being a
deactivating process. Rather it seems that events, initiated at
the cell surface, continue during a portion of the intracellular
itinerary of the receptor. In support of this view are studies
of a protein (PYP55) which was phosphorylated on tyrosine
residues in an EGF-dependent manner and closely associated with
the EGF receptor both at the cell surface and in ENs (Wada et
al., 1992). Though the functional consequences of this remain

unclear we found that the level of PYP55 tyrosine phosphorylation increased subsequent to its concentration within ENs in a manner parallel to that observed for the ERK.

Finally, it has recently been shown that the majority of membrane-associated p60$^{c-src}$ is associated with ENs (Kaplan et al., 1992) rather than with plasma membranes as originally suggested (Courtneidge et al., 1980). Taken together the above observations are consistent with a role for the endosomal apparatus in facilitating transmembrane signalling and thus mediating biological effects of insulin and growth factors.

## References

Ajioka RS, Kaplan J (1987) Characterization of endocytic compartments using the horseradish peroxidase-diaminobenzidine density shift technique. J Cell Biol 104:77-85

Al-Awqati Q (1986) Proton-translocationg ATPases. Ann Rev Cell Biol 2:179-199

Bergeron JJM, Cruz J, Khan MN, Posner BI (1985) Uptake of insulin and other ligands into receptor-rich endocytic components of target cells: the endosomal apparatus. Ann Rev Physiol 47:383-403

Burgess JW, Wada I, Ling N, Khan MN, Bergeron JJM, Posner BI (1992) Decrease in ß-subunit phosphotyrosine correlates with internalization and activation of the endosomal insulin receptor kinase. J Biol Chem 267:10077-10086

Chou CK, Dull, TJ, Russell DS, Gherzi R, Lebwohl D, Ullrich A, Rosen OM (1987) Human insulin receptors mutated at the ATP-binding site lack protein tyrosine kinase activity and fail to mediate postreceptor effects of insulin. J Biol Chem 262:1842-1847

Cohen S, Fava RA (1985) Internalization of functional epidermal growth factor receptor/kinase complexes in A-431 cells. J. Biol Chem 260:12351-12358

Courtneidge SA, Levinson AD, Bishop JM (1980) The protein encoded by the transforming gene of avian sacoma virus (pp60$^{src}$) and a homologous protein in normal cells (pp60 $^{proto-src}$) are associated with the plasma membrane. Proc Natl Acad Sci USA 77:3783-3787

Diment S, Stahl P (1985) Macrophage endosomes contain proteases which degrade endocytosed protein ligands. J Biol Chem 260:15311-15317

Diment S, Leech MS, Stahl P (1988) Cathepsin D is membrane-associated in macrophage endosomes. J Biol Chem 263:6901-6907

Diment S, Martin KJ, Stahl PD (1989) Cleavage of parathyroid hormone in mactrophage endosomes illustrates a novel pathway for intracellular processing of proteins. J

Biol Chem  264:13403:13406
Doherty II JJ, Kay DG, Lai WH, Posner BI, Bergeron JJM (1990)
    Selective degradation of insulin within rat liver
    endosomes.  J Cell Biol 110:35-42
Faure R, Baquiran G, Bergeron JJM, Posner BI (1992)    The
    dephosphorylation of insulin and epidermal growth
    factor receptors:  role  of  endosome-associated
    phosphotyrosine  phosphatase(s).     J   Biol   Chem
    267:11215-11221
Felder S, Miller K, Moehren G, Ullrich A, Schlessinger J,
    Hopkins CR (1990)    Kinase activity controls the
    sorting of the epidermal growth factor receptor within
    the multivesicular body.  Cell 61:623-634
Hamel FG, Posner BI, Bergeron JJM, Frank BH, Duckworth WC
    (1988)  Isolation of insulin degradation products from
    endosomes derived from intact rat liver.  J Biol Chem
    263:6703-6708
Hari J, Roth RA (1987)    Defective internalization of insulin
    and its receptor in cells expressing mutated insulin
    receptors lacking kinase activity.    J Biol Chem
    262:15341-15344
Honegger AM, Dull TJ, Felder S, Van Obberghen E, Bellot F,
    Szapary D, Schmidt A, Ullrich A, Schlessinger J (1987)
    Point mutation at the ATP binding site of EGF recptor
    abolishes protein-tyrosine kinase activity and alters
    cellular routing.  Cell 51:199-209
Kaplan KB, Swedlow JR, Varmus HE, Morgan DO (1992)
    Association of p60^{c-src} with endosomal membranes in
    mammalian fibroblasts.  J Cell Biol 118:321-333
Kay DG, Lai WH, Uchihashi M, Khan MN, Posner BI, Bergeron JJM
    (1986)  J Biol Chem 261:8473-8480
Khan MN, Savoie S, Bergeron JJM, Posner BI (1986)
    Characterization of rat liver endosomal fractions: in
    vivo activation of insulin-stimulable kinase in these
    structures.  J Biol Chem 261:8462-8472
Khan MN, Baquiran G, Brule C, Burgess J, Foster B, Bergeron
    JJM, Posner BI (1989)  Internalization and activation
    of the rat liver insulin receptor kinase in vivo. J
    Biol Chem 264:12931-12940
Klausner RD, Ashwell G, van Renswoude J, Harford JB, Bridges
    KR (1983) Binding of apotransferrin to K562 cells:
    explanation of the transferrin cycle.  Proc Natl Acad
    Sci USA 80:2263-2266
Klein HH, Freidenberg GR, Kladde M, Olefsky JM (1986)  Insulin
    activation of insulin receptor tyrosine kinase in
    intact rat adipocytes: an in vitro system to measure
    histone kinase activity of insulin receptors activated
    in vivo.  J Biol Chem 261:4691-4697
Klein HH, Freidenberg GR, Matthaei S, Olefsky JM (1987)
    Insulin receptor kinase following internalization in
    isolated rat adipocytes.  J Biol Chem 262:10557-10564
Lai WH, Cameron PH, Doherty II JJ, Posner BI, Bergeron JJM
    (1989a)  Ligand mediated autophosphorylation activity
    of the EGF receptor during internationalization.   J
    Cell Biol 109:2751-2760
Lai WH, Cameron PH, Wada I, Doherty II JJ, Kay DG, Posner BI,
    Bergeron JJM (1989b) Ligand mediated internalization,

recycling, and down regulation of the epidermal growth
factor receptor in vivo. J Cell Biol 109:2741-2749

Marsh M, Schmid SL, Kern H, Harms E, Male P, Mellman I,
Helenius A (1987) Rapid analytical and preparative
isolation of functional endosomes by free flow
electrophoresis. J Cell Biol 104:875-886

Mellman I, Fuchs R, Helenius A (1986) Acidification of the
endocytic and exocytic pathways. Ann Rev Biochem
55:663-700

Mostov KE, de Bruyn Kops A, Deitcher DL (1986) Deletion of
the cytoplasmic domain of the polymeric immunoglobulin
receptor prevents basolateral localization and
endocytosis. Cell 47:359-364

Pease RJ, Smith GD, Peters TJ (1985) Degradation of
endocytosed insulin in rat liver is mediated by low-
density vesicles. Biochem J 228:137-146

Posner BI, Patel B, Verma AK, Bergeron JJM (1980) Uptake of
insulin by plasmalemma and Golgi subcellular fractions
of rat liver. J Biol Chem 255:735-741

Posner BI, Khan MN, Bergeron JJM (1982) Endocytosis of peptide
hormones and other ligands. Endocr Revs 3:280-298

Posner BI, Khan MN, Bergeron JJM (1987) Internalization of
insulin and its receptor. In Raizada MK, Phillips MI,
LeRoith D (eds) Insulin, insulin-like growth factors
and their receptors in the central nervous system.
Plenum Press, New York, pp.1-10

Rosen OM, Herrera R, Olowe Y, Petruzzelli LM, Cobb MH (1983)
Phosphorylation activates the insulin receptor
tyrosine protein kinase. Proc Natl Acad Sci USA
80:3237-3240

Russell DS, Gherzi R, Johnson EL, Chou CK, Rosen OM (1987)
The protein-tyrosine kinase activity of the insulin
receptor is necessary for insulin-mediated receptor
down regulation. J Biol Chem 262:11833-11840

Schaudies RP, Gorman RM, Savage CR, Poretz RD (1987)
Proteolytic processing of epidermal growth factor
within endosomes. Biochem Biophys Res Commun 143:710-
715

Schmid SL, Fuchs R, Male P, Mellman I (1988) Two distinct
subpopulations of endosomes involved in membrane
recycling and transport to lysosomes. Cell 52:73-83

Wada I, Lai WH, Posner BI, Bergeron JJM (1992) Association of
the tyrosine phosphorylated epidermal growth factor
receptor with a 55-kD tyrosine phosphorylated protein
at the cell surface in endosomes. J Cell Biol
116:321-330

# THE TUBULAR EARLY ENDOSOME

J. Tooze and M. Hollinshead

European Molecular Biology Laboratory
Meyerhofstrasse 1
6900 Heidelberg
Germany

Endosomes are pleiomorphic organelles that are often difficult if not impossible to identify in thin sections under the electron microscope in the absence of either an electron dense endocytic tracer in their lumen or an immunocytochemical marker. As a consequence, horseradish peroxidase has been widely used as a nonspecific fluid phase endocytic tracer in electron microscopic studies of endosomes and endocytosis. The horseradish peroxidase (HRP) method involves incubating cells in medium containing HRP for times ranging from a few minutes to several hours. The cells are then fixed and incubated with hydrogen peroxide and DAB. As a result of the peroxidase reaction the DAB is polymerised and rendered insoluble in situ. Finally, the cells are treated with osmium tetroxide to osmicate them, and render electron dense the poly DAB, before being embedded.

The success of this method depends upon the presence of peroxidase active sites in endosomes after fixation. It is therefore advantageous to use preparations of HRP of high specific enzymatic activity so that the number of enzymatically active molecules in the endosomes prior to fixation is high. Of course high specific activity HRP is expensive; it is also advantageous to use mild fixation conditions so that many active

NATO ASI Series, Vol. H 74
Molecular Mechanisms of Membrane Traffic
Edited by D. J. Morré, K. E. Howell, and J. J. M. Bergeron
© Springer-Verlag Berlin Heidelberg 1993

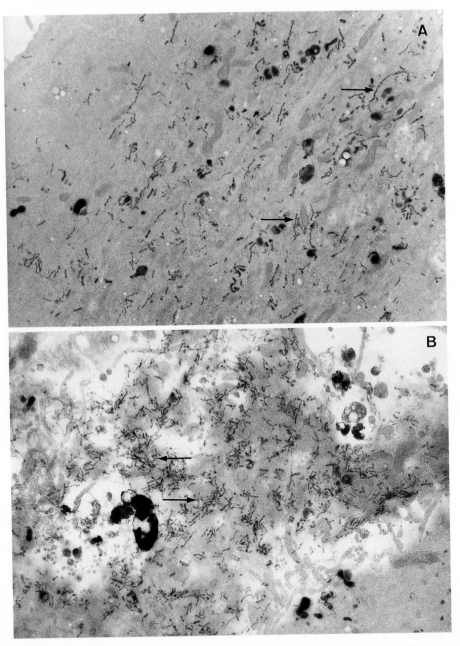

Figure 1. These two micrographs show tubular early endosomes (arrows) in AtT20 cells incubated with 10 mg/ml HRP for 60 min at 37°C (A) or 60 min at 37°C followed by 60 min at 20°C (B).

sites survive fixation. Having taken these measures to optimise the chances of HRP active sites being present, it is also advisable to incubate the fixed cells with peroxide and DAB for 30 min rather than 3 or 5 min, to allow the enzymatic reaction to produce large amounts of poly DAB.

Once the peroxidase reaction and osmication have been done, the cells are embedded in plastic, sectioned and under the electron microscope the electron dense HRP reaction product reveals the endosomes. Electron microscopists customarily cut very thin sections. They do this to obtain optimal resolution and to avoid the superimposition of the projected images of structures at different levels within the section; the thinner the section the less the superimposition. The downside of very thin sections is, first, a reduction in contrast and second, long thin tubular structures with a diameter close to the section thickness will appear as circular or oval images except in rare instances when the long axis of the tube is parallel to the section plane. Electron microscopists, because they examine thin sections, tend to overestimate the number of vesicular organelles and not to see tubular ones.

These technical considerations are not particularly profound but our success in identifying the tubular early endosomes present in many cell types (Tooze and Hollinshead, 1991) stemmed from using high specific activity HRP and examining sections up to 500 nm thick.

**Morphological characteristics of tubular endosomes**

Tubular endosomes have a diameter of about 50 nm and lengths of

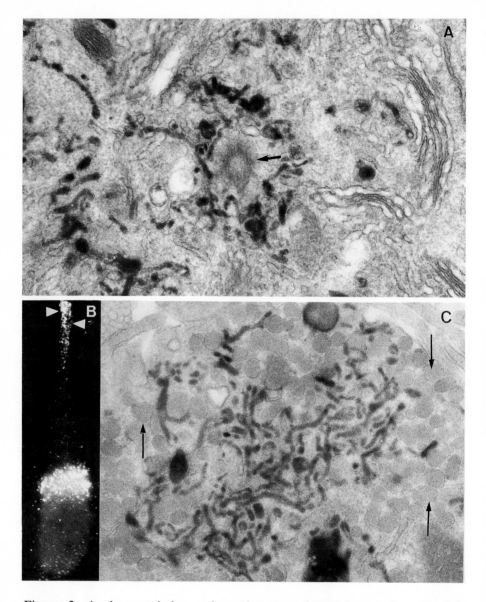

Figure 2. A shows tubular early endosomes clustered around a centriole (arrow) in an AtT20 cell. B shows immunofluorescence labelling of ACTH in an AtT20 cell; the secretory protein is abundant in the Golgi region and at the tips (growth cones) of neurites (arrowheads). C shows HRP labelled tubular endosomes in the tip of an AtT20 cell neurite, note the abundant ACTH containing secretory granules (arrows).

several μm. As Fig. 1 shows, some are extended straight tubes while others are multibranching to form discrete patches of tubular endosomal network. As best we can determine from examination of serial sections in the electron microscope, and whole cells labelled with antibody to transferrin receptor under the fluoresence microscope, these patches of network are discrete. They are not all joined in one large reticulum extending throughout the cytoplasm.

Tubular endosomes appear to be inherently tubular; their tubular form is not dependent on microtubules or other cytoskeletal elements. They do, however, appear to be stretched out on microtubules because nocodazole and mitosis both alter the intracellular distribution of tubular endosomes and cause the patches of network to become more compact. The fact that tubular endosomes are particularly abundant in ruffling edges and growth cones of neurites, but also invariably occur close to the centrioles (Fig. 2), also implies their intracellular distribution depends on microtubules. Tubular endosomes in this respect resemble the endoplasmic reticulum which uses microtubules to extend from the centre of the cell to the periphery (Terasaki et al., 1986).

Tubular endosomes are not uniformly distributed in the thickness of a cultured cell. They are concentrated near the basal plasma membrane that is in contact with the substratum, possibly because that is where the majority of microtubules are located. This intracellular distribution has an important practical consequence for anyone desirous of observing tubular endosomes in the electron microscope. The cells should be flat embedded in situ and sectioned from the bottom surface.

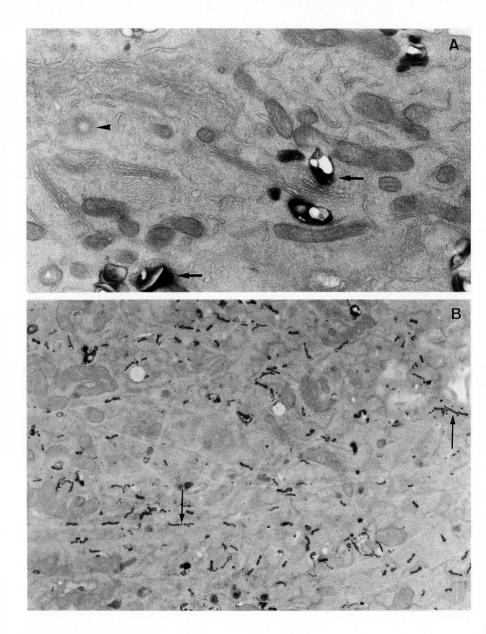

Figure 3. A shows the centriolar region (arrowhead) of an AtT20 cell following labelling with HRP for 60 min and then 60 min chase. All the HRP chases out of tubular endosomes. Labelled late endosomes are arrowed. B shows labelled tubular early endosomes in an HeLa cell incubated for 10 min with transferrin-HRP.

The results repay the extra effort.

## Tubular endosomes are early endosomes

Tubular endosomes receive HRP within the first minutes of its addition to the medium. Delivery of HRP is not inhibited by either nocodazole or cytochalasin D and continues at 20°C. A pulse of endocytic tracer delivered to tubular endosomes can be completely chased out (Fig. 3A). Furthermore, tubular endosomes are heavily labelled by transferrin-HRP used at very low concentrations, at which this ligand is accumulated by receptor mediated endocytosis (Fig. 3B); transferrin competitively inhibits this labelling. Tubular endosomes therefore contain abundant transferrin receptor. This set of properties means that tubular endosomes must be classified as early endosomes. They are, however, morphologically distinct from the cisternal-vesicular early endosome described in many cell types in previous studies; the latter and tubular early endosomes coexist and might, of course, be interconvertible. In the cells we have studied, tubular early endosomes were not observed in physical continuity with either cisternal-vesicular early endosomes or late endosomes.

Biochemical studies have shown that in PC12 cells synaptophysin occurs in early endosomes (Johnston et al., 1988; Linstedt and Kelly, 1991; Régnier-Vigoroux et al., 1991). We find that in PC12 cells and AtT20 cells, which both express synaptophysin, the tubular endosomes contain synaptophysin in their membrane. Clearly synaptophysin is not a ubiquitous component of early endosomes, but in cells that express this protein some is in tubular early endosomes.

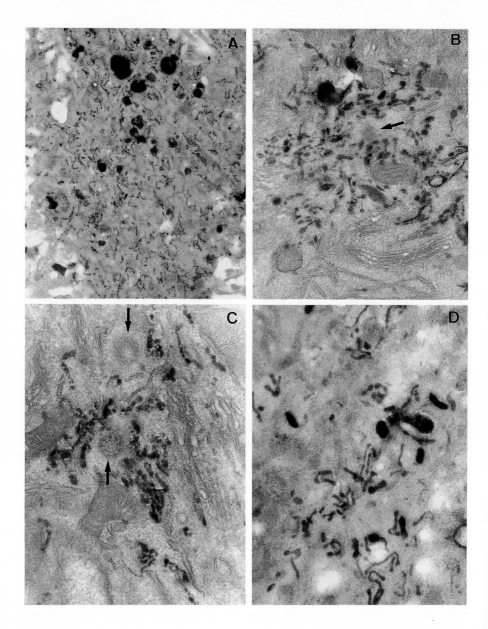

Figure 4. A and B show tubular early endosomes at the periphery (A) and around a centriole (arrow in B) of PC12 cells labelled with HRP. C shows tubular endosomes around the centrioles (arrows) of an MDCK II cell grown on plastic and D shows HRP labelled tubular endosomes in primary human foreskin fibroblasts.

## Tubular endosomes occur in many cell types in culture

The cells of the majority of the stable lines that we have examined - more than a dozen - contain tubular endosomes (Fig. 4A-C). They are particularly abundant in the neuroendocrine cells AtT20 and PC12, and in HeLa and Hep2 cells. They are virtually absent, however, from BHK21 and 3T3 cells. We have also observed abundant tubular endosomes in human foreskin fibroblasts in primary culture (Fig. 4D) and Parton et al. (1992) observed them in large numbers in the dendrites, but not the axon, of rat hippocampal neurons in primary culture. Tubular endosomes therefore occur in post-mitotic cells and are not restricted to stable cell lines. Until, however, they have been detected in cells in tissues the possibility remains that they develop as a response to the tissue culture environment.

## Tubular endosomes are sensitive to Brefeldin A

Brefeldin, A as well as causing disassembly of the Golgi apparatus, has recently been shown to induce early endosomes to become tubular (Wood et al., 1991; Lippincott-Schwartz et al., 1991; Hunziker et al., 1991; Klausner et al., 1992). In AtT20, PC12, Hep2 and HeLa cells we have studied the effects of BFA on the abundant pre-existing tubular early endosomes (Tooze and Hollinshead, 1992). The drug causes two striking morphological changes, which can be observed by electron microscopy following endocytosis of either transferrin-HRP or HRP, and by immunofluorescence microscopy with antibody against transferrin receptor. First BFA causes the discrete patches of tubular endosomal network to fuse together to form a single continuous network

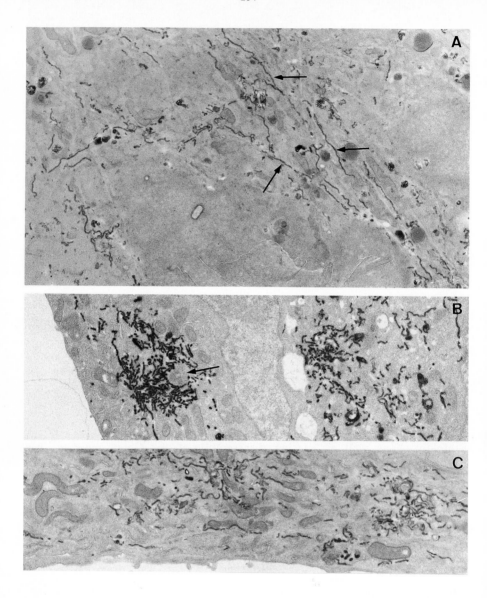

Figure 5. A shows tubular endosomes fused together into long tubes (arrows) in a HeLa cell labelled for 60 min with transferrin-HRP, then washed and incubated with 5 µg/ml BFA for 5 min. Compare this figure with figure 3. B shows a centriole (arrow) surrounded by a dense network of tubular endosomes in an AtT20 cell incubated with BFA for 30 min the then BFA and HRP for only 5 min. C shows AtT20 cells incubated in BFA for 30 min and then with 20 mg/ml HRP for only 1 min.

throughout the cytoplasm (Fig. 5A). Second in the presence of BFA the tubular endosomes become more linear, less branching and align along microtubules as straight tubules running for many tens of μm without branching (Fig. 5A). They form an open reticulum at the cell periphery but close to the centrioles the tubular endosomes in the presence of BFA branch repeatedly to form a dense three-dimensional cage enclosing the microtubule organising centre (Fig. 5B).

BFA appears to alter two endocytic properties of tubular endosomes; first they can be labelled by fluid phase endocytic tracer much more rapidly than in control cells, no doubt because fluid phase tracer is free to diffuse throughout the lumen of the entire network (Fig. 5C); second, they can be labelled by fluid phase tracer used at low concentrations, e.g. in the range 0.1 to 1.0 mg/ml, which are too low to label tubular endosomes in controls. It appears to be the case that in the presence of BFA tubular endosomes can concentrate HRP (Fig. 6).

Some of the bulbous ends of branches of tubular endosomes are coated both in the presence and absence of BFA (Tooze and Hollinshead, 1992). It is unlikely, therefore, that this coat is a βCOP Golgi coat (Donaldson et al., 1990) From the morphology of the coat seen under the electron microscope we believe that it may well contain clathrin but immunocytochemical labelling is needed either to confirm or rule out this suggestion. If it is clathrin, then we would have to conclude that at least some vesicles budding from an early endocytic compartment have clathrin coats.

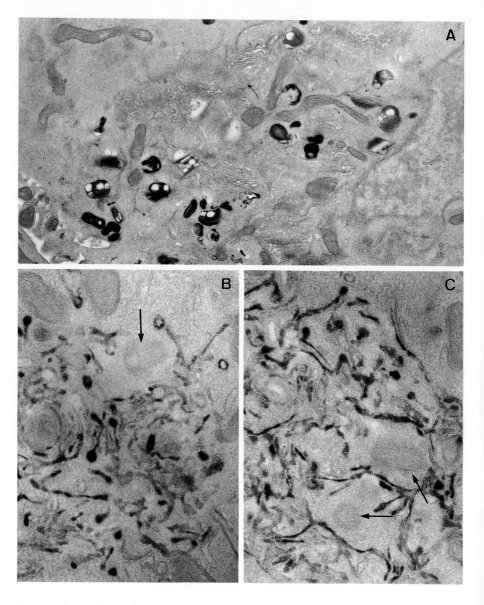

Figure 6. A shows the Golgi-centriolar region of an AtT20 cell incubated for 60 min with 0.5 mg/ml HRP. Late endosomes in which HRP accumulates are heavily labelled, but tubular endosomes do not contain the threshold level of HRP and are not labelled. B and C show AtT20 cells incubated with BFA for 30 min and then BFA and 0.1 mg/ml HRP (B) or 0.25 mg/ml HRP (C) for 60 min. The pericentriolar tubular endosomes contain HRP. Arrows indicate centrioles.

## Tubular endosomes in mitotic cells

During mitosis endocytosis is inhibited. Therefore, to observe the distribution of tubular early endosomes in mitotic cells it is necessary to load the cells with endocytic tracer while they are still in interphase prior to mitosis. We simply incubate cultures with HRP for 60-180 min, fix and prepare them for electron microscopy, and then search for the small number of mitotic cells in each section.

In all the mitotic cells of several lines that we have examined (including AtT20, PC12, Hep2, HeLa and RK13) the tubular endosomes remain tubular. They are primarily located at the periphery of metaphase and anaphase cells outside the spindle, but some tubular endosomes also occur within the volume of the spindle; however, they do not align along the spindle or aster microtubules (Fig. 7).

In telophase cells some tubular endosomes are located close to the reforming Golgi stacks near the nucleus, but large numbers are also seen extending up to the midbody (Fig. 8A), where in AtT20 cells secretory granules also accumulate (Tooze and Burke, 1987). This distribution implies that some tubular endosomes migrate anterogradely to the plus ends of microtubules. In short, at telophase we observe the beginnings of the interphase distribution of tubular endosomes with many at the cell periphery, the region of microtubule plus ends, but some around the centrioles close to the MTOC and Golgi stacks, the site of minus ends of microtubules.

When cells are pre-incubated with BFA and then incubated with BFA

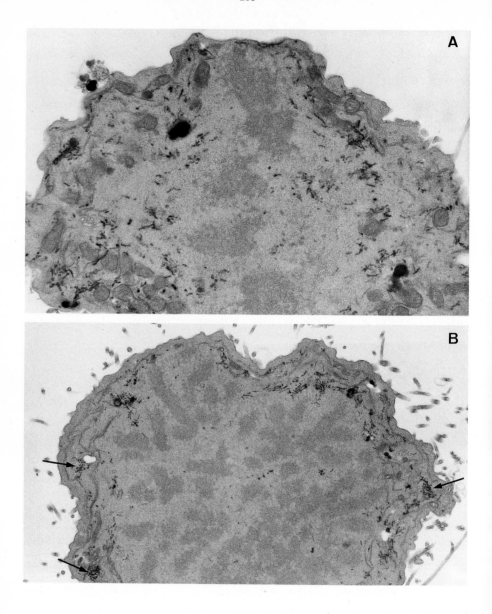

Figure 7. A shows a mitotic PC12 cell with tubular endosomes labelled with HRP. B shows a mitotic HeLa cell with HRP labelled tubular endosomes, many of which are in clusters at the periphery (arrows).

and HRP for 60 min, the distribution of tubular endosomes in the mitotic cells is essentially similar to that in mitotic cells in the absence of BFA. The very long tubular endosomes seen in BFA treated interphase cells are not seen in mitotic cells in the same cultures, they must be broken up into smaller profiles at mitosis, as the interphase microtubules depolymerise.

As expected, since during mitosis endocytosis and recycling to the medium are inhibited (Warren et al., 1984; Warren, 1985 and refs therein), HRP cannot be chased from the tubular early endosomes in cells held in mitosis by the depolymerisation of microtubules with nocodazole. In the interphase cells in the same cultures the HRP is chased out of the tubular endosomes (Fig.  ).

**Tubular endosomes and mitotic cell Golgi clusters**

In mitotic HeLa cells Lucocq et al. (1987, 1988) identified clusters of vesicular, and some tubular, profiles about 50 nm in diameter; they interpreted these structures as remnants of the Golgi apparatus which vesiculates on mitosis. Their evidence for this interpretation was primarily that a minority of the vesicles in these globular Golgi clusters could be immunogold labelled with antibody against galactosyltransferase. In conventionally fixed and epon embedded mitotic HeLa cells we also observed these clusters of 50 nm vesicles and tubules (Fig. 9), the so-called Golgi clusters.

When we examined carefully mitotic HeLa cells that had been loaded with fluid phase HRP, by incubating the cultures with the tracer for 60

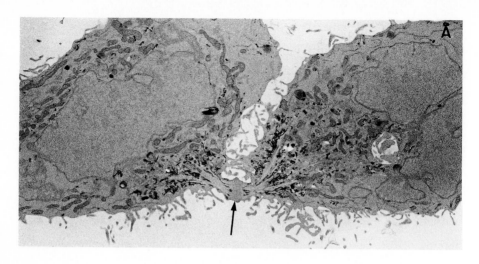

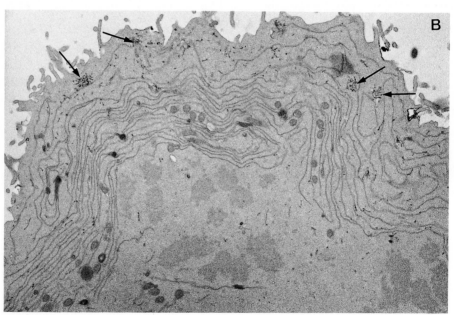

Figure 8. A. Telophase HeLa cell in which the HRP labelled tubular endosomes are aligning along midbody microtubules and some extend to the midbody (arrow). B. Part of a mitotic HeLa cell showing the layers of ER cisternae surrounding the mitotic spindle and, close to the periphery, clusters of tubular early endosomes (arrows) labelled with transferrin HRP.

or 120 min prior to fixation, we failed to detect Golgi clusters. In their place we found clusters of HRP labelled tubular endosomes (Fig. 8B). When the cultures were incubated for only 15 or 30 min with HRP before fixation and processing we observed Golgi clusters without HRP, clusters with some profiles containing HRP and others lacking it, and clusters fully labelled with HRP. These results indicate that as the endocytic pathway is saturated with HRP the Golgi clusters also become saturated with the endocytic tracer.

In mitotic HeLa cells in cultures pretreated with BFA to disassemble the Golgi apparatus, and then incubated with BFA and HRP, we observed HRP labelled clustered tubular endosomes but no unlabelled Golgi clusters. This result makes it highly unlikely that those Golgi cisternae which are susceptible to BFA contribute membrane to clustered tubular endosomes, alias Golgi clusters. We conclude that the structures identified as globular Golgi clusters in mitotic HeLa cells by Lucocq et al. (1987, 1988) are in fact clusters of tubular early endosomes. We cannot as yet, however, explain why they should be labelled by antibody against galactosyltransferase.

## Is the tubular early endosome a novel compartment?

Over the years the endocytic compartments in a wide variety of cell types have been investigated using as endocytic tracers either HRP, transferrin HRP or antibodies against endocytic receptors. However, the presence of the tubular early endosomes we have described and illustrated here has largely gone unnoticed. Why? The answer, we believe, is because most previous investigations have been made with

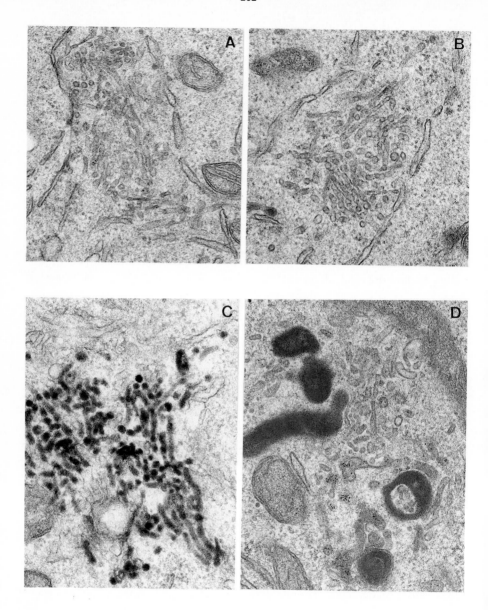

Figure 9. A and B show thin sections of clustered tubular early endosomes in a mitotic HeLa cell following conventional glutaraldehyde fixation and epon embedding. C shows HRP labelled clustered tubular endosomes in a mitotic cell. D shows clustered endosomes in a mitotic HeLa cell labelled with BSA-gold.

low specific activity HRP, often followed by short peroxidase reaction times, and then thin sections were studied. The first two factors mitigate the obtention of the detectable threshold level of HRP reaction product, while the third mitigates the visualization of tubular structures. In our experience at least 5 mg/ml of high specific activity HRP (about 1000 enzyme units per mg) must be used, together with a 30 min peroxidase reaction, and the sections should be above 150 nm thick. These factors notwithstanding, tubular early endosomes were, we believe, visualized in several cell types in earlier studies, to cite a few examples by Miller et al. (1986), Johnston et al. (1989) and Yamashiro et al. (1984), but because thin sections were examined they were often described as vesicular structures. Tubular early endosomes, including those which are invariably present around the centrioles, may also well be the same compartment as the CURL described by Geuze and his colleagues (1987). The relationship between the tubular early endosomes we have described in HeLa and Hep2 cells and the tubular early endosomes seen in vivo in Hep2 cells by Hopkins et al. (1990) remains to be clarified, but both the in vivo studies of Hopkins and his colleagues (1990) and our electron microscopy (Tooze and Hollinshead, 1991) serve to emphasize that in the cells of many lines the early endosomal compartment is much more tubular than previously credited.

**The function of tubular early endosomes**

In HeLa and AtT20 cells, the two cell types we have studied most extensively, it is clear that tubular early endosomes and cisternal-vesicular early endosomes are morphologically distinct compartments.

Both are early endosomes. We do not know how, if at all, they are inter-related. We do not know if they are inter-convertible. We do not know whether the differences in the morphology reflect differences in function. It is difficult to envisage that cells contain two functionally distinct early endosomes but it is clear that all the early endosomes in a cell at any one time do not have the same morphology. Since electron microscopy provides only a static snap-shot of what must in vivo be a set of dynamic organellar systems, there may be an equilibrium between tubular shaped early endosomes and the classic cisternal-vesicular early endosomes. In different cell types the equilibrium may be shifted in favour of one or other of these morphologies, which would explain why tubular endosomes are much more abundant in PC12, AtT20, HeLa and Hep2 cells, for example, than BHK or 3T3 cells. We also believe, however, from our survey of cell types that the total surface area of early endosomal membrane per cell varies greatly between, for example, a 3T3 cell at the low end and a HeLa or AtT20 cell at the high end of the range.

The most striking feature of tubular endosomes is the high surface area to volume ratio. Tubular endosomes might therefore act as intracellular stores of plasma membrane proteins that are ready to be moved to the plasma membrane by the endocytic recycling pathway. Both glucose transporter (Haney et al., 1991 and references therein) and plasma membrane galactosyltransferase (Eckstein and Shur, 1989) are rapidly delivered to the plasma membrane from intracellular storage compartments on receipt of the appropriate environmental signal; it is conceivable that tubular early endosomes are early endosomal domains into which such proteins are sorted and stored. The presence of synaptophysin in the tubular endosomes of AtT20 and PC12 cells is consistent with this notion.

One of the unexpected properties of tubular endosomes is their intracellular distribution. Many patches of tubular endosomes are accumulated at the cell periphery and especially in ruffling edges and the growth cones of neurites; in general they are more abundant close to the basal surface of cells than elsewhere. This is not surprising since ruffling edges in particular and the basal surface in general may be the sites of most exocytic and endocytic activity in cells in culture. But what is surprising is that in the same cells patches of tubular endosomes are found at the very minus ends of microtubules abutting the boundary of the microtubule organising centre (MTOC) and closer to it than the Golgi stacks. This distribution, while implying that tubular endosomes can move in both directions along microtubules, poses the question, why at steady state should early endosomes surround the MTOC. It is hard to believe that this distribution does not reflect some function of tubular endosomes, but what function? The MTOC is the crossroads of the cytoplasm and recent evidence indicates that some newly synthesized lysosomal proteins reach early endosomes before they reach late endosomes and lysosomes (Ludwig et al., 1992). The pericentriolar tubular early endosomes may be involved in this pathway, as well as in recycling of endocytosed molecules to the trans Golgi network; they are certainly in an ideal location.

# References

Donaldson JG, Lippincott-Schwartz J, Bloom GS, Kreis TE, Klausner RD (1990) Dissociation of a 110 kD peripheral membrane protein from the Golgi apparatus is an early event in brefeldin A action. J Cell Biol 111: 2295-2306

Eckstein DJ, Shur BD (1989) Laminin induces the stable expression of surface galactosyltransferase on lamellipodia of migrating cells. J Cell Biol 108: 2507-2517

Geuze HJ, Slot JW, Schwartz AL (1987) Membranes of sorting organelles display lateral heterogeneity in receptor distribution. J Cell Biol 104: 1715-1724

Haney PM, Slot JW, Piper RC, James DE, Mueckler M (1991) Intracellular targeting of the insulin-regulatable glucose transporter (GLUT4) is isoform specific and independent of cell type. J Cell Biol 114: 689-699

Hopkins, CR, Gibson A, Shipman M, Miller K (1990) Movement of internalised ligand-receptor complexes along a continuous endosomal reticulum. Nature 346: 335-339

Hunziker W, Andrew-Whitney J, Mellman I (1991) Selective inhibition of transcytosis by Brefeldin A in MDCK cells. Cell 67: 617-627

Johnston PA, Cameron PL, Stukenbruk H, Jahn R, de Camilli P, Südhof TC (1989) Synaptophysin is targeted to similar microvesicles in CHO and PC12 cells. EMBO J 8: 2863-2872

Klausner RP, Donaldson JG, Lippincott-Schwartz J (1992) Brefeldin A, insights into the content of membrane traffic and organelle structure. J Cell Biol 116: 1071-1080

Linstedt AD, Kelly RB (1991) Synaptophysin is sorted from endocytic markers in neuroendocrine PC12 cells but not transfected fibroblasts. Neuron 7: 309-317

Lippincott-Schwarz J, Yuan LC, Bonifacino JS, Klausner RD (1989) Rapid redistribution of Golgi proteins into the ER in cells treated with Brefeldin A, evidence of membrane cycling from Golgi to ER. Cell 56: 801-813

Lucocq JM, Pryde JG, Berger EG, Warren G (1988) A mitotic form of the Golgi apparatus identified using immunoelectronmicroscopy. *In* Cell-Free Analysis of Membrane Traffic, pp 431-440, Ed DJ Morrée, Alan R Liss Inc, New York

Lucocq JM, Pryde JG, Berger EG, Warren G (1987) A mitotic form of the Golgi apparatus in HeLa cells. J Cell Biol 104: 865-874

Ludwig T, Griffiths G, Hoflack B Distribution of newly synthesised lysosomal enzymes in the endocytic pathway of normal rat kidney cells. J Cell Biol 115: 1561-1572

Parton RG, Simons K, Dotti CG (1992) Axonal and dendritic endocytic pathways in cultured neurons. (submitted)

Régnier-Vigoroux A, Tooze SA, Huttner WB (1991) Newly synthesized synaptophysin is transported to synaptic-like microvesicles via constitutive secretory vesicles and the plasma membrane. EMBO J 10: 3589-3601.

Terasaki M, Chen LB, Fujiwara K (1986) Microtubules and the endoplasmic reticulum are highly interdependent structures. J Cell Biol 103: 1557-1568

Tooze J, Burke B (1987) Accumulation of ACTH secretory granules in the midbody of telophase AtT20 cells, evidence that secretory granules move anterogradely along microtubules. J Cell Biol 104: 1047-1057

Tooze J, Hollinshead M (1991) Tubular early endosomal networks in AtT20 and other cells. J Cell Biol 115: 635-653

Tooze J, Hollinshead M (1992) In AtT20 and HeLa cells Brefeldin A induces the fusion of tubular endosomes and changes their distribution and some of their endocytic properties. J Cell Biol, in the press

Warren, G (1985) Membrane traffic and organelle division. Trends Biochem Sci 10: 439-443

Warren G, Davoust J, Cockcroft A (1984) Recycling of transferrin receptors in A431 cells is inhibited during mitosis. EMBO J 3: 2217-2225

Yamashiro DJ, Tycko B, Fluss SR, Maxfield FR (1984) Segregation of transferrin to a mildly acidic (pH 65) para-Golgi compartment in the recycling pathway. Cell 37: 789-800

# DIRECT COMPARISION OF THE ENDOCYTIC ROUTES OF FLUORESCENTLY-LABELLED LIPIDS AND TRANSFERRIN

Satyajit Mayor and Frederick R. Maxfield.
Department of Pathology, College of Physicians and Surgeons
Columbia University
630 West 168th Street
New York, NY 10032
USA

Intracellular trafficking of proteins during endocytosis has been extensively characterized. This process is mediated by vesicular carriers such as coated vesicles, sorting endosomes and late endosomes, all of which are composed primarily of membrane proteins and lipid. A central question in the endocytic process concerns the sorting of recycling components from lysosomally-directed components. An iterative fractionation model (Dunn and Maxfield, 1992; Dunn et al., 1989) and other current models of endocytic sorting (Linderman and Lauffenburger, 1988) propose that the geometry of the sorting endosome directs bulk membrane, including membrane-associated receptors, towards the recycling pathway while the volume content, consisting of the acid-released ligands, is lysosomally targetted. However, it is not known whether recycling membrane receptors follow bulk membrane flow or if these proteins are actively sorted from the lysosomally-directed material because of specific cytoplasmic tail sequences. The kinetics of movement of the bulk carrier, membrane lipids, will directly address this issue; a difference in the kinetics of lipid-traffic in relation to protein-traffic will provide evidence for mechanisms of selective protein transport or absence of difference will support the hypothesis that proteins follow 'bulk' membrane traffic.

Dunn and co-workers (Dunn et al., 1989) have previously shown that in Chinese Hamster Ovary cells transfected with the human transferrin receptor (TRVb-1 cells) fluorescently-labelled transferrin (F-Tf) and low density lipoprotein (LDL) enter the cells in the same endocytic compartment called sorting endosomes (See Fig. 1). Consequently F-Tf rapidly sorts from LDL, and accumulates in a separate compartment called the recycling compartment, localized to a peri-nuclear region of the cell. In addition F-Tf rapidly reaches steady state levels in sorting endosomes (within 2 min) while LDL continues to accumulate in these endosomes over a 10 min period.

NATO ASI Series, Vol. H 74
Molecular Mechanisms of Membrane Traffic
Edited by D. J. Morré, K. E. Howell, and J. J. M. Bergeron
© Springer-Verlag Berlin Heidelberg 1993

Using similar techniques of quantitative fluorescence microscopy (Dunn and Maxfield, 1990; Maxfield and Dunn, 1990) in Figs. 2 and 3 we show that $N$-[7-(4-nitrobenzo-2-oxa-1,3-diazole)]-6-aminocaproic acid-labelled-sphingomyelin ($C_6$-NBD-SM) and F-Tf also enter TRVb-1 cells in sorting endosomes, however, unlike LDL, $C_6$-NBD-SM reaches steady state with similar kinetics as F-Tf and also acumulates in the recycling compartment. (Under the conditions of the experiment the labelled lipid remains exofacial and upon internaliztion, and the lipid is recycled via endocytic pathways, co-localized with a rapidly recyling membrane receptor, the transferrin receptor (see Fig. 2 and Koval and Pagano, 1989).) The half time for exit from sorting endosomes (into peri-nuclear-localized recycling endosomes) is also similar for NBD-SM and F-Tf (S. Mayor and F. R. Maxfield, manuscript in preparation). Similarity of the endocytic pathways and kinetics of trafficking of transferrin and a membrane lipid marker, $C_6$-NBD-SM, provides strong evidence that recycling of receptors follow 'bulk membrane flow', consistent with an iterative fractionation/geometric model of protein and membrane sorting in the endocytic process.

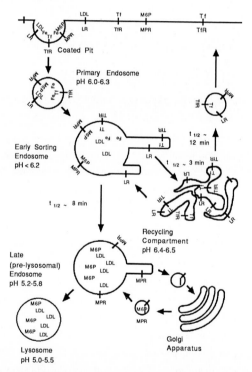

**Figure 1. Schematic of Endocytic Process**
This schematic illustrates various fates of internalized molecules, diferric transferrin (FeTfFe), LDL and mannose 6-phosphate-containing molecules (M6Ps), and their receptors, LR, TfR and MPR, respectively. LDL is released from LR in the acidic milieu of early sorting endosomes. LR is then recycled back to the surface. Internalized LDL is eventually delivered to lysosomes.

FeTfFe releases its bound iron in an early endosomal compartment. Apotransferrin (Tf), at the acidic pH of endosomes, remains associated with TfR and is recycled back to the plasma membrane via a peri-nuclear localized recycling compartment. This sorting from non-recycled molecules takes place in tubulo-vesicular endosomal compartment, referred to as an early sorting endosome. In this scheme the late endosome represents a compartment that that contains endocytosed molecules destined for the lysosome but does not recycling LRs or TfRs. Endocytosed M6Ps are released from their receptor (MPR) into a pre-lysosomal late endosomal compartment to which newly synthesized Man-6-P-proteins are delivered from the trans-Golgi network. MPRs are retrieved and recycled back to the trans-Golgi while the M6Ps are eventually delivered to the lysosome. This late comartment is a second sorting compartment that also receives lysosomally destined endocytosed molecules.

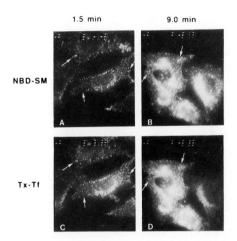

**Figure 2. Intracellular accumulation of Texas Red-Transferrin (Tx-Tf) and C6-NBD-SM in TRVb-1 cells.**

TRVb-1 cells were incubated with C6-NBD-SM vesicles {1 μM total lipid; C6-NBD-SM : dioleylphosphatidylcholine (1:1, mole/mole)} and 20 μg/ml Tx-Tf at 4°C for 30 min in Hepes-buffered (pH 7.3) Hams F-12 media containing 2 mg/ml ovalbumin (HF-Ova) and washed extensively in ice cold HF-Ova. The cells were then warmed up to 37 °C for the indicated periods, fixed and back-exchanged in 5% BSA (6 x 5 min changes) to remove surface C6-NBD-SM (Koval and Pagano, 1989). NBD-fluoresence images (A, B) were recorded using a Leitz fluorescence microscope equipped with a 63X, NA 1.4 objective and a 530-560 nm band pass excitation filter, a 580 nm dichroic mirror and a 580-nm long pass emission filter and Texas Red fluoresence images (C, D) were recorded using the same equipment but a Texas Red filter set (540-580 nm band pass excitation filter, a 595 nm dichroic mirror and a 620-nm long pass emission filter) was used to visualize the fluoresence.. Images were recorded on a JVC CR6650U video cassette recorder with a Videoscope KS1380 image intensifier. Neutral density filters (5%, 13%, 33% transmission) were used to keep the fluoresence intensities from exceeding the camera's linear range. Images were recorded only from one focal plane and digitized as described (Dunn et al., 1989; Maxfield and Dunn, 1990). Digitized Images were photographed via a Freeze Frame Video Recorder coupled to a 35 mm camera and printed at the same settings for all pairs of images.. Neutral density filters (5%, 13%, 33% transmission) were used to keep the fluoresence intensities from exceeding the camera's linear range.and photographed. The images of Tx-Tf and C6-NBD-SM show an almost complete colocalization of fluoresence patterns (compare A with C and B with D). Note the delivery of both the labelled species to punctate spots (2 and 9 min) and also to the perinuclear region (para-Golgi recycling endosomes) at later times (9 min). There appears to be no appreciable change in endosome (punctate spot) brightness for the two fluoresent labels over the 9 min period of incubation. See Fig. 3 for quantitative analysis of a time course Tx-Tf and $C_6$-NBD-SM internalization into sorting endosomes.

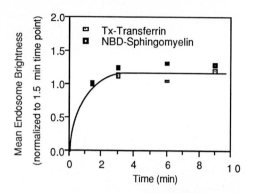

**Figure 3.  Time course of accumulation of $C_6$-NBD-SM and Tx-Tf in sorting endosomes.**
TRVb-1 cells were incubated with $C_6$-NBD-SM and Tx-Tf for the indicated times, fixed and the fluorescence images were recorded as described in Fig. 2. The digitized images were background corrected and the total intensities in size (4-50 pixel) selected images were obtained as described (Dunn et al., 1989). The fluoresence in the para-Golgi region of Tf and NBD-SM images was eliminated by image processing. The data points shown are the geometric mean of measured endosome brigntness of 8-12 fields of 3-8 cells each, corresponding to 2000-5000 endosomes/data set. Standard errors from the mean are in each case smaller than the size of the symbol. The lipid concentration of $C_6$-NBD-SM in the labelling solution was chosen so that the mean endosome brightness of the 10 min endosomes increased proportionately with increasing concentration of $C_6$-NBD-lipid in the labelling solution. The data show that $C_6$-NBD-SM and Tf rapidly (within 2 min) reach a steady state level of mean endosome brightness. Under the same conditions DiO-LDL continues to accumulate over the same period in sorting endosomes (see also Dunn et al., 1989).

## References

Dunn KW, Maxfield FR (1990) Use of Fluorescence microscopy in the study of receptor-mediated endocytosis. In: Herman B and Jackson K (eds) Optical microscopy for biology, Wiley-Liss, New York, p 153

Dunn KW, Maxfield FR (1992) Delivery of ligands from sorting endosomes to late endosomes occurs by maturation of sorting endosomes. J Cell Biol 117: 301-310

Dunn KW, McGraw TE, Maxfield FR (1989) Iterative fractionation of recycling receptors from lysosomally destined ligands in an early sorting endosome. J Cell Biol 109: 3303-3314

Koval M, Pagano RE (1989) Lipid recycling between the plasma membrane and intracellular compartments: transport and metabolism of fluorescent sphingomyelin analogues in cultured fibroblasts. J Cell Biol 108: 2169-2181

Linderman JJ, Lauffenburger DA (1988) Analysis of intracellular receptor/ligand sorting in endosomes. J Theor Biol 132: 203-245

Maxfield FR, Dunn KW (1990) Studies of endocytosis using image intensification fluorescence microscopy and digital image analysis. In: Herman B, Jackson K (eds) Noninvasive techniques in cell biology, Wiley-Liss, New York, p 357

# LINKAGE OF PLASMA MEMBRANE PROTEINS WITH THE MEMBRANE SKELETON: INSIGHTS INTO FUNCTIONS IN POLARIZED EPITHELIAL CELLS.

W. James Nelson
Department of Molecular and Cellular Physiology
Stanford University School of Medicine
Stanford, CA 94305-5426
USA

## *Introduction*

Interactions between integral membrane proteins and components of the cytoskeleton, termed the membrane skeleton, are thought to be important in many cellular functions in a wide variety of cell types (Bennett, 1990, Nelson, et al., 1990). A limited number of the proteins involved in these interactions have been identified and characterized, including: interactions between the actin-based cytoskeleton and cell-cell (Ozawa, et al., 1990, Takeichi, 1991, Nelson, et al., 1990, McCrea and Gumbiner, 1991) and cell-substratum adhesion proteins (Burridge, et al., 1988), and ion transport proteins (Nelson and Veshnock, 1987, Morrow, et al., 1989, Srinivasan, et al., 1988). These interactions may be important in restricting protein distributions in the plane of the lipid bilayer, regulating the assembly of protein complexes, and modulating the response of cells to their external environment. Although the identity of membrane proteins that interact with the cytoskeleton is limited at present, detailed analyses of those interactions are providing insight into functions of the membrane skeleton complex.

A role for interactions between membrane proteins and the cytoskeleton are readily apparent in specialized transporting epithelial cells (eg. kidney, intestine). These cells form a tight cell monolayer which separates two biological compartments in the organism, and which regulates the ionic composition of these compartments (Rodriguez-Boulan and Nelson, 1989). The formation of a transporting epithelium requires specific interactions between cells, and between

NATO ASI Series, Vol. H 74
Molecular Mechanisms of Membrane Traffic
Edited by D. J. Morré, K. E. Howell, and J. J. M. Bergeron
© Springer-Verlag Berlin Heidelberg 1993

the cells and the substratum. In addition, ion transport proteins must be restricted in their distributions between domains of the plasma membrane that face the two biological compartments, termed apical and basal-lateral, in order to regulate vectorial transport of ions and solutes (Rodriguez-Boulan and Nelson, 1989). This short overview will focus on interactions between specific proteins of the actin-based cytoskeleton, cell adhesion proteins and ion transport proteins in the context of the establishment and maintenance of polarized transporting epithelial cells.

## *The Membrane Skeleton: Similarities in the Basic Building Block Between Erythrocytes and Non-Erythroid (Epithelial) Cells.*

Specific protein-protein interactions in the membrane skeleton were first described and characterized in the human erythrocyte (reviewed in Bennett, 1990). In erythrocytes the principle cytoskeletal protein remaining in the cell following extraction with a non-ionic detergent (Triton X-100) is spectrin, a tetramer composed of two nonidentical, high molecular weight proteins (Shotton, et al., 1979). Detailed binding studies demonstrated that spectrin bound to the plasma membrane through a high affinity binding site on another high molecular weight peripheral membrane protein, termed ankyrin (Bennett and Stenbuck, 1979). Ankyrin in turn was shown to bind with high affinity to the cytoplasmic domain of an integral membrane protein, the anion transporter (Bennett and Stenbuck, 1980). These interactions define the basic building block of the membrane skeleton in the erythrocyte (see Figure 1). Direct visualization of the erythrocyte membrane skeleton in the electron microscope revealed that spectrin tetramers form an hexagonal array on the cytoplasmic surface of the erythrocyte membrane; further binding studies demonstrated that a ternary complex containing protein 4.1, adducin and short actin oligomers probably interlinks spectrin tetramers at the vertices of this hexagonal array (Ungewickell, et al., 1979, Gardner and Bennett, 1986, Gardner and Bennett, 1987).

Several studies of congenital hemolytic anemias in mice, in which one or more of the components of the membrane skeleton are deleted or mutated, have shown directly that the membrane skeleton plays an important role in maintaining the shape of the erythrocyte, and in restricting the mobility of the anion transporter in the plane of the membrane (for example see, Sheetz, et al., 1980).

Extensive searches for protein homologs of ankyrin and spectrin in nonerythroid cells has shown that these proteins exist as several isoforms in all tissues and organs analyzed (Bennett, 1990, Nelson and Lazarides, 1984). Analysis of the coding sequences of erythroid and non-erythroid ankyrin and spectrin (fodrin) indicate that these proteins are structurally highly related to their homologs in the erythrocyte (Wasenius, et al., 1989, Lux, et al., 1990). Furthermore, the subcellular distribution and extractability of ankyrin and fodrin from non-erythroid cells indicate that these proteins associate with the plasma membrane and are only partially extracted with buffers containing Triton X-100, indicating the presence of a membrane-associated protein complex similar to that described in erythrocytes (Nelson and Veshnock, 1986, Morrow, et al., 1989).

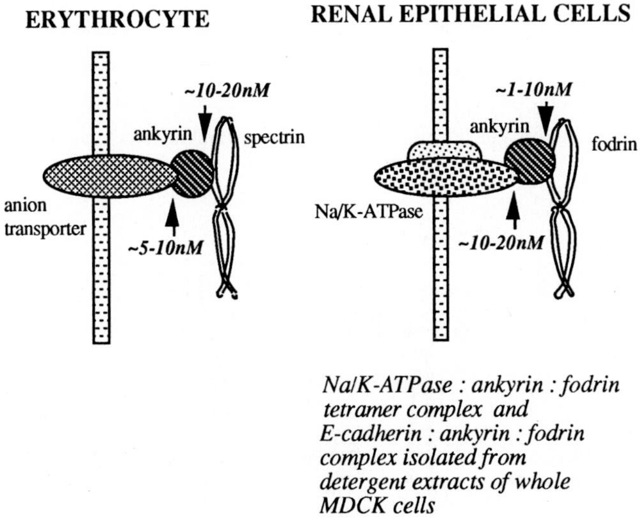

**FIGURE 1.** Schematic representation of membrane skeleton complexes identified in erythrocytes and in renal epithelial cells. Proteins were identified with specific antibodies. The affinities of each protein-protein interaction are given, and were determined by direct measurement of the binding of purified proteins *in vitro*.

Several membrane proteins have been shown to bind to ankyrin in non-erythroid cells. These include Na/K-ATPase (Nelson and Veshnock, 1987, Morrow, et al., 1989) and the voltage-sensitive Na channel(Srinivasan, et al., 1988). Binding between these proteins and ankyrin has been demonstrated directly with purified proteins, and by isolation of protein complexes extracted from whole cells. The binding affinities are similar to that of ankyrin to anion transporter (Bennett and Stenbuck, 1980). In addition, a complex of these proteins isolated from whole MDCK cell extracts (Nelson and Hammerton, 1989) has an apparent structural organization consistent with a complex similar to the unit structure of the erythrocyte ankyrin/spectrin/anion transporter complex (see Figure 1).

## Restricted Subcellular Distribution of Membrane Skeleton Complexes in Polarized Epithelial Cells

In the erythrocyte, the membrane skeleton is uniformly distributed on the plasma membrane where it may be involved in maintaining cell shape and the lateral mobility of bound integral membrane proteins (see above). These properties of the membrane skeleton have, thus far, not been extensively studied in non-erythroid cells. However, studies of the subcellular distribution of the membrane skeleton in specialized epithelial cells have provided insight into possible functions.

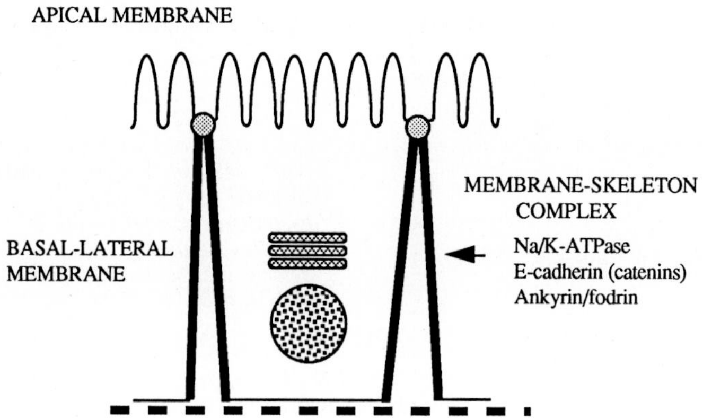

**FIGURE 2.** Schematic representation of the distributions of Na/K-ATPase, E-cadherin, ankyrin and fodrin in a polarized MDCK cell. Protein distributions were determined by immunocytochemistry (see text).

These distributions have been studied in renal and intestinal epithelial cells at the light and electron microscope levels (Drenckhahn, et al., 1985, Nelson and Veshnock, 1986, Morrow, et al., 1989, Nelson, et al., 1990). In polarized transporting epithelial cells, the distributions of ankyrin and fodrin are restricted to a specific domain of the plasma membrane. In MDCK epithelial cells, ankyrin and fodrin are restricted to the basal-lateral membrane; there appears to be little or none of these proteins associated with the apical membrane domain (Morrow, et al.,

1989, Nelson, et al., 1990). In epithelial cells which contain a brush border at the apical membrane (eg proximal tubule epithelia, enterocytes), ankyrin and fodrin have been localized also to the apical membrane domain (Davis, et al., 1989). However, at least in the case of fodrin, the proteins may be associated with the terminal web, a protein complex that interlinks the microvillar actin rootlets in the subcortical cytoplasm ( reviewed in Mooseker, 1985).

The restricted distribution of ankyrin and fodrin in MDCK cells and in renal epithelial cells *in situ* coincides with the distribution of several integral membrane proteins. In polarized transporting epithelial cells, integral membrane proteins are distributed between functionally distinct plasma membrane domains, termed apical and basal-lateral; differences in the membrane distributions of these proteins is the basis for vectorial transport of ions and solutes between the two biological compartments separated by the epithelium (Rodriguez-Boulan and Nelson, 1989). The basal-lateral membrane domain faces the serosa and contains proteins involved in cell-cell and cell-substratum interactions, growth hormone receptors and a variety of ion transport proteins (Rodriguez-Boulan and Nelson, 1989). Na/K-ATPase is a major protein of the basal-lateral membrane and is critically involved in the establishment and maintenance of a transepithelial gradient of Na ions that facilitates ion and solute uptake and transport to the serosa (Sweadner, 1989). Detailed comparison of the distributions of Na/K-ATPase, ankyrin and fodrin has confirmed that these proteins have identical distributions on the basal-lateral membrane of renal epithelial cells (Nelson and Veshnock, 1986, Morrow, et al., 1989, Nelson, et al., 1990). It is interesting to note that in retinal pigmented epithelial cells (Gundersen, et al., 1991) and in choroid plexus (Wright, 1972) Na/K-ATPase is localized to the apical membrane. Significantly, ankyrin and fodrin are also restricted to the apical membrane domain in these cell types (Gundersen, et al., 1991, Marrs, Mays and Nelson, in preparation). Taken together, these results strongly suggest that Na/K-ATPase, ankyrin and fodrin are co-localized in a membrane skeleton protein complex in a wide variety of cell types.

## *Regulation of Membrane Skeleton Distribution in Polarized Epithelial Cells: the Role of E-Cadherin-Mediated Cell-Cell Contacts*

Early studies indicated that the development of cell surface polarity of Na/K-ATPase and the membrane skeleton in MDCK cells was initiated upon the induction of cell-cell contact (Nelson and Veshnock, 1986, Nelson and Veshnock, 1987). In single MDCK cells, Na/K-ATPase, ankyrin and fodrin were uniformly distributed on the cell surface. The proteins were readily extracted from the cells in isotonic salt buffers containing Triton X-100, indicating that they were not assembled into an insoluble complex characteristic of the membrane skeleton. Significantly, analysis of the extracted proteins on sucrose gradients followed by non-denaturing polyacrylamide gels (Nelson and Hammerton, 1989) indicated that Na/K-ATPase, ankyrin and fodrin were in a protein complex similar to the UNIT complex described in Figure 1. This result has been interpreted as evidence that UNIT complexes of the membrane skeleton are pre-assembled on the plasma membrane, and require a signal to assemble into higher-ordered complexes (Nelson and Hammerton, 1989). It was proposed that assembly of these higher-ordered complexes at only sites of cell-cell contact, and not elsewhere in the cell, would result in the accumulation of these proteins in a polarized cell surface distribution (Nelson, 1989).

The principle cell adhesion protein in MDCK cells is the $Ca^{++}$-dependent adhesion protein, E-cadherin (uvomorulin) (Gumbiner and Simons, 1986, Gumbiner, et al., 1988). Cell-cell adhesion and initiation of the development of cell surface polarity can be induced by E-cadherin-mediated cell-cell contact. To determine directly whether E-cadherin is involved in initiating membrane skeleton assembly and the development of Na/K-ATPase polarity, fibroblasts transfected with E-cadherin were used as a model system (Nagafuchi, et al., 1987). Fibroblasts do not express E-cadherin, but constitutively express Na/K-ATPase, ankyrin and fodrin. In untransfected cells, Na/K-ATPase and fodrin were distributed uniformly over the cell surface. Expression of full-length E-cadherin in confluent cultures of fibroblasts resulted in the accumulation of E-cadherin, Na/K-ATPase and fodrin at sites of cell-cell contacts (McNeill, et al., 1990); if E-cadherin-mediated cell adhesion was disrupted by removal of extracellular $Ca^{++}$, E-cadherin, Na/K-ATPase and fodrin became uniformly distributed over the surface of the cells. These results provide direct evidence that E-cadherin-mediated cell-cell adhesion induced the remodelling of the cell surface distribution of Na/K-ATPase to cell-cell contacts (McNeill, et al., 1990).

How is the redistribution of Na/K-ATPase to sites of cell-cell contact driven by E-cadherin-mediated cell adhesion? Both membrane proteins have been shown to interact directly with components of the actin-based cytoskeleton. In the case of E-cadherin, 3 cytoplasmic proteins have been shown to be tightly bound to the cytoplasmic domain of E-cadherin. These cytoplasmic proteins are termed $\alpha$-, $\beta$-, and $\gamma$-catenin (Ozawa, et al., 1990, McCrea and Gumbiner, 1991, Takeichi, 1991). In addition, a small fraction of MDCK E-cadherin (~30%) has been recovered from whole cell extracts in a complex with ankyrin and fodrin (Nelson, et al., 1990). At present, it is not known whether ankyrin or fodrin interact with the cytoplasmic domain of E-cadherin directly or with one of the catenins. Nevertheless, these results indicate that E-cadherin is associated with the same class of cytoskeletal proteins as Na/K-ATPase. The role of cytoplasmic linkage of E-cadherin to Na/K-ATPase through these cytoskeletal proteins in driving the distribution of Na/K-ATPase to sites of cell-cell contact was also tested in the fibroblast system described above. Fibroblasts were transfected with E-cadherin that comprised a deletion of the C-terminal cytoplasmic domain containing the binding site(s) to these cytoskeletal proteins (McNeill, et al., 1990). Under these conditions, neither Na/K-ATPase or fodrin were localized to cell-cell contacts, presumably due to the loss of linkage of the cytoskeletal proteins to the cytoplasmic domain of E-cadherin.

### Regulation of Na/K-ATPase Distribution in Polarized MDCK Cells

The results discussed thus far raise the question as to the consequence of membrane skeleton assembly on the basal-lateral membrane on the organization of Na/K-ATPase in polarized epithelial cells. Extensive previous studies had shown that in polarized MDCK cells the majority of membrane proteins are delivered vectorially from the Golgi complex to either the apical or basal-lateral membrane (Rodriguez-Boulan and Nelson, 1989). If this was the case for Na/K-ATPase, protein arriving at the basal-lateral membrane would become incorporated into the assembled membrane skeleton complex. To test this hypothesis, the delivery of newly-synthesized Na/K-ATPase at the cell surface was monitored by domain-specific protein biotinylation (Hammerton, et al., 1991). Surprisingly, the results showed that between 4 to 96 hours after induction of cell-cell contact newly-synthesized Na/K-ATPase was delivered in approximately equal amounts to both the apical and basal-lateral membranes. However, analysis of the steady state distribution of Na/K-ATPase, either by confocal immunofluorescence microscopy or by cell surface biotinylation, showed that the distribution of Na/K-ATPase was restricted to the basal-lateral membrane within 48-72 hours following induction of cell-cell contact (Hammerton, et al., 1991). These results showed that despite delivery of newly-synthesized Na/K-ATPase to both cell surface domains, protein accumulated, and became restricted to the basal-lateral membrane domain.

To investigate the mechanism for accumulation of Na/K-ATPase on the basal-lateral membrane we compared the residence times of newly-synthesized Na/K-ATPase on the apical and basal-lateral membranes. The results showed that Na/K-ATPase delivered to the apical membrane had a very short residence time ($t_{1/2}$ 1-2hrs). In contrast, Na/K-ATPase delivered to the basal-lateral membrane had an increasingly long residence time ($t_{1/2}$ >36hrs at 144 hours after the induction of cell-cell contact). This 30-40 - fold difference in the residence times of Na/K-ATPase on the apical and basal-lateral membranes provides an explanation of how the distribution of this protein becomes restricted to the basal-lateral membrane in the face of protein delivery to both cell surface domains (Hammerton, et al., 1991).

# A Function of the Membrane Skeleton in Polarized (MDCK) Epithelial Cells: Specific Retention of Na/K-ATPase in the Basal-Lateral Plasma Membrane

What are the functions of the membrane skeleton in polarized MDCK cells? Three results described above support a role for the membrane skeleton as a retention system for membrane proteins, such as Na/K-ATPase, in the membrane:

1) Na/K-ATPase, ankyrin and fodrin are co-localized on the basal-lateral membrane of polarized MDCK cells.

2) Na/K-ATPase, ankyrin and fodrin form a defined, high affinity protein complex (UNIT), which is insoluble in buffers containing Triton X-100 (an indication of the formation of a higher-ordered structure).

3) Na/K-ATPase has a 30-40 - fold longer residence time on the basal-lateral membrane (in association with the membrane skeleton) than on the apical membrane (in the absence of the membrane skeleton).

A simple model that includes these observations and data is that cell-cell contact through the cell adhesion protein E-cadherin induces that assembly of the membrane skeleton on the lateral membrane in association with the contact site (Nelson, et al., 1990). Other membrane proteins (eg. Na/K-ATPase) that are bound to the membrane skeleton are included in the growing membrane skeleton complex. Over a period of time the constituent proteins become insoluble to extraction with buffers containing Triton X-100 as they become assembled into a higher-ordered structure on the lateral membrane. Newly-synthesized proteins that are delivered to the basal-lateral membrane are incorporated into the membrane skeletal complex through their high affinity interaction with ankyrin. The formation of a higher-ordered membrane skeleton complex excludes membrane proteins from internalization resulting in the long residence times of the constituent membrane proteins.

In contrast, there is no cell-cell contact on the apical membrane. Hence, membrane skeleton assembly is not induced. Due to the absence of the membrane skeleton on the apical membrane, newly synthesized Na/K-ATPase that is delivered there is not retained in the membrane, as in the case on the basal-lateral membrane, and is subject to rapid, constitutive internalization. It is interesting to note that the time required to clear a fluid-phase marker protein from either the apical or basal-lateral membranes of polarized MDCK cells was shown to be 1-2 hours (Balcarova, et al., 1984); this time is similar to the residence of newly-synthesized Na/K-ATPase on the apical membrane. Thus, the development of cell surface polarity of Na/K-ATPase, ankyrin and fodrin may be a consequence of differences in the sites of assembly of the membrane skeleton and the residence times of proteins between the apical and basal-lateral membranes.

*Acknowledgements.*

Work from the author's laboratory was supported by grants from the National Institutes of Health, the National Science Foundation, and the March of Dimes. The author is the recipient of an American Heart Association Established Investigator Award.

*References*

Balcarova SJ, Pfeiffer SE, Fuller SD, Simons K (1984) Development of cell surface polarity in the epithelial Madin-Darby canine kidney (MDCK) cell line. EMBO 3: 2687-2694.

Bennett V (1990) Spectrin-based membrane skeleton: A multipotential adaptor between plasma membrane and cytoplasm. Physiol Rev 70: 1029-1065.

Bennett V (1990) Spectrin: a structural mediator between diverse plasma membrane proteins and the cytoplasm. Curr. Op. Cell Biol 2: 51-56.

Bennett V, Stenbuck PJ (1979) Identification and partial purification of ankyrin, the high affinity membrane attachment site for human erythrocyte spectrin. J. Biol. Chem 254: 2533-2541.

Bennett V, Stenbuck PJ (1980) Association between ankyrin and the cytoplasmic domain of band 3 isolated from human erthrocyte membranes. J. Biol. Chem. 255: 6424-6432.

Burridge K, Fath K, Kelly T, Nuckolls G, Turner C (1988) Focal adhesion: transmembrane junctions between the extracellular matrix and the cytoskeleton. Annu. Rev. Cell Biol. 4: 487-525.

Davis J, Davis L, Bennett V (1989) Diversity in membrane binding sites of ankyrin. J. Biol. Chem. 264: 6417-6426.

Drenckhahn D, Schulter K, Allen D, Bennett V (1985) Colocalization of band 3 with ankyrin and spectrin at the basal membrane of intercalated cells in the rat kidney. Science 230: 1287-1289.

Gardner K, Bennett V (1986) A new erythrocyte membrane-associated protein with calmodulin binding activity: identification and purification. J. Biol. Chem. 261: 1339-1348.

Gardner K, Bennett V (1987) Modulation of spectrin-actin assembly by erythrocyte adducin. Nature 328: 359-362.

Gumbiner B, Simons K (1986) A functional assay for proteins involved in establishing an epithelial occluding barrier: identification of a uvomorulin-like polypeptide. J. Cell Biol. 102: 457-468.

Gumbiner B, Stevenson B, Grimaldi A (1988) The role of the cell adhesion molecule uvomorulin in the formation and maintenance of the epithelial junctional complex. J. Cell Biol. 107: 1575-1587.

Gundersen D, Orlowski J, Rodriguez-Boulan E (1991) Apical polarity of Na,K-ATPase in retinal pigment epithelium is linked to a reversal of the ankyrin-fodrin submembrane cytoskeleton. J. Cell Biol. 112: 863-872.

Hammerton RW, Krzeminski KA, Mays RW, Ryan TA, Wollner DA, Nelson WJ (1991) Mechanism for regulating cell surface distribution of Na/K-ATPase in polarized epithelial cells. Science 254: 847-850.

Lux SE, John KM, Bennett V (1990) Analysis of cDNA for human erythrocyte ankyrin indicates a repeated structure with homology to tissue-differentiation and cell-cycle control proteins. Nature 344: 36-42.

McCrea PD, Gumbiner B (1991) Purification of a 92kDa cytoplasmic protein tightly associated with the cell-cell adhesion molecule E-cadherin (Uvomorulin). J. Biol. Chem. 266: 4514-4520.

McNeill H, Ozawa M, Kemler R, Nelson WJ (1990) Novel function of the cell adhesion molecule uvomorulin as an inducer of cell surface polarity. Cell 62: 309-316.

Mooseker MS (1985) Organization, chemistry, and assembly of the cytoskeletal apparatus of the intestinal brush border. Annu. Rev. Cell Biol. 1: 209-242.

Morrow JS, Cianci CD, Ardito T, Mann AS, Kashgarian M (1989) Ankyrin links fodrin to the alpha subunit of Na/K-ATPase in Madin-Darby canine kidney cells and in intact renal tubule cells. J. Cell Biol. 108: 455-465.

Nagafuchi A, Shirayoshi Y, Okazaki K, Yasuda K, Takeichi M (1987) Transformation of cell adhesion properties by exogenously introduced E-cadherin cDNA. Nature 329: 340-343.

Nelson WJ (1989) Topogenesis of plasma membrane domains in polarized epithelial cells. Curr. Opp. Cell Biol. 1: 660-668

Nelson WJ, Hammerton RW (1989) A membrane-cytoskeletal complex containing Na/K-ATPase, ankyrin, and fodrin in Madin-Darby canine kidney (MDCK) cells: implications for the biogenesis of epithelial cell polarity. J. Cell Biol. 108: 893-902.

Nelson WJ, Hammerton RW, Wang AZ, Shore EM (1990) Involvement of the membrane-cytoskeleton in the development of epithelial cell polarity. Sem. Cell Biol. 1: 359-371.

Nelson WJ, Lazarides E (1984) Assembly and establishment of membrane-cytoskeleton domains during differentiation: spectrin as a model system. Mod. Cell Biol. 2: 219-246.

Nelson WJ, Shore EM, Wang AZ, Hammerton RW (1990) Identification of a membrane-cytoskeletal complex containing the cell adhesion molecule uvomorulin (E-cadherin), ankyrin, and fodrin in Madin-Darby canine kidney epithelial cells. J. Cell Biol. 110: 349-357.

Nelson WJ, Veshnock PJ (1986) Dynamics of membrane skeleton (fodrin) organization during development of polarity in Madin-Darby canine kidney epithelial cells. J. Cell Biol. 103: 1751-1765.

Nelson WJ, Veshnock PJ (1987) Ankyrin binding to Na/K-ATPase and implications for the organization of membrane domains in polarized cells. Nature 328: 533-536.

Nelson WJ, Veshnock PJ (1987) Modulation of fodrin (membrane skeleton) stability by cell-cell contact in Madin-Darby canine kidney epithelial cells. J. Cell Biol. 104: 1527-1537.

Ozawa M, Ringwald M, Kemler R (1990) Uvomorulin-catenin complex formation is regulated by a specific domain in the cytoplasmic region of the cell adhesion molecule. Proc. Natl. Acad. Sci. U S A 87: 4246-4250.

Rodriguez-Boulan E, Nelson WJ (1989) Morphogenesis of the polarized epithelial cell phenotype. Science 245: 718-725.

Sheetz MP, Schindler M, Koppel D (1980) Lateral mobility of integral membrane proteins is increased in spherocytic erythrocytes. Nature 285: 510-512.

Shotton DM, Burke BE, Branton D (1979) The molecular structure of erythrocyte spectrin. J. Mol. Biol. 131: 303-329.

Srinivasan Y, Elmer L, Davis J, Bennett V, Angelides K (1988) Ankyrin and spectrin associate with voltage-dependent sodium channels in brain. Nature 333: 177-180.

Sweadner KJ (1989) Isoenzymes of the Na/K-ATPase. Biochem. Biophys. Acta 988: 185-220.

Takeichi M (1991) Cadherin cell adhesion receptors as a morphogenetic regulator. Science 251: 1451-1457.

Ungewickell E, Bennett PM, Calver R, Ohanian V, Gratzer WB (1979) In vitro formation of a complex between cytoskeletal proteins of the human erythrocyte. Nature 380: 811-814.

Wasenius VM, Saraste M, Salver P, Erammaa M, Holm L, Lehto VP (1989) Primary structure of the brain alpha-spectrin. J. Cell Biol. 108: 79-93.

Wright EM (1972) Mechanisms of ion transport across the choroid plexus. J. Physiol. 226: 545-571.

# CELL SPECIFICITY AND DEVELOPMENTAL VARIATION IN THE TARGETING PATHWAYS OF FRT CELLS

Chiara Zurzolo
Department of Cell Biology and Anatomy
Cornell University Medical college
New York, NY 10021, USA

The polarity of epithelial cells is established gradually with time and different markers polarize at different rates (Balcarova et al., 1984). Recent data suggest that the molecular mechanisms by which proteins are delivered to the plasma membrane varies in cells derived from different tissues (Lisanti et al.1989; Le Bivic et al.1990 a,b; Matter et al.1990; Bartles et al.1990). It was previously shown that DPPIV follows an indirect route to the apical surface in liver cells (Bartles et al.1990) and in the human intestinal cell line Caco-2 (Matter et al.1990), but when transfected into MDCK cells, DPPIV is vectorially targeted to the apical surface (Casanova et al. 1991; Low et al. 1991).

To determine to what extent tissue specific sorting variation and developmental variation contribute to the establishment of the polarized phenotype we studied the biogenetic routes to the cell surface of DPPIV, an apical marker, and of Ag 35/40 kD, a basolateral marker at different times of cell culture in FRT cells.

The Fischer rat thyroid (FRT) cell line displays a highly polarized epithelial phenotype (Nitsch et al. 1985; Zurzolo et al. 1991). Like as MDCK cells (Rodriguez-Boulan et al. 1978), influenza virus buds from the apical surface while vesicular stomatitis virus (VSV) assembles at the basolateral surface, which correlates with restricted apical or basolateral distributions of, respectively, influenza hemagglutinin and VSV G protein (Zurzolo et al.1992a) Furthermore the surface distribution of several apical (DPPIV) and basolateral markers (alpha and beta Na,K-ATPase, uvomorulin, Ag 35/40 kD, transferrin receptor and ZO-1) is identical to that displayed by these antigens in MDCK and CaCo 2 cells (Zurzolo et al. 1992b).

NATO ASI Series, Vol. H 74
Molecular Mechanisms of Membrane Traffic
Edited by D. J. Morré, K. E. Howell, and J. J. M. Bergeron
© Springer-Verlag Berlin Heidelberg 1993

Using domain selective biotinylation and immunoprecipitation with specific antibodies we found that, in FRT cells, the apical polarity of DPPIV increased from 65% on the first day of culture to ~ 85-90% on the seventh day. The basolateral distribution of Ag 35/40 kD was well established on the first day (> 80% on the basolateral surface) and increased to 99% by day 7.

To follow the surface appearance of the newly synthesized DPPIV and Ag 35/40 kD, we used radioactive pulse-chase and surface-domain selective biotinylation, at different times of the chase, of FRT monolayers grown on filters. After 7 days at confluency, DPPIV was initially detected on the apical surface after 30 min of chase and accumulated there with a half-time of ~90 min, reaching a plateau by 240 min. No transient peak was observed on the basolateral membrane where a small amount of DPPIV (~15% of total surface delivery) was progressively detected with exactly the same kinetics as the apically delivered protein, also plateauing after 240 min. The initial surface appearance of Ag 35/40 kD was on the basolateral surface and was also detected at 30 min of chase, however, this protein reached a plateau much faster than DPPIV, after only 60 min of chase. Very little Ag 35/40 kD (<2%) was detected at any time of chase at the apical surface.

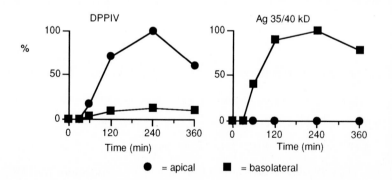

Fig.1

A very different situation was observed in FRT monolayers confluent for just one day. In these cells, delivery of DPPIV was equal to apical and basolateral surfaces until the 120 min time point (Fig 1). Delivery to the apical surface continued to increase between 120 and 240 min whereas delivery to the basolateral surface decreased continuously after 120 min (Fig 1). This indicates that a pool of DPPIV passed through the basolateral membrane and then accumulated on the apical membrane. In contrast, delivery of Ag 35/40 kD was highly polarized to the basolateral surface (Fig 1).

The results of the targeting experiments suggested that DPPIV might be vectorially targeted to the apical surface at day 7 but may utilize a transcytotic route at day 1. To directly determine the fate of the basolateral pool of DPPIV, we carried out a biotin transcytosis assay at 1 and 7 days of culture. A striking difference was observed in the levels of transcytosis of DPPIV at days 1 and 7 . At day 1, large amounts of basolateral DPPIV (>50 %) were transcytosed to the apical surface in four hours. No transcytosis of DPPIV was detected in 7-day old monolayers. Thus, FRT cells transcytose DPPIV at early times in the formation of a polarized monolayer but abandon this targeting route in favour of direct delivery as the monolayer matures.

Our results show that DPPIV is directly delivered to the apical membrane of fully polarized FRT cells, differently from liver and intestinal cells, indicating a cell specificity in the sorting of the same protein in different epithelial cell. Furthermore we demonstrate a developmental change in the targeting profile of FRT cells. A transcytotic route for the apical marker DPPIV is progressively abandoned, with time of culture, in favour of a direct apical pathway. Thus, epithelial cells not only display tissue-specific variation of targeting pathways but also developmental variations. The biological reasons for this second type of variation is unknown but we speculate that it may play a role in the initial passage from an unpolarized to a polarized stage during establishment of the epithelial monolayer.

Unlike DPPIV, Ag 35/40 kD was directly targeted at all times. This indicates that apical and basolateral sorting machineries develop along independent lines and that they are distinct from each other. Interestingly, both machineries increased their efficiency as the monolayer matured, which resulted in an enhanced surface polarization of both markers. This may result from improved sorting in the Golgi apparatus, or from improved accuracy in the targeting of post-Golgi vesicles to the cell surface. The latter may be due to enhanced polarization of surface receptors for apical and basolateral carrier vesicles, or may refect the organization of microtubules along an apical-basal axis .

References:

Balcarova-Stander, J , S E  Pfeiffer, S. D. Fuller, and K. Simons (1984)
        Development of cell surface polarity in the epithelial Madin-Darby
        canine  kidney (MDCK) cell line. EMBO J  3:2687-2694
Bartles, J.R. and A. L. Hubbard (1988) Plasma membrane protein sorting in
        epithelial cells: Do secretory proteins hold the key? T I B S 13:181-
Casanova, J.E., Y. Mishumi, Y.Ikeara, A. L. Hubbard, and K. E. Mostov (1991)
        Direct  apical sorting of Rat liver dipeptidylpeptidase IV from the
        TGN in  MDCK cells. J Biol Chem  266:24428-24435
Le Bivic, A., Y. Sambuy, K. Mostov, and E. Rodriguez-Boulan (1990a) Vectorial
        targeting of an endogenous apical membrane sialoglycoprotein
        and  uvomorulin in MDCK cells. J  Cell Biol  110:1533-1539
Le Bivic, A., A. Quaroni, B. Nichols, and E. Rodriguez-Boulan (1990b) Biogenetic
        pathways of plasma membrane proteins in Caco-2, a human
        intestinal epithelial cell line. J  Cell Biol  111:1351-1361
Lisanti, M., A. Le Bivic, M. Sargiacomo, and E. Rodriguez-Boulan (1989) Steady
        state distribution and  biogenesis of endogenous MDCK
        glycoproteins: evidence for intracellular sorting and polarized cell
        surface delivery. J  Cell Biol  109:2117-2128
Matter, K., M. Brauchbar, K. Bucher, and H. P. Hauri (1990) Sorting of
        endogenous plasma membrane proteins occurs from two sites in
        cultured human intestinal epithelial cells (Caco-2). Cell 60:429-437.
Nitsch, L., D. Tramontano, F. S. Ambesi-Impiombato, N. Quarto, and S.
        Bonatti(1985) Morphological and functional polarity in an
        epithelial thyroid cell line. Eur  J  Cell Biol 38:57-66
Rodriguez-Boulan, E. and D. D. Sabatini (1978) Asymmetric budding of viruses
        in epithelial monlayers: a model system for study of epithelial
        polarity. Proc Natl Acad Sci USA 75:5071-5075
Zurzolo, C., R. Gentile, A. Mascia, C. Garbi, C. Polistina, L. Aloj, V. E.
        Avvedimento, and L. Nitsch (1991) The polarized epithelial
        phenotype is dominant in hybrids between polarized and
        unpolarized rat thyroid cell lines. J  Cell Science 98:65-73
Zurzolo, C., C. Polistina, M. Saini, R. Gentile, G. Migliaccio, S. Bonatti, and L.
        Nitsch (1992a) Opposite polarity of virus budding and of viral
        envelope glycoprotein distribution in epithelial cells derived from
        different tissues. J Cell Biol 117:551-564
Zurzolo, C., A. Le Bivic, A.Quaroni, L. Nitsch, and E. Rodriguez-Boulan (1992b)
        Modulation of transcytotic and direct targeting pathways in a
        polarized thyroid cell line. Embo J 6:2337-2344

# PHOSPHOGLUCOMUTASE IS A CYTOPLASMIC GLYCOPROTEIN IMPLICATED IN THE REGULATED SECRETORY PATHWAY

Denise Auger, Pam Bounelis, and Richard B. Marchase
Department of Cell Biology, BHSB 690
The University of Alabama at Birmingham
Birmingham, AL 35294-0005
USA

Phosphorylation can have a profound effect on a protein's functional capabilities. One strategy for identifying proteins whose behaviors change in response to an external stimulus is to determine those in which changes in phosphorylation correlate with the presentation of the stimulus. In *Paramecium tetraurelia*, which exhibits a regulated secretory pathway, this strategy led to the isolation of parafusin (Gilligan and Satir, 1982), a 63 kDa cytoplasmic phosphoprotein that loses its phosphate upon stimulation with secretagogue. It then regains its phosphate within seconds thereafter (Ziesness and Plattner, 1985).

Recent data (Satir *et al.*, 1990) suggest that the change in parafusin's state of phosphorylation does not directly involve ATP, a protein kinase, or a phosphatase. Rather, the dephosphorylation appears to be due to the removal of $\alpha$Glc-1-P from the protein's oligosaccharide, followed by its rapid replacement via transfer from UDP-Glc. In this ciliate this cytoplasmic glycosylation event thus appears to be reversible and dependent upon external stimuli. The enzyme catalyzing the replacement of parafusin's phosphate, UDP-Glc: glycoprotein glucose-1-phosphotransferase (Glc phosphotransferase), has previously been shown in mammalian tissue homogenates to add $\alpha$Glc-1-P to O-linked mannose present on a cytoplasmic acceptor of about 63 kDa (Srisomsap *et al.*, 1988). We have recently determined that the turnover of $\alpha$Glc-1-P on the 63 kDa acceptor in rat synaptosomes and PC-12 cells is also correlated with the presentation of a secretory stimulus (in preparation). It therefore was of interest to determine the identity of the 63 kDa glycoprotein that served as the primary acceptor in the Glc phosphotransferase reaction.

## Phosphoglucomutase Is the Primary Acceptor in the Glc Phosphotransferase Reaction

As previously demonstrated (Srisomsap *et al.*, 1989), the acceptor for this reaction resides in the 100,000 x g supernatant of rat liver homogenates, while a

NATO ASI Series, Vol. H 74
Molecular Mechanisms of Membrane Traffic
Edited by D. J. Morré, K. E. Howell, and J. J. M. Bergeron
© Springer-Verlag Berlin Heidelberg 1993

second required constituent, presumably the Glc phosphotransferase, resides in the high-speed pellet. Furthermore, acceptor-rich fractions, following preparative isoelectric focusing of the supernatant, are found near a pH of 6.0. Supernatant proteins having isoelectric points near 6.0 were combined with an aliquot of the high-speed pellet and incubated with the $^{35}$S-labeled β-phosphorothioate analogue of UDP-Glc (Marchase *et al.*, 1987). SDS-polyacrylamide gel electrophoresis (SDS-PAGE) confirmed that this preparation contained a single labeled macromolecule with a mobility corresponding to approximately 63 kDa. Following removal of excess labeled precursor by molecular sieve chromatography, the proteins were subjected to a series of conventional chromatographic separations including hydrophobic interaction chromatography on Butyl-Sepharose. At each step fractions containing the labeled phosphoglycoprotein were pooled for further fractionation. Finally, the material was subjected to reversed phase high performance liquid chromatography in an increasing linear gradient of acetonitrile. The radioactively labeled macromolecule was one of the last proteins to be eluted. SDS-PAGE of the fractions showed a Commassie blue-staining protein of 63 kDa whose distribution across the gradient corresponded precisely to the radioactivity detected in autoradiographs. This material was then subjected to semi-preparative SDS-PAGE and transferred electrophoretically to Immobilon-P. The labeled protein was analyzed by sequential Edman degradation. This analysis generated the following amino acid sequence: NH$_2$-Val-Lys-Ile-Val-Thr-Val-Lys-Thr-Gln-Ala-Tyr-Pro. A comparison to protein data bases determined a correspondence at 11 of the 12 positions to the amino terminal of rabbit muscle phosphoglucomutase (Ray *et al.*, 1983), which also has a molecular weight near 63 kDa. At the same time, a parallel analysis was being carried out on the labeled acceptor from yeast. Digestion by the endoproteinase Asp N generated two peptides that were also homologous to rabbit muscle phosphoglucomutase. From the correspondence of molecular weights, the near-identity of these three peptide sequences, and other data using yeast and rat liver (in preparation), it was apparent that the protein that had been purified in both of these systems was phosphoglucomutase.

In order to confirm that phosphoglucomutase was in fact a cytoplasmic glycoprotein, commercial rabbit muscle phosphoglucomutase was subjected to acid hydrolysis. The resulting hydrolysate was then analyzed for the presence of both glucose and mannose utilizing linked enzymatic protocols (Bergmeyer,

1974). Approximately one mole of glucose (0.90 ± 0.16; n=4) and one mole of mannose (1.03 ± 0.21; n=4) were found per mole of phosphoglucomutase hydrolyzed. These and other data (in preparation) strongly suggest that the phosphoglycoprotein that has been described previously by our laboratory is the well-studied enzyme phosphoglucomutase.

## What Is the Significance of the Phosphooligosaccharide Present on Phosphoglucomutase?

Phosphoglucomutase is important in metabolism, interconverting Glc-1-P and Glc-6-P. However, the turnover of its phosphooligosaccharide in response to a secretory stimulus appears to be accompanied by its transient association with cellular membranes in both paramecia (Satir *et al.*, 1988) and mammalian cells (unpublished data). While this membrane association could be important solely as a regulatory mechanism of metabolism, it appears at least as likely that this highly conserved enzyme serves a second, independent function perhaps related to vesicle trafficking or fusion. This concept, referred to as "gene sharing" (Piatigorsky and Wistow, 1991), appears to apply to several housekeeping enzymes including, for instance, glyceraldehyde-3-phosphate dehydrogenase (Meyer-Siegler *et al.*, 1991).

Cytoplasmic glycosylation may function in a very different manner from glycosylation within the endoplasmic reticulum and Golgi apparatus. Since cytoplasmic glycoproteins are constitutively localized to the same subcellular compartment as the antagonistic glycosidases and transferases, the possibility of cyclic, stimulus-dependent turnover of their sugar residues should be considered. There are now several other examples of cytoplasmic glycosylation events, the best studied of which gives rise to O-linked GlcNAc (Haltiwanger *et al.*, 1992). The data presented here may thus reflect but one example of a group of regulated posttranslational modifications in which sugars are used to modulate a cytoplasmic protein's behavior.

## References:

Bergmeyer HU (1974) Methods of enzymatic analysis. Academic Press, New York

Gilligan DM, Satir BH (1982) Protein phosphorylation/dephosphorylation and stimulus-secretion coupling in wild type and mutant *Paramecium*. J Biol Chem 257:13903-13906

Haltiwanger RS, Kelly WG, Roquemore EP, Blomberg MA, Dong LYD, Kreppel L, Chou TY, Hart GW (1992) Glycosylation of nuclear and cytoplasmic proteins is ubiquitous and dynamic. Biochem Soc Trans 20:264-269

Marchase RB, Hiller AM, Rivera AA, Cook JM (1987) The ($\beta^{35}$S) phosphorothioate analogue of UDP-Glc is effeciently utilized by the glucose phosphotransferase and is resistant to hydrolytic degradation. Biochim Biophys Acta 91:157-162

Meyer-Siegler K, Mauro DJ, Seal G, Wurzer J, deRiel JK, Sirover MA (1991) A human nuclear uracil DNA glycosylase is the 37-kDa subunit of glyceraldehyde-3-phosphate dehydrogenase. Proc Natl Acad Sci USA 88:8460-8464

Piatigorsky J, Wistow G (1991) The recruitment of crystallins: new functions precede gene duplication. Science 252:1078-1079

Ray WJ, Hermodson MA, Puvathingal JM, Mohoney WC (1983) The complete amino acid sequence of rabbit muscle phosphoglucomutase. J Biol Chem 258:9166-9174

Satir BH, Busch G, Vuoso A, Murtaugh TJ (1988) Aspects of signal transduction in stimulus exocytosis-coupling in *Paramecium*. J Cell Biochem 36:429-443

Satir BH, Srisomsap C, Reichman M, Marchase RB (1990) Parafusin, an exocytic-sensitive phosphoprotein, is the primary acceptor for the glucosylphosphotransferase in *Paramecium tetraurelia* and rat liver. J Cell Biol 111:901-907

Srisomsap C, Richardson KL, Jay JC, Marchase RB (1988) Localization of the glucose phosphotransferase to a cytoplasmically accessible site on intracellular membranes. J Biol Chem 263:17792-17797

Srisomsap C, Richardson KL, Jay JC, Marchase RB (1989) An αglucose-1-phosphate phosphodiesterase is present in rat liver cytosol. J Biol Chem 264:20540-20546

Zieseniss E, Plattner H (1985) Synchronous exocytosis in paramecium cells involves very rapid (<1 s), reversible dephosphorylation of a 65-kD phosphoprotein in exocytosis-competent strains. J Cell Biol 101:2028-2035

# PROTEIN TARGETING AND THE CONTROL OF Cl⁻ SECRETION IN COLONIC EPITHELIAL CELLS

A.P. Morris, S.A. Cunningham, D.J. Benos and R.A. Frizzell
Department of Physiology and Biophysics
University of Alabama at Birmingham
Birmingham, AL 35294-0005

Human colonic epithelial cells can secrete chloride in response to cAMP-dependent stimulation following cellular polarization. This is correlated with the presence of 8 pS cAMP-activated Cl⁻ channels within the apical membranes of these cells and their absence from the plasma membranes of unpolarized cells (Morris *et al.*, 1992). The protein responsible for this cAMP-activated Cl⁻ conductance is now recognized to be the Cystic Fibrosis Transmembrane Conductance Regulator (CFTR, Rommens *et al.*, 1989; Bear *et al.*, 1992). We have, therefore, postulated that the apical membrane expression of CFTR, and the ability of these cells to secrete Cl⁻ in response to cAMP, is manifest when this protein targets to the apical membrane during epithelial cell differentiation (Morris *et al.*, 1992).

This hypothesis has been evaluated by comparing CFTR gene transcription and post-translational CFTR processing with cAMP-activated Cl⁻ secretion in both polarized and unpolarized HT-29 cell lines. Comparable levels of CFTR mRNA were expressed in polarized and unpolarized cells. During the period when the polarized (Cl.19A) cell line reached confluency and formed a cAMP-responsive Cl⁻-secreting monolayer, no change in the level of CFTR gene expression was observed (Fig. 1). CFTR protein expression likewise remained constant throughout this period. Thus, cellular polarization and the attainment of cAMP-dependent Cl⁻ secretory function was not associated with changes in either CFTR message or protein levels. CFTR immunoprecipitated from both cell lines ran with an apparent molecular mass (Mr) of ~170 kDa on SDS gels. The glycosylation status of CFTR was

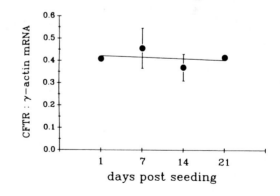

**FIGURE 1.** Ratio of CFTR to γ-actin mRNA expression as a function of day's post-seeding.

NATO ASI Series, Vol. H 74
Molecular Mechanisms of Membrane Traffic
Edited by D. J. Morré, K. E. Howell, and J. J. M. Bergeron
© Springer-Verlag Berlin Heidelberg 1993

found to be complex: CFTR oligosaccharide in both cell lines was insensitive to Endo H but was sensitive to N-glycanase. Since no difference was observed in the carbohydrate content of CFTR isolated from either cell line, the lack of a cAMP-activated Cl⁻ conductance in unpolarized cells was not due to incomplete N-linked glycoprocessing of the protein.

To determine if the glycosylation status of CFTR can affect cAMP-dependent Cl⁻ secretion, polarized (Cl.19A) monolayers were pre-incubated for 24 h with inhibitors of key enzymes found within the N-linked glycoprocessing pathway: swainsonine, deoxymannojirimycin and deoxynojirimycin. All inhibitors affected the Mr of CFTR. In contrast, no change in forskolin-evoked Isc or Gt was observed. Thus, changes in the N-linked oligosaccharide content of CFTR do not affect its function as a cAMP-regulated Cl⁻ channel. Since equivalent Cl⁻ current was recorded in the presence of each inhibitor, it is also unlikely that the apical membrane targeting of CFTR was altered under these conditions. This postulate has implications for the targeting hypothesis of CF, which argues that some mutant variants of CFTR result in deficient cAMP-mediated Cl⁻ secretion, not because the protein is dysfunctional, but because CFTR does not reach the plasma membrane (Cheng *et al.*, 1990). Our findings support the experimental basis for this hypothesis, i.e. that incomplete mutant CFTR glycoprocessing reflects a failure of these proteins to progress through the subcellular compartments of the N-linked glycoprocessing pathway where carbohydrate modification occurs.

To perturb the targeting of newly synthesized CFTR protein, Cl.19A monolayers were pre-incubated with the fungal metabolite Brefeldin A (BFA) for various times. BFA caused a time-dependent decrease in FSK-stimulated Isc and Gt across polarized monolayers (Fig. 2) and across the apical membrane of monolayers in which the basolateral membranes were permeabilized with nystatin. CFTR immunoprecipitated from BFA-treated cells ran at two different molecular masses: 170 kDa and 145 kDa. The lower Mr band, like that isolated from

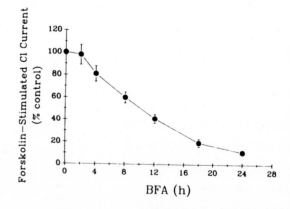

**FIGURE 2. The effect of Brefeldin A (BFA) pre-treatment on forskolin-stimulated Cl⁻ secretion by polarized Cl.19A monolayers.**

DMJ-treated monolayers, was Endo H-sensitive. Thus, the 145 kDa band is a high mannose form of CFTR trapped at its site of synthesis within the endoplasmic reticulum of these cells, while the 170 kDa form represents mature CFTR retrieved from the apical membrane domain (see below).

The localization of CFTR within control and BFA-treated HT-29 colonocytes was determined by immunofluorescence, performed using the Genzyme monoclonal antibody (M13) and FITC-conjugated secondary antibodies. Polarized colonocytes were found to contain CFTR predominantly localized to the apical membrane domain, visualized as a diffuse reticulated band. In unpolarized HT-29 cells, or in the Cl.19A clone prior to polarization, CFTR was localized to the perinuclear region with no apparent plasma membrane staining. Thus, CFTR location in HT-29 epithelial cells changes when cells polarize, and its location correlates with the ability of cAMP to stimulate the plasma membrane Cl⁻ conductance. BFA treatment of polarized monolayers attenuated cAMP-stimulated Cl⁻ secretion and also caused CFTR to accumulate in large coalesced intracellular vesicles which co-localized with the cell's microtubule organizing centers (MTOC's). Following BFA removal, both apical membrane domain CFTR staining and cAMP-mediated Cl⁻ secretion recovered (within 24-36 h) to near control levels. The slow recovery of both parameters probably reflects the delayed functional, as opposed to structural, recovery of the N-glycan processing pathway (Sampath et al., 1992). Thus, BFA inhibits the targeting of CFTR containing vesicles towards the apical membrane of polarized cells, and this is correlated with inhibition of cAMP-mediated Cl⁻ secretion. In unpolarized colonocytes, BFA has no effect on CFTR location. The unpolarized cell contains mature (fully processed) CFTR, yet this protein does not target to the plasma membrane because the apical membrane has not yet formed.

These results indicate that a mechanism responsible for N-linked glycoprotein targeting is present at sites distal to the Golgi cisternae, and that this vesicular secretory pathway develops with the differentiation process that leads to epithelial cell polarization. The MTOC- associated trans-Golgi network (TGN) is postulated to be the site of protein sorting (Griffiths & Simmons, 1986). Our finding that CFTR is found within vesicles associated with this structure in polarized, BFA-treated monolayers and that unpolarized HT-29 cells contain an underdeveloped TGN (Laboisse, 1990) suggests that the TGN is the site of this peripheral, apically-directed targeting pathway. In the absence of this pathway, fully processed CFTR remains intracellular and no cAMP-activated plasma membrane Cl⁻

conductance is detectable. It will be interesting to determine whether this peripheral targeting pathway contributes to the inappropriate localization of mutant forms of CFTR found in CF epithelia.

## REFERENCES

Bear CE, Canhui L, Kartner N, Bridges RJ, Jensen TJ, Ramjeesing M, Riordan JR (1992) Purification and functional reconstitution of the cystic fibrosis transmembrane conductance regulator. Cell 68:809-818

Cheng SH, Gregory RJ, Marshall J, Paul S, Souza DW, White GA, O'Riordan CR, Smith AE (1990) Defective intracellular transport and processing of CFTR is the molec ular basis of most cystic fibrosis. Cell 63:827-834

Griffiths G, Simmons K (1986) The trans golgi network: sorting at the exit site of the golgi complex. Science 234:438-443

Laboisse CL (1990) In: Augenlicht LH (ed) Molecular Biology of Colon Cancer. Raven Press, New York, pp 1389-1418

Morris AP, Cunningham SA, Benos DJ, Frizzell RA (1992) Cellular differentiation is required for cAMP but not $Ca^{2+}$-dependent Cl⁻ secretion in colonic epithelial cells expressing high levels of cystic fibrosis transmembrane conductance regulator. J Biol Chem 267:5575-5583

Rommens JM, Iannuzzi MC, Kerem B-S, Drumm ML, Melmer G, Dean M, Rozmahel R, Cole JL, Kennedy D, Hidaka M, Zsiga M, Buckwald M, Riordan JR, Tsui L-C, Collins FS (1989) Identification of the cystic fibrosis gene: chromosome walking and jumping. Science 245:1059-1065

Sampath D, Varki A, Freeze HH (1992) The spectrum of incomplete N-linked oligosaccharides synthesized by endothelial cells in the presence of brefeldin A. J Biol Chem 267:4440-4455

# APICAL MEMBRANE PROTEIN SORTING IS AFFECTED BY BREFELDIN A

C. B. Brewer and M. G. Roth
Department of Biochemistry
U. T. Southwestern Medical Center
5323 Harry Hines Blvd.
Dallas, TX 75235
USA

Polarized epithelial cells create and maintain plasma membrane domains with distinct protein compositions. The molecular mechanisms used by such cells to segregate newly-synthesized proteins and to deliver them to their appropriate places are not well understood. In the model Madin Darby canine kidney (MDCK) cell line, it appears that apical and basal membrane proteins travel through the same biosynthetic compartments as far as the trans Golgi network (TGN) and are then separated and delivered directly to their respective plasma membrane domains (Rindler et al., 1984; Matlin and Simons, 1984; Fuller et al., 1985).

The drug brefeldin A (BFA) has recently shown promise as a tool for studying morphology and mechanisms of membrane-bounded organelles involved in protein transport. BFA has been shown to affect endoplasmic reticulum to Golgi transport and, in addition, to alter the structure and function of the TGN and endocytic compartments. (See Pelham, 1991, for review.) In MDCK cells, Low et al. (1991) found that a low concentration of BFA had an inhibitory effect on total apical protein secretion but not on total basal secretion nor on Golgi morphology. Later, Hunziker et al. (1991), also using a low BFA concentration, showed an inhibitory effect on basal-to-apical transcytosis of the polyimmunoglobulin receptor.

We have examined the effect of a low concentration of BFA on the polarity of surface delivery of an apical membrane protein, influenza hemagglutinin (HA) and a basolateral membrane protein, the HA mutant called Tyr 543 (Lazarovits and Roth, 1988). In our initial pulse-chase assays on filter-grown MDCK cells, we found that treatment with 1 μg/ml BFA (3.6 μM) causes a dramatic reduction in the rate of delivery of HA apically, along with an increase in the rate of its missorting to the basal surface. In contrast, Tyr 543 retains its 96-100% basal polarity after the same treatment. Figure 1 shows a time course of HA surface delivery in the absence and in the presence of BFA. In the absence of BFA, the ratio of apical to basal HA is roughly 10:1, but with BFA treatment, basal HA exceeds apical.

NATO ASI Series, Vol. H 74
Molecular Mechanisms of Membrane Traffic
Edited by D. J. Morré, K. E. Howell, and J. J. M. Bergeron
© Springer-Verlag Berlin Heidelberg 1993

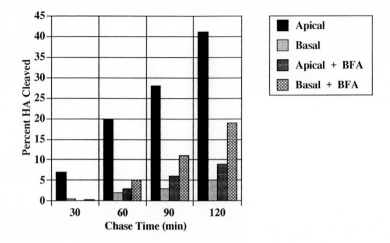

Figure 1. Effects of BFA on surface arrival of HA. MDCK cells were grown for 5 days on porous supports, then infected with influenza virus. Four and one-half hours post-infection, cells were pulse-labeled for 6 minutes. Labeled proteins were chased to the surface at 37⁰ for the time periods shown, then apical or basal HA's were marked by cleavage with cold trypsin on ice. The graph shows for each sample the cleaved HA as a percent of the total HA, corrected for the small percent cleaved in a sample with no trypsin added to the medium.

The fact that correct apical sorting is inhibited while basal sorting is not, for membrane proteins as well as for secreted proteins, suggests that it will be possible to use BFA to help in elucidating the mechanism of sorting. A first step is to determine in what exocytic compartment the BFA exerts its effect so that, ultimately, in vitro methods may be used to dissect the molecular components of sorting.

In agreement with published reports which show that low concentrations of BFA have no apparent effect on Golgi morphology in MDCK cells (Low et al., 1991; Hunziker et al., 1991), we find that the rate at which pulse-labeled HA acquires resistance to endoglycosidase H is not affected by 1 μg/ml BFA, even after a 30-min pretreatment. From this result, we conclude that the selective inhibitory effect of BFA on apical sorting occurs at a site which follows the medial Golgi in the biosynthetic pathway of glycoproteins. To compare the site of BFA action kinetically with the site of sulfation in the trans Golgi, we used an assay in which cells were labeled with $H_2{}^{35}SO_4$ in the absence or presence of BFA (no pretreatment), then surface HA's were marked using trypsin on ice. To order the site of BFA action with respect to the movement of HA from the TGN, we used a pulse-label and a 2 hr 20º block to accumulate labeled HA in the TGN, all in the absence of BFA, then added BFA during short chases to the surface at 37º. Conclusions from these experiments are summarized in the following table.

| Compartment | Marker | BFA Effect |
| --- | --- | --- |
| Medial Golgi | Endo H resistance | After |
| Trans Golgi | Sulfation | After |
| TGN | 20º block | In or after |

A model that is consistent with all of these results is based on the idea that apical and basal proteins are intermingled when they arrive in the TGN but then are segregated into membrane patches which have different proteins associated with them on the cytoplasmic face. The effect of BFA treatment may be to promote dissociation of a cytoplasmic protein that is an essential part of apically-destined membrane patches, while leaving basally-destined patches unaltered. With prolonged BFA treatment, HA may be hindered in leaving the TGN via the disabled apical patches and may undergo increasing leakage into basal patches, and thence to the basal surface.

References

Fuller SD, Bravo R, Simons K (1985) An enzymatic assay reveals that proteins destined for the apical or basolateral domains of an epithelial cell line share the same late Golgi compartments. EMBO J 4:297-307

Hunziker W, Whitney JA, Mellman I (1991) Selective inhibition of transcytosis by brefeldin A in MDCK cells. Cell 67:617-627

Lazarovits J, Roth M (1988) A single amino acid change in the cytoplasmic domain allows the influenza virus hemagglutinin to be endocytosed through coated pits. Cell 53:743-752

Low SH, Wong SH, Tang BL, Tan P, Subramaniam VN, Hong W (1991) Inhibition by brefeldin A of protein secretion from the apical cell surface of Madin-Darby canine kidney cells. J Biol Chem 266:17729-17732

Matlin KS, Simons K (1984) Sorting of an apical plasma membrane glycoprotein occurs before it reaches the cell surface in cultured epithelial cells. J Cell Biol 99:2131-2139

Pelham HRB (1991) Multiple targets for brefeldin A. Cell 67:449-451

Rindler MJ, Ivanov IE, Plesken H, Rodriguez-Boulan E, Sabatini DD (1984) Viral glycoproteins destined for the apical or basolateral membrane domains traverse the same Golgi apparatus during their intracellular transport in double infected Madin-Darby canine kidney cells. J Cell Biol 98:1304-1319

# STRUCTURAL ANALYSIS OF THE CLATHRIN TRISKELION

Inke S. Näthke[1] and Frances M. Brodsky
Departments of Pharmacy and Pharmaceutical Chemistry, School of Pharmacy,
University of California
San Francisco, CA 94143

Much has been learned about the biochemistry of coat proteins surrounding membrane vesicles involved in intracellular protein traffic, but very little is known about how their structure mediates their function. To address this issue we characterized the molecular features of clathrin that are responsible for its unique shape and self-assembly into a vesicle coat (reviewed in Brodsky, 1988).

Clathrin has a three-legged structure, called a triskelion. Each triskelion consists of three 192kD heavy chains. To each heavy chain one clathrin light chain is bound. In mammalian species, there are two types of clathrin light chains, $LC_a$ and $LC_b$ (Brodsky, 1988). These light chains apparently provide the regulatory domains for clathrin to interact with other cytoplasmic proteins (Brodsky et al., 1991). Clathrin triskelia polymerize into a polyhedral lattice that forms the outer coat of clathrin-coated vesicles and clathrin-coated pits, organelles responsible for receptor-mediated endocytosis (Brodsky, 1988).

The goal of the studies described here was to establish a structural model of how clathrin subunits fold into a triskelion and how triskelia form the polyhedral lattice. These studies resulted in identification of a region of the heavy chain sequence important for trimerization, as well as the description and localization of a site required for binding of heavy chains to light chains and the mapping of a site involved in regulation of clathrin assembly and disassembly. This work is reported elsewhere (Näthke et al., 1992) in detail and was the result of a collaboration between ourselves and John Heuser, Andrei Lupas, Jeff Stock, and Christoph W. Turck.

To identify the region involved in the trimerization of clathrin heavy chains, clathrin was digested with trypsin or chymotrypsin and the trimerization state of resulting heavy chain fragments was established using a sizing column. We found that fragments with molecular weights of 56 and 59kD eluted from the column at a position corresponding to proteins with a molecular weight of >150kd, suggesting that they are trimeric. Fragments with molecular weights 44 and 38kd eluted at a position corresponding to their molecular weight indicating that they are monomeric. N-terminal sequencing of the fragments indicated that they were

[1]Present address: Department of Molecular and Cellular Physiology, Stanford University Medical School, Stanford, CA 94305-5426.

NATO ASI Series, Vol. H 74
Molecular Mechanisms of Membrane Traffic
Edited by D. J. Morré, K. E. Howell, and J. J. M. Bergeron
© Springer-Verlag Berlin Heidelberg 1993

derived from the C-terminal third of the heavy chain. The smaller, monomeric fragments shared N-termini with the larger, trimeric fragments and the larger fragments contained 100 additional amino acids at their C-terminus. Furthermore, the larger fragments were missing 100 amino acids from the C-terminus of the heavy chain. This analysis mapped the trimerization site to a 100 amino acid region that is 100 amino acids removed from the C-terminus of the clathrin heavy chain (Figure 1). This is consistent with the previous finding by Lemmon et al. (1991) that the C-terminal 57 amino acids in the clathrin heavy chain of yeast are not required for trimerization.

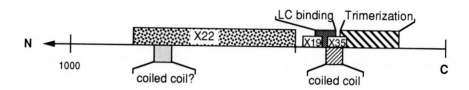

Figure 1: Linear arrangement of functional sites in the C-terminal third of clathrin heavy chain. The arrangement is based on proteolytic mapping studies described in the text. The regions labeled by X19, X22, and X35 indicate the boundaries of these monoclonal antibody binding sites. LC binding indicates the light chain binding site. The coiled coil regions were predicted by computer analysis (Lupas et al., 1991).

The trimeric heavy chain fragments co-eluted with smaller fragments that were shown to be derived from clathrin light chains by antibody reactivity and N-terminal sequencing. This suggested that light chains were protected from proteolysis in these triskelion fragments and may bind to a region in heavy chains near the trimerization site. Sequence analysis of heavy chain fragments bound to clathrin light chain affinity columns established that a region in the heavy chain sequence, directly adjacent to the trimerization site, is required for light chains to bind.

To identify regions in the heavy chain that are involved in lattice formation we mapped the binding sites of monoclonal antibodies specific for the clathrin heavy chain. Two of these antibodies (X19 and X35) inhibit lattice formation whereas the third (X22) does not (Blank and Brodsky, 1986). These antibodies were used to immunoprecipitate proteolytic fragments of clathrin heavy chain and the N-terminal sequence of the bound fragments was determined, localizing their binding sites in the linear sequence of clathrin heavy chain (Figure 1). Since X19 and X35 interfere with lattice formation, their binding site marks a region in the heavy chain sequence that is important for lattice assembly. The co-localization of the light chain binding site and the binding site for these antibodies (Figure 1) is supported by previous observations that X19 and X35 bind better to clathrin from which light chains have been removed and that X35 interferes with rebinding of purified light chains to clathrin heavy chains (Blank and Brodsky, 1987). Binding of the third antibody, X22, was mapped to a relatively large 38kD region which represents the smallest proteolytic fragment we were able to isolate.

To correlate the linear arrangement of functional sites in the clathrin heavy chain with the physical location of these sites on triskelion legs, immunoelectron microscopy was used to visualize monoclonal antibodies bound to triskelia. The binding sites for X35, X19, and X22 were respectively 4.6, 9.4, and 14.5nm from the triskelion vertex, spanning the central or "hub" region, which extends from the vertex to the bend in the leg (approximately 15nm away from the vertex). Considering the physical location of the antibody binding sites, their location in the protein sequence and the dimensions of triskelion legs, a folding model for the C-terminal third of the clathrin heavy chain within the triskelion hub was generated (Figure 2). The folding of clathrin light chains within the triskelion hub and the structural basis for their interaction with clathrin heavy chains are also predicted in the model in Figure 2. Based on the presence of heptad repeats in the region of clathrin light chains involved in binding clathrin heavy chains (Brodsky et al., 1987; Scarmato and Kirchhausen, 1990), the interaction between light chains and heavy chains is thought to be mediated by alpha helical coiled coils (Jackson et al., 1987; Kirchhausen et al., 1987b). When the heavy chain sequence was analyzed for the presence of coiled coils, residues in the light chain binding region mapped by proteolysis were strongly predicted to form coiled coils (Figure 1). This prediction of coiled coil structures was generated using a newly developed algorithm (Lupas et al., 1991) and was found for the clathrin heavy chain sequence of rat, yeast and dictyostelium, strengthening the prediction. A second region, previously suggested to be involved in light chain–heavy chain interaction (Kirchhausen et al., 1987a), was weakly predicted to form coiled coils (indicated by "coiled coil?" in Figures 1 and 2). This region could also be involved in light chain binding, but, based on our proteolysis data, is not sufficient. The coiled coil region in light chains is interrupted in the middle, dividing the heavy chain binding domain into two coiled coil regions. These are of similar length as the predicted coiled coil region in the heavy chains that is required for light chain binding. This information suggests that light chains fold back on themselves to interact with the heavy chain to form a 3 or 4 helix bundle.

To determine whether the light chains fold back on themselves, possible interaction between opposite termini of the heavy chain binding region was investigated. The binding of X45, an antibody directed against the C-terminal region of the heavy chain binding domain of light chains was inhibited by an antibody that reacts with a sequence at the N-terminus of the heavy chain binding region, 70 amino acids away from the X45 binding site. In combination with the effects of other antibodies on X45 binding, this data suggested a close proximity of the N- and C-terminal regions of the heavy chain binding domain in light chains and supported the proposed arrangement of light chains folded back on themselves in the triskelion. This arrangement is also supported by our observation (unpublished) that an antibody (LCB.1) against the N-terminus of clathrin light chains (residues 1-24) decorates triskelia at the vertex as does an antibody to a C-terminal region (residues 188-208) of light chains (Kirchhausen et al., 1983). Additionally, the uncoating protein, which binds to a region near the N-terminus of light chains (residues 47-71), also binds to the vertex of the triskelion (Heuser and Steer, 1989).

The folding model for clathrin heavy chains and light chains within the triskelion hub and the physical localization of functional sequences on the clathrin triskelion established several structural features of the triskelion (Näthke et al., 1992). The trimerization site was mapped near the vertex, 100 amino acids from the C-terminus of the clathrin heavy chain. Clathrin light chains were predicted to fold back on themselves and bind to a region directly adjacent to the trimerization site. This region is also important for the assembly of clathrin triskelia into lattices and thus represents a key regulatory domain for clathrin interaction with itself during assembly and with the uncoating protein during disassembly. Now that relatively small functional regions in the sequence of clathrin heavy chains have been identified, mutagenesis can be used to understand the molecular basis for clathrin assembly into triskelia and lattices in more detail.

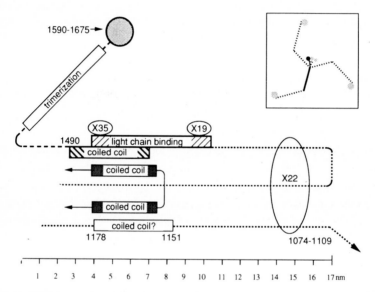

Figure 2: Folding of clathrin heavy and light chains in the triskelion hub. The sequence boundary of the trimerization site determined by proteolysis is indicated by the dashed line and consists of residues 1490-1587. The domain that may form the globular protrusion on top of triskelia at the vertex (residues 1590-1675) is indicated by the shaded circle. The predicted region that could mediate trimerization (residues 1522-1572) is indicated by an open box. The main predicted coiled coil region in heavy chains (residues 1460-1489) is marked by the box with dark hatchmarks, and the region required for light chain binding, as mapped by proteolysis, is indicated by the box with light hatchmarks. Antibody binding sites for X35, X19, and X22, as mapped by electron microscopy and proteolysis, are indicated in ovals. The dotted line represents a possible folding pattern for the heavy chain that would allow predicted coiled coil regions in the heavy chains (residues 1460-1489 and residues 1151-1178 [lightly stippled box]) to form a helix bundle with the equivalent light chain regions. We have found that residues 1151-1178 are not sufficient for heavy chains to associate with light chains, and although the coiled coil forming potential for these residues is much weaker than for residues 1460-1489 (at least in rat and yeast), we cannot eliminate the possibility that this region is necessary for light chain binding. The position of the light chain and its coiled coil regions are represented by the solid line and the darkly stippled boxes. The solid line in the inset indicates which portion of the triskelion is illustrated in this model. (Reproduced with permission from Cell Press from Näthke et al., 1992).

**References:**

Blank, G S , Brodsky, F M (1986) Site-specific disruption of clathrin assembly produces novel structures. EMBO J *5*, 2097-2095.

Blank, G S , Brodsky, F M (1987) Clathrin assembly involves a light chain-binding region. J Cell Biol *105*, 2011-2019.

Brodsky, F M (1988) Living with clathrin: Its role in intracellular membrane traffic. Science *242*, 1396-1402.

Brodsky, F M , Galloway, C J , Blank, G S , Jackson, A P , Seow, H -F , Drickamer, K , Parham, P (1987) Localization of clathrin light-chain sequences mediating heavy-chain binding and coated vesicle diversity. Nature *326*, 203-205.

Brodsky, F M , Hill, B L , Acton, S L , Näthke, I , Wong, D H , Ponnambalam, S , Parham, P (1991) Clathrin light chains: arrays of protein motifs that regulate coated vesicle dynamics. TIBS *16*, 208-213.

Heuser, J , Steer, C J (1989) Trimeric binding of the 70-kD uncoating ATPase to the vertices of clathrin triskelia: a candidate intermediate in the vesicle uncoating reaction. J Cell Biol *109*, 1457-1466.

Jackson, A P , Seow, H -F , Holmes, N , Drickamer, K , Parham, P (1987) Clathrin light chains contain brain-specific insertion sequences and a region of homology with intermediate filaments. Nature *326*, 154-159.

Kirchhausen, T , Harrison, S C , Chow, E P , Mattaliano, R J , Ramachandran, K L , Smart, J , Brosius, J. (1987a) Clathrin heavy chain: molecular cloning and complete primary sequence. Proc Natl Acad Sci USA *84*, 8805-8809.

Kirchhausen, T , Harrison, S C , Parham, P , Brodsky, F M (1983) Location and distribution of the light chains in clathrin trimers. Proc Natl Acad Sci USA *80*, 2481-2485.

Kirchhausen, T , Scarmato, P , Harrison, S C , Monroe, J J , Chow, E P , Mattaliano, R J , Ramachandran, K L , Smart, J E , Ahn, A H , Brosius, J (1987b) Clathrin light chains LCa and LCb are similar, polymorphic, and share repeated heptad motifs. Science *236*, 320-324.

Lupas, A , Van Dyke, M , Stock, J (1991) Predicting coiled coils from protein sequences. Science *252*, 1162-1164.

Näthke, I S , Heuser, J , Lupas, A , Stock, J , Turck, C W , Brodsky, F M (1992) Folding and trimerization of clathrin subunits at the triskelion hub. Cell *88*, 899-910.

Scarmato, P , Kirchhausen, T (1990) Analysis of clathrin light chain-heavy chain interactions using truncated mutants of rat liver light chain LCB3. J Biol Chem *265*, 3661-3668.

# POLARITY IN NEURONS AND EPITHELIAL CELLS : DISTRIBUTION OF ENDOGENOUS AND EXOGENOUS ION PUMPS AND TRANSPORTERS

G.Pietrini
Department of Cellular and Molecular Physiology
Yale University School of Medicine
333 Cedar St.
06510 CT
USA

The surface membranes of neurons and epithelial cells are divided into distinct domains characterized by markedly different protein compositions. Similar mechanisms may function to generate and maintain this polarity in both systems. Viral glycoproteins which are sorted to the epithelial apical and basolateral surfaces, respectively, have been shown to accumulate in the axonal and dendritic domains of infected neurons (Dotti and Simons, 1990) Thus, the signals and mechanisms responsible for apical or basolateral sorting in epithelia may result in axonal or dendritic targetting in neurons. In order to test this model directly we have examined the sorting behavior of two isoforms of the Na, K-ATPase and of the transporter for GABA in primary culture of hippocampal neurons and in polarized epithelial cells. The Na, K-ATPase is an integral membrane protein complex responsible for establishing the electrochemical gradients of sodium and potassium across the plasma membrane of the mammalian cells. The enzyme is composed of two subunits, alpha and beta, both of which exist as multiple isoforms. Three alpha isoforms, encoded by three different genes, have been cloned from the rat. Polarized epithelial cells express the alpha 1 isoform. In brain, alpha 1 and alpha 2 are expressed both in neuronal and in non-neuronal cells, while alpha 3 is the neuron-specific isoform (Sweadner et al. 1990). The GABA-transporter belongs to the noradrenaline/dopamine transporter gene family whose members are thought to terminate synaptic activity by sodium and cloride dependent reaccumulation of neurotrasmitters into presynaptic termini (Guastella et al., 1990). Indirect immunofluorescence of fully polarized hippocampal neurons carried out with specific monoclonal and polyclonal antibodies raised against two isoforms of the Na,K-ATPase (alpha 1 and alpha 3) and the GABA-transporter polypeptides were

NATO ASI Series, Vol. H 74
Molecular Mechanisms of Membrane Traffic
Edited by D. J. Morré, K. E. Howell, and J. J. M. Bergeron
© Springer-Verlag Berlin Heidelberg 1993

performed. We find that the GABA-transporter is localized exclusively in the axolemma of the cultured neurons while the alpha 1 and alpha 3 isoforms of the sodium pump are expressed in both axons and dendrites. We have expressed the cDNA encoding the GABA-transporter and the neuronal isoform of the sodium pump (alpha 3) in polarized epithelial cells (MDCK or LLC-PK1). Data obtained by immunofluorescence and by GABA transport assay reveal that the functional GABA transporter is accumulated in the apical plasma membrane of MDCK cells. Istead, both the exogenously expressed alpha3 and the endogenous alpha 1 polypeptides exhibit a clear basolateral localization in LLC-PK1 cells, as revealed by confocal analysis. It is known that the sodium pump is associated with cytoskeletal elements that, in epithelial cells, underly the basolateral membrane. When we determined the distribution of the ankyrin in hippocampal neurons we found that a non-isoform specific antibody decorated both axons and dendrites. This result is consistent with the hypothesis that the polarized distribution of the sodium pump is established by a retention mechanism acting at the plasma membrane level. In conclusion our results suggest that these cell types utilize similar sorting mechanisms. In the case of the GABA-transporter the axonal sorting information leads to apical targeting in polarized epithelial cells. Furthermore, the putative retention signals of the sodium pump may lead to the stabilization of the protein basolaterally in the epithelial cells and in both surfaces in neuronal cells.

References

Dotti CG, Simons K (1990) Polarized sorting of viral glycoproteins to the axons and dendrites of hippocampal neurons in culture. Cell 62:63-72

GuastellaJ , Nelson N, Nelson H, Czyzyk L, Keynan S, Miedel MC, Davidson N, Lester HA, Kanner BI (1990) Cloning and expression of a rat brain GABA transporter. Science (Wash DC) 249: 1303-1306

Hammerton RW, Kremiscki KA, Mays RW, Ryan TA, Wollner DA, Nelson WJ (1991) Mechanism for regulating cell surface distribution of Na, K-ATPase in polarized epithelial cells. Science ( Wash DC) 254: 847-850

Sweadner KJ (1989) Isozymes of the Na, K-ATPase. Bioch Biophys Acta 988:185-220

# BIOCHEMICAL PROPERTIES AND EXPRESSION OF TAU PROTEINS IN THE ENDOCYTIC COMPARTMENT DURING LIVER REGENERATION

M.Vergés, *W.H.Evans and C.Enrich
Departamento de Biología Celular
Facultad de Medicina.Universidad de Barcelona.
08028-Barcelona. Spain.

Microtubules are cytoskeletal structures involved in several important cellular functions including cell division and the movement of organelles in the cell. Most of these functions are underlined with the dynamic polymerization or depolymerization capacity of tubulin and many other proteins associated with tubulin, including tau proteins known to actively participate in these processes.

Tau proteins, the low molecular type of microtubule-associated proteins (MAPs), were partially purified from rat liver endocytic fractions. These tau proteins were shown to be low affinity calmodulin-binding proteins (Kakiuchi and Sobue, 1983), heat-stable and soluble in 2.5% perchloric acid. Tau proteins showed the characteristic high degree of charge and molecular weight when analysed by two-dimensional gel electrophoresis (Cleveland et al, 1977). In the present studies, five different antibodies were used to definitively identify the hepatic tau proteins.

Liver regeneration is an excellent model to study the transition of hepatocytes from the quiescent to a proliferative state. In fact, following a partial hepatectomy, the hepatocytes are recruited to G1 phase of the cell cycle and after 18 h the DNA synthesis is initiated; 30 h later the first wave of mitosis is observed.

Changes in the expression of several proteins of the endosomes and in the overall organization of the endocytic compartment have been observed during liver regeneration (Enrich et al. 1992). The expression of tau proteins in the endocytic compartment and in domain specific plasma membranes was studied at different time periods after a partial hepatectomy. Endosomes from 6, 12, 24, 36 and 48 hours of the regenerative period were isolated and their

---

*Natl Inst Med Res
Mill Hill. London
England.

NATO ASI Series, Vol. H 74
Molecular Mechanisms of Membrane Traffic
Edited by D. J. Morré, K. E. Howell, and J. J. M. Bergeron
© Springer-Verlag Berlin Heidelberg 1993

protein profiles were shown to be similar to endosomes prepared from livers of non-operated and sham-operated animals.

Western-blotting, using an anti-tau antibody of these endosome fractions resolved by SDS-PAGE, showed a progressive increase in the amount of tau proteins with a peak expression (a 4-fold increase) occurring 12 h after a partial hepatectomy (pre-replicative phase of the regenerative process). No changes occurred in the endosomes isolated from sham-operated rats. Two-dimensional gel electrophoresis was applied to study in greater detail the changes of tau proteins during liver regeneration using a non-equilibrium pH gel electrophoresis in the first dimension (NEPHGE). The results also showed not only the increased expression of tau protein but also a slightly change in pI, with the tau proteins being more acidic at 12 and 24 hours. This charge change may reflect a different degree of phosphorylation of these tau proteins in regenerating liver. Interestingly, when lateral or sinusoidal plasma membrane fractions from control and regenerating livers were also compared by immunoblotting, no changes in the expression of tau proteins were detected. Thus, during the pre-replicative (G1) and S phases of the hepatocyte's proliferative activation tau proteins in the endocytic compartment are specifically increased. This event is related temporally to a general rearrangement of the endocytic compartment and the general arrest of membrane traffic that will occurs later during the mitosis (Enrich et al.,1992).

Cleveland DW, Hwo S-Y, Kirschner MW (1977) Physical and chemical properties of purified tau factor and the role of tau in microtubule assembly. J Mol Biol 116:227-247

Enrich C, Vergés M, Evans WH (1992) Reorganization of the endocytic compartment during liver regeneration. Exp Cell Res (in press) July 1992.

Kakiuchi S, Sobue K (1983) Control of cytoskeleton by calmodulin and calmodulin-binding proteins. Trends Biochem Sci 2: 59-62

# IDENTIFICATION OF HEPATIC ENDOCYTIC PROTEINS POTENTIALLY INVOLVED IN MEMBRANE TRAFFIC

C.Enrich, M.Vergés and W.H.Evans*
Departamento de Biología Celular
Facultad de Medicina. Universidad de Barcelona.
08028-Barcelona. Spain.

To understand the mechanisms that control membrane traffic along cellular endocytic pathways the isolation of distinct classes of vesicles and the functional characterization of their membrane proteins is necessary. Using a combination of sucrose and Nycodenz density gradients, we have isolated three highly purified endocytic fractions from rat liver homogenates that kinetically corresponds to "early" and "late" endosomes and a "receptor-enriched" fraction (Evans and Flint,1985; Enrich et al. 1990). Since the microtubules are credited to be involved in the transport of endocytic vesicles (Gruenberg and Howell, 1989), the association of various cytoskeletal-related proteins with liver endosomes was thoroughly investigated and the results compared with those obtained with domain identified plasma membranes.

Using Western-blotting and specific antibodies we demonstrated by SDS-PAGE that liver plasma membranes contained actin, α-spectrin, myosin, vinculin, tubulin and MAP-2, whereas these components were not detected in the endocytic membranes. In contrast, endosomes were highly enriched in tau proteins, the low molecular weight type of microtubule-associated proteins.

The properties of these tau proteins in hepatic endosomes were analysed in detail. Three monoclonal and two polyclonal antibodies, that recognised tau proteins in neuronal tissues, were used to immunologically identify these proteins in hepatic endosomes. Furthermore, these tau proteins were shown to be calmodulin-binding, heat-stable and soluble in 2.5% perchloric acid. When analysed by two-dimensional gel electrophoresis, their pI and molecular weight heterogeneity were identical to the tau proteins from brain (Cleveland et al. 1977).

---

*Natl Inst Med Res
Mill Hill. London
England.

NATO ASI Series, Vol. H 74
Molecular Mechanisms of Membrane Traffic
Edited by D. J. Morré, K. E. Howell, and J. J. M. Bergeron
© Springer-Verlag Berlin Heidelberg 1993

The immunolocalization of tau proteins was studied using freshly prepared hepatocytes, Hep G2 and NRK cells and liver tissue. The tau antibodies stained 400-600 nm vesicle-like structures radially arranged in Hep G2 cells and displaying a cytoplasmic punctate staining in NRK cells and hepatocytes. The degree of co-localization of tau proteins and two endocytic markers, the asialoglycoprotein and the mannose 6-phosphate receptors, was studied in Hep G2 cells and in liver tissue, using indirect immunofluorescence and double immuno-gold labeling at the electron microscopy level. The results demonstrate a high degree of co-localization of tau proteins with endocytic structures. The staining of tau proteins was mainly concentrated in the characteristic tubular proyections of endosomes. Although these tau proteins cannot be considered as endosome-resident or endosome-specific proteins, the high concentration in endocytic structures in rat liver and in Hep G2 cells is marked. A role as linkers between microtubules and cellular organelles can be postulated on the basis of the present results.

Cleveland DW, Hwo S-Y, Kirschner MW (1977) Physical and chemical properties of purified tau factor and the role of tau in microtubule assembly. J Mol Biol 116:227-247

Enrich C, Tabona P, Evans WH (1990) Two-dimensional electrophoretic analysis of the proteins and glycoproteins of liver plasma membrane domains and endosomes. Implications for endocytosis and transcytosis. Biochem J 271: 171-178.

Evans WH, Flint N (1985) Subfractionation of hepatic endosomes in Nycodenz gradients and by free-flow electrophoresis. Separation of ligand-transporting and receptor-enriched membranes. Biochem J 232:25-32

Gruenberg J, Howell KE (1989) Membrane traffic in endocytosis: insights from cell-free assays. Ann Rev Cell Biol 5:453-481

# STUDIES ON SEQUENCE REQUIREMENTS FOR BASOLATERAL TARGETING OF THE POLYMERIC IMMUNOGLOBULIN RECEPTOR IN MDCK CELLS

B. Aroeti[1] and K. Mostov

Departments of Anatomy, and Biochemistry and Biophysics, and Cardiovascular Research Institute

University of California

San Francisco, CA 94143-0452

Accurate sorting and targeting mechanisms are essential for the development and maintenance of the spatial asymmetry characterizing most of eukaryotic cells. Proteins are assumed to harbour sorting signals that play critical roles in their sorting and targeting to the final location in the cell. The properties of these signals and the cellular machinery that recognizes them are still largely unknown. A significant progress has been achieved in identifying and characterizing the signals that mediate sorting of plasma membrane proteins into the endocytic pathway (Trowbridge, 1991). These signals, which consist of a Tyr in the context of ß-turn conformation (Bansal & Gierasch, 1991; Eberie et al., 1991) were localized to the cytoplasmic domain of many membrane proteins that enter the cell via coated-pit mediated endocytosis. Cytosolic proteins belonging to the adaptor family were shown to interact with the Tyr signal. These proteins are therefore likely to comprise the cellular machinery that specifically sorts membrane proteins into coated-pits and the endocytic pathway.

Sorting events have been extensively studied in epithelial cells, particularly in the Madin-Darby canine kidney (MDCK) cell line (Mostov et al., 1992). When cultured on permeable filter supports, MDCK cells form a well-polarized monolayer with distinct apical and basolateral plasma membranes. We use the polymeric immunoglobulin receptor (pIgR) expressed in MDCK cells as a model system to study sorting mechanisms. This receptor first travels from the trans-Golgi network to the basolateral surface where it binds its natural ligand, dIgA. The ligand-receptor complex is then endocytosed and transcytosed to the apical surface. It has been also shown that the membrane-proximal 17 residues in the 103 amino-acid carboxy-terminal cytoplasmic domain is responsible for the basolateral targeting of the receptor. Deletion of this segment, leaving the remainder of the cytoplasmic tail fused in frame, resulted in an impaired basolateral phenotype: the receptor was primarily sorted to the apical surface. Strikingly, the 17-mer segment could be transplanted into a normally apical protein and redirect it to the basolateral surface. This implies that the putative basolateral signal is autonomous and dominant. Taken together, these results decrease the likelihood that additional redundant basolateral signals exist in the rest of the tail. Recent studies revealed some degree of homology between sequences in the cytoplasmic tail of the low density lipoprotein receptor (LDLR) and residues in the 17-mer basolateral signal of the pIgR (Yokode et al., 1992). Except for this case, the 17-mer domain shares no obvious homology or structural features in common with cytoplasmic tails of other membrane proteins. Therefore, it is difficult to discern what the basolateral signals are. Site-directed mutagenesis studies have indicated that in at least five cases the Tyr endocytosis signal overlapped with the basolateral targeting signal. Conversely, in the

---

[1]B. Aroeti was supported by a fellowship from the Human Frontiers of Science Program Organization.

NATO ASI Series, Vol. H 74
Molecular Mechanisms of Membrane Traffic
Edited by D. J. Morré, K. E. Howell, and J. J. M. Bergeron
© Springer-Verlag Berlin Heidelberg 1993

case of LDLR, transferrin receptor, and pIgR, mutation of the Tyr internalization signal that blocked endocytosis had no effect on basolateral sorting. One possibility is that, the Tyr endocytosis signal resembles structurally to the basolateral sorting signal. Alternatively, we may simply not found the mutations that separate the two functions.

The major goal of our studies is to identify the exact residues that comprise the basolateral targeting determinant in the cytoplasmic domain of the pIgR. The identification of these residues can directly lead to the analysis of the secondary structure of the signal and to the elucidation of similar sorting signals in other membrane proteins. Our findings encouraged us to systematically mutate the 17-mer segment and resolve the precise residues required for basolateral targeting of the pIgR. An adaptation of both, successive truncations of residues from the C-terminus and alanine-scanning mutagenesis, were utilized to provide high-resolution analysis of the basolateral targeting signal in the 17-mer segment. In the first approach, the basolateral delivery of the pIgR was quantitatively measured upon deletion of single residues from the C-terminus of the 17-mer domain. If the deletion of a residue has no effect on the basolateral targeting phenotype, then it is reasonable to assume that the examined residue is not contained within the sorting epitope. However, deletion of a residue results in a movement of the negatively-charged C-terminus. The basolateral targeting determinant may be sensitive to a negatively charged microenvironment. Thus, even if the deleted amino acid is not part of the sorting determinant, impaired basolateral targeting may still be observed, simply because of the new position of the charged C-terminus relative to the signal. In the second approach, the targeting of the pIgR was determined following systematic replacements of individual amino acids to an Ala. Alanine was chosen as the replacement residue because it eliminates the side chain beyond the ß carbon without imposing severe constraints on secondary structure and tertiary conformation (as can glycine or proline), nor does it impose extreme electrostatic or steric effects. Furthermore, alanine is the most abundant amino acid in proteins and is found frequently in both buried and exposed positions and all variety of secondary structures (Cunningham & Wells, 1989). Therefore, if substitution of a residue for an Ala results in a disturbance of the basolateral targeting phenotype, one can conclude that the R group of the examined residue is a critical entity of the basolateral targeting epitope.

Since each of the mutagenesis techniques may suffer of certain deficiencies, high resolution mapping of the basolateral targeting determinant in the pIgR should be examined by a variety of mutagenesis strategies. Combining of mutagenesis with structure studies (e.g. circular dichroism and 2D NMR) is essential as it will allow us to obtain a more detailed and accurate picture of the features of the basolateral targeting signal.

References
Bansal A & Gierasch LM (1991) The NPXY internalization signal of the LDL receptor adopts a reverse-turn conformation. Cell 67: 1195-1201.
Cunningham BC & Wells JA (1989) High-resolution epitope mapping of hGH-receptor interactions by alanine-scanning mutagenesis. Science 244: 1081-1085.
Eberle W, Sander C, Klaus W, Schmidt B, von Figura K & Peters C (1991) The essential Tyrosine of the internalization signal in lysosomal acid phosphatase is part of a ß turn. Cell 67: 1203-1209.
Mostov K, Apodaca G, Aroeti B, & Okamoto C (1992) Plasma membrane protein sorting in polarized epithelial cells. J. Cell Biol. 116: 577-583.
Trowbridge IS (1991) Endocytosis and signals for internalization. Current Opinion in Cell Biology 3: 634-641.
Yokode M, Pathak RK, Hammer RE, Brown MS, Goldstein JL & Anderson GW (1992) Cytoplasmic sequence required for basolateral targeting of LDL receptor in livers of transgenic mice. J. Cell Biol. 117: 39-46.

# ENDOCYTOSIS BY *TRYPANOSOMA BRUCEI*: PROTEASES AND THEIR POSSIBLE ROLE IN THE DEGRADATION OF ANTI-VSG ANTIBODIES.

Lonsdale-Eccles JD, Grab DJ, Russo DCW, and Webster, P.
ILRAD
PO Box 30709
Nairobi
Kenya

African trypanosomes can evade the immune system of their host by changing their variable surface glycoproteins (VSGs). Although healthy parasites do not normally release VSG (Black *et al.* 1982; Lonsdale-Eccles & Grab, 1986), they can remove surface-bound, VSG-specific antibodies (Seyfang, 1990). We have, therefore, looked for alternative explanations for the disappearance of antibody-VSG complexes such as their possible uptake into the endocytic network of the parasites (Webster, 1989).

The endosomal-lysosomal system of *Trypanosoma brucei* was studied using chemical and temperature perturbations. Chemical agents such as tyloxapol have helped us to distinguish between different organelles but have caused other problems. For example, the lysosomes of trypanosomes isolated from infected rats treated with 300 mg tyloxapol/kg appeared to be destabilised. Furthermore, the activity of the trypanosome cysteine proteases (trypanopains) was induced, by more than 100%. Analysis of the subcellular fractions of

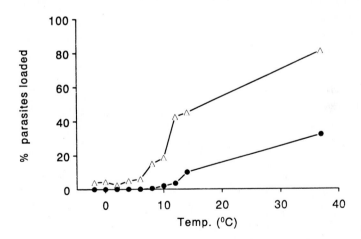

Fig 1 *T. brucei* (600 μl, 2 x 10^7/ml) were incubated at the appropriate temperature for 2 min in RPMI, 5% rat serum and supplements (Lonsdale-Eccles and Grab, 1986). Bovine serum albumin-colloidal gold (BSA-Au; 35 μl of 25 O.D. units) was added. After 5 min 600 μl 1% gluteraldehyde was added. The samples were analysed by electron microscopy (Webster, 1989) and the percentage of cells containing gold in their flagellar pockets (-Δ-) or intracellularly (-●-) was determined.

NATO ASI Series, Vol. H 74
Molecular Mechanisms of Membrane Traffic
Edited by D. J. Morré, K. E. Howell, and J. J. M. Bergeron
© Springer-Verlag Berlin Heidelberg 1993

these tyloxapol-treated trypanosomes showed even greater increases in trypanopain activity; granular (6.2-fold) microsomal (9-fold) and supernatant (20-fold). These results are indicative of a complex interplay with inhibitors and/or activators.

Because of the complications associated with the chemical perturbation, we focussed more on the temperature changes (Fig 1). Below 6°C *T. brucei* exhibited little or no endocytosis of BSA-Au. Between 6°C and 10°C there was a marked increase in the uptake with a more gradual uptake thereafter. BSA-Au is observed in the flagellar pocket before it is seen in the main cell body. These differences have been used in the characterisation of distinct *T. brucei* subcellular organelles separated by centrifugation in Percoll and free flow electrophoresis (Grab *et al.* 1987; Grab *et al.*, unpublished). They have also been exploited to study the uptake of proteins and hydrolysis of peptides (Lonsdale-Eccles *et al.* 1989) as well as the endocytosis of surface bound immunoglobulins.

When *T. brucei* is treated with anti-VSG antibodies at 0°C, antibody is found over the whole surface of the parasites. Upon warming to 37°C the antibodies disappear rapidly, with double exponential decay kinetics, from the surface. The faster rate ($r_1$) probably corresponds to the rate of clearance from the cell body while the slower rate ($r_2$) to the clearance from the flagellum. For Fab, $r_1/r_2 = 18.8$ while for F(ab)$_2$, $r_1/r_2 = 25.6$. Little or none of the antibody that is lost is found inside the parasites. However, if protease inhibitors (50 µg/ml each of leupeptin, E-64, antipain and chymostatin) are included in the incubation medium, massive accumulation of antibody is seen in the parasite endosomal-lysosomal region. We conclude that *T. brucei* can endocytose many molecules, including anti-VSG/VSG complexes, which may be broken down by intracellular proteolysis. This may prove to be an important mechanism by which the parasites mitigate the effect of their host immune responses.

Black SJ, Hewett RS and Sendashonga CN (1982) *Trypanosoma brucei* variant-surface antigen is released by degenerating parasites but not by actively dividing parasites. Parasite Immunol 4:233-244

Grab DJ, Webster P, Ito S, Fish WR, Verjee Y and Lonsdale-Eccles JD (1987) Subcellular localisation of a variable surface glycoprotein phosphatidylinositol-specific phospholipase-C in African trypanosomes. J Cell Biol 105:737-746

Lonsdale-Eccles JD and Grab DJ (1986) Purification of African trypanosomes can cause biochemical changes in the parasites. J Protozool 34:405-408

Lonsdale-Eccles JD, Mpimbaza GWN, Verjee Y and Webster P (1989) Proteolysis and endocytosis in *Trypanosoma brucei*. In "Protein traffic in parasites and mammalian cells" Lonsdale-Eccles JD ed. ILRAD, Nairobi, pp 94-97

Mbawa ZR, Gumm ID, Shaw E and Lonsdale-Eccles JD (1992) Characterisation of a cysteine protease from bloodstream forms of *Trypanosoma congolense*. Eur J Biochem 204: 371-379

Seyfang A, Mecke D and Duszenko M (1990) Degradation, recycling and shedding of *Trypanosoma brucei* variant surface glycoprotein. J Protozool 37:546-552

Webster P (1989) Endocytosis by African trypanosomes. I Three-dimensional structure of the endocytic organelles in *Trypanosoma brucei* and *T. congolense*. Eur J Cell Biol 49:295-302

# RECEPTOR MEDIATED ENDOCYTOSIS OF EGF IN A CELL FREE SYSTEM

T. E. Redelmeier, C. J. Lamaze and S. L. Schmid
Department of Cell Biology
The Scripps Research Institute
La Jolla, CA. 92037.

## Introduction

Endocytosis of many essential nutrients and growth factors is mediated by transmembrane receptors which bind to coated pits and are internalized via clathrin coated vesicles (for a recent review see Smythe and Warren). The receptors can be classified on the basis of whether ligand binding is required to promote efficient internalization. For example, the nutrient hormone receptors LDL and Tfn are internalized efficiently in the absence of ligands whereas EGF stimulates the endocytosis of the EGF receptor. These observations imply that the factors responsible for the efficient endocytosis of receptors maybe different for different classes of receptors.

Recent work in our laboratory has indicated that adaptors are involved in the sequestration of the Tfn receptor into clathrin coated pits (Smythe et al., 1992). Using cell free assays which measure the ATP, temperature and cytosol dependent internalization of the Tfn receptor into plasma membrane derived elongated coated pits, Smythe and co-workers have established that purified adaptors stimulate the sequestration of labelled Tfn into structures which are inaccessible to added anti-Tfn antibodies but accessible to the low molecular weight reducing agent Mesna. These results provide further evidence to support the hypothesis that adaptors mediate the binding of receptors (perhaps via internalization signals residing in the cytoplasmic domain of the receptor) to clathrin coated pits (Pearse and Robinson, 1990).

It is presently less clear whether adaptors also mediate the binding of the second class of receptors to clathrin coated pits. We report here a modified approach to measuring endocytosis of ligands into cells. This assay uses an ELISA based detection system for monitoring the sequestration of ligands into elongated coated pits (accessible to Mesna but inaccessible to Avidin) or coated vesicles which are inaccessible to both reagents. We use this assay to compare the internalization of EGF and Tfn in semi intact cells.

NATO ASI Series, Vol. H 74
Molecular Mechanisms of Membrane Traffic
Edited by D. J. Morré, K. E. Howell, and J. J. M. Bergeron
© Springer-Verlag Berlin Heidelberg 1993

## Results and Discussion

Figure 1: ATP and Cytosol Dependent Sequestration of BSS-EGF into Elongated Coated Pits

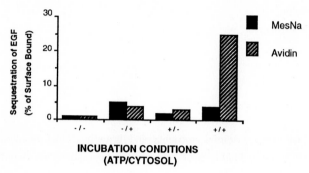

BSS-EGF internalization was assayed as described in Smythe *et. al.*, 1991.

The results presented in Figure 1 demonstrate that the ELISA based assay can be readily used to measure internalization of biotinylated EGF in semi-intact cells. Sequestration into deeply invaginated coated pits (Avidin) is temperature, time, cytosol and ATP dependent. The EGF receptors are not efficiently internalized into coated vesicles i.e. structures inaccessible to the low molecular weight reducing agent MesNa. This is consistent with the earlier suggestions by Smythe and co-workers (1991) that only receptors present in the elongated coated pits at the time of semi-intact cell preparation will efficiently be incorporated into coated vesicles. However, in contrast to Tfn internalization (Smythe *et al.*, 1992) EGF internalization is not stimulated by bovine brain adaptors. We are currently investigating the molecular basis for this observation.

## REFERENCES

Pearse, B.M.F. and Robinson, M.S. (1990) Annu Rev. Cell Biol. 6: 151-171.
Smythe, E.M. and Warren, G. (1991) Eur. J. Biochem 202: 689-699.
Smythe, E.M., Redelmeier, T.E. and Schmid, S.L. (1992) Methods in
    Enzymology (in press)
Smythe, E.M., Carter, L.L. and Schmid, S.L. (submitted to J.C.B.)

# Transferrin on the Basolateral Surface Regulates Apical $^{59}$Fe Uptake and Transport Across Intestinal Epithelial Cells

X. Alvarez-Hernandez and J. Glass
Department of Medicine and Center for Excellence in Cancer Research
Louisiana State University Medical Center
1501 Kings Hwy. P.O. Box 33932
Shreveport LA 71130-3932

Although the physiology of iron uptake across the intestinal epithelium has been well described, many of the molecular mechanisms of iron transport remain to be elucidated. Measurements of iron uptake using whole animals or various gut sac preparations have defined three phases of iron uptake: an uptake phase in which iron enters the apical or lumenal surface; an intracellular phase in which iron is either transported to the basolateral surface for subsequent delivery to the plasma or is stored to be lost when the epithelial cells are sloughed; and a transfer phase by which iron is released from the epithelium to the plasma (Conrad, M.E.). Regulation of iron absorption is at the point of entry, the intestinal cell; there is no excretory regulation. Kinetic analyses of iron uptake suggest that regulation occurs at either the uptake or intracellular phases (Nathanson, M.H. and McLaren, G.D.) or at the transfer phase as we have suggested previously (Alvarez, X. et. al.). The molecular details of the regulatory mechanisms are not known.

We have recently described the Caco-2 cell line grown in bicameral chambers as a model for intestinal iron uptake (Alvarez, X. et. al.). The utility of this model is that it enables the uptake, intracellular, and transfer phases of iron uptake to be studied with greater ease than with animal or organ models. Transport of iron in the Caco-2 system responds to the iron status of the cell with iron deficient cells increasing and iron loaded cells decreasing transport into the basolateral chamber. In the studies presented here we have extended the model to investigate the effect of additions of the plasma proteins transferrin and bovine serum albumin to the basal chamber, studying both the uptake of $^{59}$Fe and the transport of $^{59}$Fe into the basal chamber as well as the intracellular

NATO ASI Series, Vol. H 74
Molecular Mechanisms of Membrane Traffic
Edited by D. J. Morré, K. E. Howell, and J. J. M. Bergeron
© Springer-Verlag Berlin Heidelberg 1993

distribution of $^{59}$Fe. The effect of transferrin (Tf), on iron (Fe) transport including both the intracellular distribution of $^{59}$Fe and $^{59}$Fe transport across intestinal epithelial cell monolayers was analyzed. Caco-2 cells with three different Fe statuses, deficient (FeD;[Fe]<0.1uM), normal (FeN;[Fe]=1.1uM) and loaded (FeH;[Fe]=65uM) were grown on porous membranes in bicameral chambers as described previously. The handling of $^{59}$Fe, offered to the apical side as 1uM $^{59}$Fe-ascorbate, in FeD cells was markedly affected by additions to the basal chamber with $22.2 \pm 3.0 \times 10^4$, $8.2 \pm 0.6 \times 10^4$, and $2.7 \pm 0.4 \times 10^4$ atoms $^{59}$Fe/cell/min transported across the cell monolayer into the basal chamber in the presence of apoTf, BSA, and no added protein respectively. Unexpectedly and paradoxically, total $^{59}$Fe uptake (cellular and transported $^{59}$Fe) by FeD was decreased by basolateral Tf with uptake in FeD cells of $2.6 \times 10^8$ with Tf, $4.8 \times 10^8$ with BSA and $4.8 \times 10^8$ with no basal additions (atoms/cell). HPLC analysis with tandem Zorbax 450-250 columns of the transported $^{59}$Fe demonstrated $^{59}$Fe either on Tf or BSA when either were present in the basal chamber or, in the absence of additions, both on high and low MW moieties corresponding to no known Fe binding protein. Analysis of intracellular $^{59}$Fe by IEF-gels of cell lysates demonstrated $^{59}$Fe migrating both with a very basic pI and with the pIs of ferritin (Ft) at a ratio of 200:1 in FeD cells. The presence of Tf in the basolateral side further decreased the small amount of $^{59}$Fe in Ft. The $^{59}$Fe distribution in FeN and FeH cells was about 5:1 (basic $^{59}$Fe:$^{59}$Fe-Ft); in addition in the FeH cells $^{59}$Fe was also found spread at pIs more basic than Ft as if the Ft capacity had been exhausted. These studies demonstrate that basolateral Tf affects both the apical uptake of Fe and the transport of Fe across intestinal epithelium. The latter effect is seen even when cellular content of ferritin is high.

This work was supported in part by grant DK-41279 from the National Institutes of Health and the Center for Excellence in Cancer Research, Louisiana State University Medical Center.

Alvarez-Hernandez, X , Nichols, G M and Glass, J (1991) Biochim Biophys Acta 1070: 205-208.

Conrad, M E (1987) In: Physiology of the Gastrointestinal Tract. Johnson L.R. (Ed.), Raven Press, New York. 1437-1453.

Nathanson, M H and McLaren, G D (1987) J Nutr 117:1067-1075.

# INTRACELLULAR TRAFFICKING OF *Salmonella typhimurium* WITHIN HeLa EPITHELIAL CELLS

Francisco Garcia-del Portillo and B. Brett Finlay
Biotechnology Laboratory, University of British Columbia
6174 University Boulevard, Vancouver, B.C.V6T 1Z3
CANADA

*Salmonella typhimurium* is an intracellular pathogen which penetrates (invades) and multiplies within epithelial cells (Finlay and Falkow, 1988, 1989). Unlike some intracellular pathogens, *S. typhimurium* remains within a vacuole (Finlay and Falkow, 1988, 1989). Following internalization this vacuole migrates to the perinuclear region, and after a lag period of 2-3 h, bacteria initiate intracellular multiplication (Finlay and Falkow, 1989). We are currently characterizing the intracellular trafficking of bacteria-containing vacuoles.

## Localization of CI-M6PR in *S. typhimurium*-infected HeLa epithelial cells

Indirect immunofluorescence studies using a polyclonal antibody to the mannose 6-phosphate receptor (CI-M6PR) (Griffiths et al. 1988, Kornfeld and Mellman, 1989) demonstrate that CI-M6PR does not colocalize with bacteria at any time. These results would indicate that in infected epithelial cells *S. typhimurium* does not reach any compartment(s) that harbors this receptor or passes through it transiently. Colocalization of bacteria and CI-M6PR was not observed in other epithelial cell lines, such as MDCK, nor with other intracellular bacteria that reside in vacuoles, such as *Yersinia enterocolitica*.

## Presence of lysosomal membrane glycoproteins (lgps) in *S. typhimurium*-containing vacuoles

Indirect immunofluorescence assays with different anti-lgp antibodies have shown that *S. typhimurium* does not reach any lgp-containing compartment until 2 h after invasion, in contrast to the rapid lysosomal targeting of *Y. enterocolitica*.

NATO ASI Series, Vol. H 74
Molecular Mechanisms of Membrane Traffic
Edited by D. J. Morré, K. E. Howell, and J. J. M. Bergeron
© Springer-Verlag Berlin Heidelberg 1993

Interestingly, at 4 h post-infection, *S. typhimurium* appears in lgp-containing vacuoles, and long lgp-labeled tubules, with periodic swellings, emerge from the intracellular bacteria. These tubules are never seen in uninfected cells or in HeLa cells infected with *Y. enterocolitica*. We further characterized the nature of these tubules, and preliminary data indicate that they are not labelled with lucifer yellow, even after 90 min, which would label the lysosomal compartment. We have also determined that *Salmonella*-induced tubules are microtubule associated, much like the brefeldin A-induced and macrophage tubular lysosomes (Lippincot-Schwarz et al. 1991, Swanson et al. 1987).

## Conclusions

Upon internalization within HeLa epithelial cells *S. typhimurium* is not colocalized with cellular compartments containing CI-M6PR. Bacteria-containing vacuoles have no lgps (Fukuda, 1991) during the first 2 h post-infection, and at approximately 4 h *S. typhimurium* induces the formation of lgp-containing tubules, that are microtubule associated but morphologically different to BFA-induced tubular lysosomes.

## References

Finlay BB, Falkow S (1988) Comparison of the invasion strategies used by *Salmonella cholerae-suis*, *Shigella flexneri* and *Yersinia enterocolitica* to enter cultured animal cells: endosome acidification is not required for bacterial invasion or intracellular replication. Biochimie 70:1089-1099
Finlay BB, Falkow S (1989) *Salmonella* as an intracellular parasite. Mol Microbiol 3:1833-1841
Fukuda M (1991) Lysosomal membrane glycoproteins. J Biol Chem 266:21327-21330
Griffiths G, Hoflack B, Simons K, Mellman I, Kornfeld S (1988) The mannose 6-phosphate receptor and the biogenesis of lysosomes. Cell 52:329-341
Kornfeld S, Mellman I (1989) The biogenesis of lysosomes. Annu Rev Cell Biol 5:483-525
Lippincott-Schwartz J, Yuan L, Tipper C, Amherdt M, Orci L, Klausner, RD (1991) Brefeldin A's effects on endosomes, lysosomes, and the TGN suggest a general mechanism for regulating organelle structure and membrane traffic. Cell 67: 601-616
Swanson J, Bushnell A, Silverstein SC (1987) Tubular lysosome morphology within macrophages depend on the integrity of cytoplasmic microtubules. Proc Natl Acad Sci USA 84:1921-1925

# FUNCTIONAL ANALYSIS OF DYNAMIN, A GTPASE MEDIATING EARLY ENDOCYTOSIS.

H.S. Shpetner, C.C. Burgess, and R.B. Vallee
Cell Biology Group
Worcester Foundation for Experimental Biology
Shrewsbury, MA 01545

Dynamin was initially identified as a nucleotide-dissociable 100 kD microtubule-binding protein in calf brain cytosol, distinct from cytoplasmic dynein and kinesin (Shpetner and Vallee, 1989). Initial biochemical studies indicated a likely nucleotidase activity, and subsequent molecular cloning in rat indicated that the N-terminal 300 amino acids contained three consensus sequence elements for GTP-binding (Obar et al., 1990). No other homology was found with either ras proteins, the $\alpha$-subunits of G proteins, or any of the GTPases associated with either protein translation or protein translocation into the endoplasmic reticulum. However, extensive homology has been found with three other GTP-binding proteins: the mammalian Mx proteins ($M_r$ 72-80 kD; Horisberger et al., 1990), involved in mediating the interferon-induced reponse to viral infection, the product of the yeast VPS1 gene ($M_r$ 79 kD; Rothman et al, 1990), involved in protein sorting from the Golgi body to the vacuole, and the product of the yeast MGM1 gene ($M_r$ 94 kD; Jones and Fangman, 1992), involved in replication of the mitochondrial genome.

Molecular analysis of Drosophila head cDNAs identified multiple isoforms ($M_r$ = 94-99 kD) of a protein homologous with rat brain dynamin (Chen et al., 1991; van der Bliek and Meyerowitz, 1991). Genetic analysis indicated that the the fly proteins are the products of the shibire gene. Mutations in this gene were initially found to exhibit temperature-sensitive defects in synaptic transmission, due to a failure to recycle synaptic vesicles (Poodry and Edgar, 1979). Subsequently, the shibire gene product was implicated in the budding of clathrin-coated vesicles from the plasma membrane in a variety of cell types (Kosaka and Ikeda, 1983; Kessel et al., 1989). Thus, dynamin appears to play a general role in the very early events of endocytosis.

To begin to elucidate the mechanism of action of dynamin, we have begun to analyze the functional properties of the protein both in vitro and in vivo. Purified bovine brain dynamin exhibits very high steady state GTP turnover rates (0.04-0.20 sec$^{-1}$), compared to other GTPases, indicating a possible mechanochemical function (Shpetner and Vallee, 1992).

NATO ASI Series, Vol. H 74
Molecular Mechanisms of Membrane Traffic
Edited by D. J. Morré, K. E. Howell, and J. J. M. Bergeron
© Springer-Verlag Berlin Heidelberg 1993

The rate of steady-state GTP hydrolysis was found to be further accelerated up to 75-fold by microtubules, to rates of 1.5-6.1 $sec^{-1}$. Maximal activation was seen at low microtubule concentrations (~0.1 mg/ml), and neither tubulin dimers, other polyanions (f-actin, vimentin filaments, or polyglutamic acid), nor clathrin coat proteins appreciably stimulated GTPase activity, suggesting that microtubules specifically modulate dynamin GTPase activity. Nevertheless, because a role for microtubules in early endocytosis has not been identified, it is possible that microtubules are mimicking a physiological activator of dynamin GTPase activity yet to be identified.

To identify the intracellular structures with which dynamin interacts, we have transiently transfected COS-7 cells with constructs encoding wild-type or mutant forms of the protein. Cells transfected with constructs encoding the wild-type protein exhibited a diffuse, cytoplasmic dynamin immunofluorescence staining pattern, and both clathrin and α-adaptin exhibited a uniform, punctate distribution, as seen in untransfected cells. In contrast, cells transfected with mutant dynamin constructs that either contained point mutations in the GTP-binding domain, or lacked this domain entirely, exhibited punctate and apparently tubular dynamin-containing structures. In many of the mutant cells, clathrin- and α-adaptin-reactive spots were found to be strikingly clustered. Some of the clathrin- and α-adaptin-containing structures co-localized with the mutant dynamin. The distribution of γ-adaptin was unaffected in these cells, indicating that dynamin specifically interacted with plasma membrane-derived coated vesicles. These findings strongly suggest that, during its GTPase cycle, dynamin interacts with clathrin-coated structures associated with the plasma membrane, and further studies will focus on clarifying the precise role of this interaction in the mechanism of action of dynamin.

Chen, M.S., Obar, R.A., Schroeder, C.S., Austin, T.W., Poodry, C.A., Wadsworth, S.C. and Vallee, R.B. (1991) Nature 351, 583-586.

Horisberger, M.A., McMaster, G.K., Zeller, H., Wathelet, M.G., Dellis, J. and Content, J. (1990) J. Virol. 64, 1171-1181.

Jones, B. A. and Fangman. W.L. (1992) Genes & Develop. 6, 380-389.

Kessel, I. Holst, B.D. and Roth, T.F. (1989) Proc. Natl. Acad. Sci. U.S.A. 86, 4968-4972.

Kosada, T. and Ikeda, K. (1983) J. Cell Biol. 97, 499-507.

Obar, R., Collins, C.A., Hammarback, J.A., Shpetner, H.S. and Vallee, R.B. (1990) Nature 347, 256-261.

Rothman, J.H., Raymond, C.K., Gilbert, T., O'Hara, P.J. and Stevens, T.H. (1990) Cell 61, 1063-1074.

Poodry, C.A. and Edgar, L. (1979) J. Cell Biol. 81, 520-527.

Shpetner, H.S. and Vallee, R.B. (1989) Cell 59, 421-432.

Shpetner, H.S. and Vallee, R.B. (1992) Nature 355, 733-735.

van der Bliek, A.M. and Meyerowitz, E.M. (1991) Nature 351, 411-414.

# IDENTIFICATION OF A REGULATORY DOMAIN IN DYNAMIN

J.S. Herskovits, C.C. Burgess, and R.B. Vallee
Cell Biology Group
Worcester Foundation for Experimental Biology
222 Maple Avenue
Shrewsbury, MA 01545
U.S.A.

Our laboratory has identified dynamin as a 100 kD GTP-binding protein which binds to microtubules in a nucleotide-sensitive manner and possesses a potent microtubule-stimulated GTPase activity (Shpetner and Vallee, 1989; 1992). Molecular cloning of rat brain dynamin revealed two striking features in the deduced primary sequence of the molecule: (1) an N-terminal 300 amino acid GTP-binding domain and (2) a C-terminal 100 amino acid basic, pro-rich region (Obar et al, 1990). Recent work has indicated that dynamin is the apparent homologue of the product of the *Drosophila* gene *shibire*[ts] (Chen et al, 1991; van der Bliek and Meyerowitz, 1991). *shibire*[ts] flies exhibit temperature-sensitive paralysis (Griglatti et al, 1973),which is due to a block in coated and noncoated vesicle budding from the plasma membrane (Kosada and Ikeda, 1983). Therefore, dynamin is thought to play a key role in the endocytic pathway.

To define the microtubule-binding region of dynamin, we exposed it to a number of proteases. Papain digestion of purified calf brain dynamin generated a stable 90 kD fragment which was unable to bind microtubules, while an accompanying ~8 kD fragment retained microtubule-binding activity. This fragment was localized to the C-terminus of dynamin using a polyclonal antipeptide antibody which we raised against the 21 C-terminal residues of the rat brain sequence. Consistent with the removal of the microtubule-binding domain, papain digestion destroyed the microtubule-stimulated GTPase activity, but had no effect on the basal GTPase level.

Intriguingly, we find that both rat and fly C-terminal domains contain sequences that resemble a GTP-binding element important in the modulation of *ras* and EF-Tu GTPase activity via guanine nucleotide release proteins (GNRPs) (Bourne et al, 1991). We conclude that the C-terminal region of dynamin is an important regulatory domain, though, in view of the apparent role of the protein in the earliest steps of endocytosis, we are uncertain whether it interacts with microtubules *in vivo*. In addition, the effect of the dynamin C-terminus on GTPase activity leads us to speculate that in the folded molecule it is in close spatial proximity to the amino terminal GTP-binding region.

## Functional Sites Near Dynamin's C-Terminus

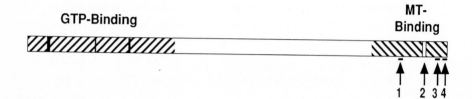

Figure 1. Schematic diagram of dynamin. The 300 N-terminal residues contain the three GTP-binding consensus elements. The basic, pro-rich C-terminus contains the microtubule(MT)-binding domain. Functional sites designated as follows: (1) the predicted location of the papain cleavage site; (2) the putative C-terminal GTP-binding element PPPAGSAL; (3) a peptide splice site found in *Drosophila*; and (4) the region used for the immunogenic peptide.

Bourne, H.R., Sanders, D.A., and McCormick, F. (1991) Nature 349, 117-127.

Chen, M.S., Qbar, R.A., Schroeder, C.S., Austin, T.W., Poodry, C.A., Wadsworth, S.C., and

Vallee, R.B. (1991) Nature 351, 583-586.

Grigliatti, T.A. Hall, L., Rosenbluth, R., and Suzuki, D.T. (1973) Molec. gen. Genet. 120, 107-114.

Kosada, T. and Ikeda, K. (1983) J. Cell Biol. 97, 499-507.

Obar, R.A., Collins, C.A., Hammarback, J.A., Shpetner, H.S. and Vallee, R.B. Nature 347, 256-261 (1990)

Shpetner, H.S. and Vallee, R.B. (1989) Cell 59, 421-432.

Shpetner, H.S. and Vallee, R.B. (1992) Nature 355, 733- 735.

van der Bliek, A.M. and Meyerowitz, E.M. (1991) Nature 351, 411-414.

# POTASSIUM DEPLETION STIMULATES CLATHRIN-INDEPENDENT ENDOCYTOSIS IN RAT FOETAL FIBROBLASTS.

Ph. Cupers, A. Kiss, P. Baudhuin and P.J. Courtoy
Cell Biology Unit, University of Louvain Medical School and
International Institute of Cellular and Molecular Pathology
Avenue Hippocrate, 75
1200 Brussels
Belgium

Evidence that pinocytosis may proceed through clathrin-coated and clathrin-independent pits and vesicles largely relies on perturbations, such as potassium depletion, which block the formation of endocytic clathrin-coated pits (discussed in Hansen et al., 1991, J. Cell Biol., 113, 731-741). The clathrin-independent pathway is then characterized by residual endocytosis. However, the possibility that this pathway is also affected by the perturbation has received little attention so far. We report here that potassium depletion enhances fluid-phase endocytosis by the clathrin-independent pathway in rat fibroblasts.

In rat foetal fibroblasts, the procedure of Madshus et al. (J. Cell Physiol., 1987, 131, 6-13) resulted in a 96 % depletion of intracellular potassium. By morphometry, both number and size of clathrin-coated round pits and vesicles were reduced by potassium depletion so that their cumulative surface and volume were decreased by ~20-fold. Treated cells also showed a 10-fold decrease of intracellular $^{125}I$-transferrin accumulation and a two-fold increase of cell surface-bound transferrin, implying an actual 20-fold inhibition of transferrin endocytosis. By ultrastructural cytochemistry, short pulses (15-30 sec) of horseradish peroxidase (HRP) uptake resulted in labelling of clathrin-coated and non clathrin-coated vesicles in control cells and a similar number of small, exclusively non clathrin-coated vesicles in potassium-depleted cells, indicating that fluid-phase uptake was not blocked by the treatment.

The usual measurements of accumulation of fluid-phase tracers underestimate the rate of endocytosis, in proportion of regurgitation (Besterman et al., 1981, J. Cell Biol., 91, 716-727). Indeed, accumulation of HRP by control and treated cells did not proceed linearly with time and decreased as early as after 4 to 5 min, owing to the existence of an early endosomal compartment open to regurgitation. Direct evaluation of the rate of fluid-phase endocytosis by HRP accumulation during short pulses (1 to 4 min) was $273 \pm 29$ nl/h/mg cell protein (n=4; Fig. 1a). Assuming that clathrin-coated pits cover ~1.2 % of the pericellular membrane and have a lifetime of ~1 min (Hansen et al., 1992, Exp. Cell Res., 199, 19-28), and on the basis of our morphometric data on rat foetal fibroblasts (pericellular surface = 2100 $\mu m^2$; average surface of a clathrin-coated particle = 0.0748 $\mu m^2$; average volume of a clathrin-coated particle = 0.0021 $\mu m^3$), it can be computed that the clathrin-coated pathway accounts for ~50 % of endocytic fluid entry in control cells. Potassium-

NATO ASI Series, Vol. H 74
Molecular Mechanisms of Membrane Traffic
Edited by D. J. Morré, K. E. Howell, and J. J. M. Bergeron
© Springer-Verlag Berlin Heidelberg 1993

depleted cells internalized HRP at the same rate as in control cells during pulses of 1-4 min (296 ± 29 nl/h/mg cell protein, n=4; Fig. 1a).

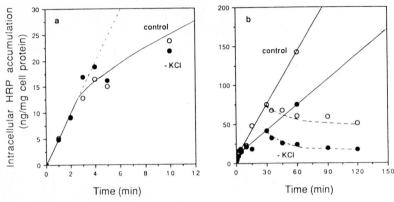

**Figure 1 : Kinetics of peroxidase accumulation**
Control (open symbols) and potassium-depleted (closed symbols) rat foetal fibroblasts were incubated at 37°C in simplified medium containing 1 mg/ml HRP, in presence or in absence of 10 mM KCl. Cells were extensively washed at 4°C and peroxidase activity was assayed in the lysates. For short pulses (a), the initial slopes indicate a similar rate of fluid-phase endocytosis in control and in potassium-depleted cells. Longer pulses (b) demonstrate the decrease of accumulation and the acceleration of regurgitation (dotted lines) in treated cells.

For longer uptake (15-120 min) at 1 mg/ml HRP, the rate of fluid-phase endocytosis was estimated by pulse-chase experiments, as the sum of accumulation and regurgitation. In control cells, HRP accumulation amounted to 155 ± 19 ng/h/mg cell protein (n=4) and its release, estimated by the initial slope of chase, was 91 ± 20 ng/h/mg cell protein (Fig. 1b). This can be translated into a fluid-phase endocytosis rate of 246 ± 15 nl/h/mg cell protein (n=4), close to the direct measurements using short pulses. Release was no longer detectable after 1 h of chase, indicating that HRP remaining in the cell after this interval (~80 % of accumulated tracer) had been irreversibly sequestered. Regurgitation from the endosomal compartment fitted an exponential decay (half-life of 12 min). In potassium-depleted cells, HRP accumulation was decreased by half, at 80 ng/h/mg cell protein (n=2), a phenomenon easily explained by a ~two-fold enhancement of the rate of release (150 ng/h/mg cell protein, n=2; Fig. 1b). Again, this implies that fluid-phase endocytosis was in fact very close to that of control cells (230 nl/h/mg cell protein). Irreversible sequestration occurred after 30 min of chase, at about 55 % of accumulated HRP, and the half-time for endosomal clearance by regurgitation was decreased to 7 min.

In conclusion, intracellular potassium depletion in rat foetal fibroblasts blocks the formation of endocytic clathrin-coats and receptor-mediated endocytosis of transferrin to the same extent, indicating that coat formation is a rate-limiting step in transferrin entry, but has no appreciable effect on fluid-phase endocytosis. Since clathrin-coated pits normally contribute to one-half of pinocytosis in rat foetal fibroblasts, potassium-depleted cells must compensate for the loss of the clathrin-dependent pathway by stimulating their clathrin-independent pathway. In turn, this indicates the possibility that clathrin-independent endocytosis can be regulated.

# CLATHRIN INTERACTION WITH NADH DEHYDROGENASES OF RAT LIVER PLASMA MEMBRANE.

P. Navas, J.M. Villalba, J.C. Rodríguez-Aguilera, A. Canalejo and M.I. Burón

*Departamento de Biología Celular. Facultad de Ciencias. Universidad de Córdoba. 14004 Córdoba, Spain.*

Eukaryotic cells contain a variety of transplasma membrane redox activities some of which have been related to control of cell growth (Crane et al.,1985). Both iron-containing oxidants and ascorbate free radical (AFR) are reduced by the transplasma membrane electron transport and stimulate cell growth (Sun el al., 1985; Alcaín et al., 1990). These two activities do not necessarily represent the same via of electron transport. In fact, we show here different properties between NADH-AFR reductase and NADH ferricyanide (FeCN) reductase.

## Role of clathrin on NADH dehydrogenases

NADH-AFR reductase of rat liver plasma membranes obtained by two-phase partitioning was 80% inhibited by anti-clathrin antibody (Fig 1), but NADH-FeCN reductase was unaffected. Clathrin-depleted membranes (Schook et al. 1979) showed a 95% inhibition of NADH-AFR reductase (Table 1), but the bulk of NADH-FeCN reductase was only slightly sensitive to clathrin removal.

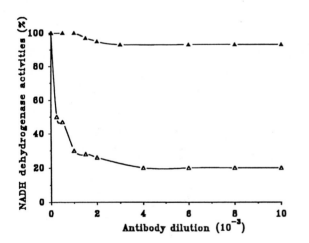

Fig.1.- Effect of anticlathrin antibody on NADH dehydrogenases.

NADH-AFR reductase was recovered in reconstituted membranes up to 80% with $Ca^{++}$, 78% with $K^+$, and only 63% with $Mg^{++}$ (Table 1).

## Peculiarities of the NADH dehydrogenases

NADH-AFR reductase activity was reduced by solubilization and was not detected in the supernatants. NADH-FeCN reductase was solubilized from membranes with up to 98% of the original activity being recovered in the supernatants (Table 1).

With WGA lectin NADH-AFR reductase was inhibited up to 95% but the NADH-FeCN reductase was insentitive.

NATO ASI Series, Vol. H 74
Molecular Mechanisms of Membrane Traffic
Edited by D. J. Morré, K. E. Howell, and J. J. M. Bergeron
© Springer-Verlag Berlin Heidelberg 1993

## Discussion

Both NADH-AFR-reductase (Navas et al, 1988) and NADH-FeCN reductase (Goldenberg et al, 1979) activities have been identified as parts of the transplasma membrane redox system. NADH-AFR reductase associated to clathrin-coated membranes could be related to energy production via generation of a membrane potential as it has been shown for coated vesicles (Morré et al 1987) and plant plasma membrane (González-Reyes et al., 1992).

In contrast, NADH-FeCN reductase does not show the same propierties than NADH-AFR reductase, being insensitive to both clathrin removal and WGA lectin and being solubilized by chaps.

We then show here a clear difference between both dehydrogenases activities, although the mechanisms implicated in the transplasma membrane electron flow remain uninvestigated.

Supported by DGICYT Spanish grant no. PB89-0337-CO2-01.

Table 1. Properties of NADH dehydrogenase activities of rat liver plasma membranes. Specific activities are expressed as nmoles of NADH oxidized/min/mg protein. Brackets: percentage of non treated membranes. N.D.: non detected. n=4.

| Fractions/ Additions | NADH-AFR reductase | NADH-FeCN redcutase |
|---|---|---|
| Plasma membrane: | $14.5 \pm 0.1$ | $600 \pm 30$ |
| - Clathrin-depleted | $0.6 \pm 0.01$ (4) | $578 \pm 20$ (96) |
| - Clathrin-enriched supernatants | N.D. | N.D. |
| Recontituted membranes: | | |
| + $Ca^{++}$ (25 mM) | $11.6 \pm 0.08$ (80) | $603 \pm 36$ (100) |
| + $K^+$ (100 mM) | $11.3 \pm 0.09$ (78) | $580 \pm 40$ (97) |
| + $Mg^{++}$ (25 mM) | $9.2 \pm 0.04$ (63) | $592 \pm 31$ (98) |
| Chaps 0.1% | | |
| a) pellets | $6.67 \pm 0.05$ (46) | $168 \pm 17$ (28) |
| b) supernatants | $0.02 \pm 0.008$ (0) | $432 \pm 25$ (72) |
| WGA lectin | $0.78 \pm 0.06$ (5) | $525 \pm 51$ (88) |

## References

Alcaín FJ, Burón MI, Rodríguez-Aguilera JC, Villalba JM, and Navas P (1990) Cancer Res 50:5887-5891

Crane FL, Sun IL, Clarck MG, Grebing C and Löw,H (1985) Biochim Biophys Acta 811:233-264

Goldenberg M, Crane FL and Morré DJ (1979) J Biol Chem 254:2491-2498

González-Reyes JA, Döring O, Navas P, Obst G and Böttger M (1992) Biochim Biophys Acta 1098:177-183

Morré DJ, Crane FL, Sun IL and Navas P (1987) Ann NY Acad Sci 498:153-171

Navas P, Estévez A, Burón MI, Villalba JM and Crane,FL, (1988) Biochem Biophys Res Commun 154:1029-1033

Schook W, Puszkin S, Bloom V, Ores C and Kochura S (1979) Proc Natl Acad Sci USA 76:116-121

Sun IL, Crane FL, Grebing C and Löw H (1985) Exp Cell Res 156:528-536

# PUTATIVE ADAPTOR PROTEINS OF CLATHRIN COATED VESICLES FROM DEVELOPING PEA

Juliet M. Butler and Leonard Beevers
Department of Botany and Microbiology
University of Oklahoma
770 Van Vleet Oval
Norman, OK 73019

Membrane bound vesicles with an external lattice of the protein clathrin appear to be a ubiquitous cellular transport motif in eukaryotic cells. In plants, roles for coated vesicles in endocytosis (Tanchak et. al 1984) and the intracellular transport of storage proteins during seed development have been described (Harley and Beevers 1989).

In addition to the protein clathrin, coated vesicles isolated from animal tissues contain two polypeptide complexes, termed assembly or adaptor proteins which, *in vivo*, appear to mediate the interaction of clathrin with specific receptors, serving to sort receptors, and thus their ligand cargo, to their cellular destinations (Pearse 1988). While the identification and characterization of the assembly/adaptor complexes of animal tissues has been extensive, no comparable studies have been made of plant tissues. Here we report the identification of several candidates for coated vesicle adaptor proteins of developing pea, *Pisum sativum L. cv. Burpeeana*.

Coated vesicles were obtained from field grown peas harvested 24-27 days post anthesis using either sucrose step (Keen 1987) or linear Ficoll/$D_2O$ (Pearse 1983) density gradient centrifugation and incubation with pancreatic ribonuclease A (1 mg/ml f.c.). Vesicles were extracted with Tris-HCL, coat proteins concentrated using an Amicon PM-10 membrane, and clathrin and adaptors purified using the procedure of Keen (1987). Partially purified adaptor proteins were phosphorylated *in vitro* as described by Keen et al (1987).

With the exception of large non-protein peaks at both ends, the Superose 6 elution profile of extracted coat proteins from vesicles is similar to that reported by Keen (1987) for bovine brain. The polypeptide composition, as determined by SDS-PAGE, of protein peak 1 and 2 fractions is shown in Figure 1. Protein peak 1, consisting of isolated triskelions, exhibits bands at 190 kD (clathrin heavy chain), and of 50 and 46 kD (identified by our laboratory as clathrin light chains). Protein peak 2, identified by Keen (1987) as adaptors, is enriched in polypeptides of 28, 48, and 70 kD with faint bands also observed at 110 kD. Autophosphorylation of partially purified adaptor proteins produced detectable bands at 48, 55, 83, and 110 kD as shown in Figure 2.

NATO ASI Series, Vol. H 74
Molecular Mechanisms of Membrane Traffic
Edited by D. J. Morré, K. E. Howell, and J. J. M. Bergeron
© Springer-Verlag Berlin Heidelberg 1993

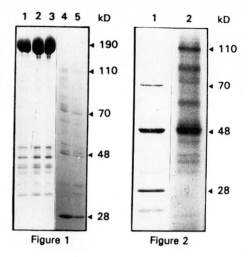

Fig. 1. SDS-PAGE of Superose 6 chromatography of Tris-HCL extracted coat proteins Lane 1-3. Protein Peak 1, Clathrin heavy and light chains. Lane 4-5. Protein Peak 2, Adaptor proteins.

Fig. 2. Autophosphorylation of concentrated adaptor proteins Lane 1. SDS-PAGE. Lane 2. Autoradiography.

Figure 1                    Figure 2

We have demonstrated that polypeptides of approximately 28, 48, 70, and 110 kD are components of the Tris-HCL extractable coat proteins of clathrin coated vesicles from peas. These molecular weights correspond somewhat with the molecular weights for Golgi derived AP-1, with the exception of the 28, and 70 kD proteins. No polypeptide corresponding to the 20 kD subunit of AP-1 has been detected. Since the molecular weights of clathrin light chain from pea also differs from mammalian tissue, the difference in subunit size is not unexpected. The autophosphorylation of the 48 and 110 kD proteins lends additional support to the candidacy of these proteins as plant adaptor polypeptides. Supported by NSF DCB-8916621 to LB.

REFERENCES

Harley SM, Beevers L (1989) Coated vesicles are involved in the transport of storage proteins during seed development in *Pisum sativum* L. Plant Physiol 91:674-678

Keen JH (1987) Clathrin assembly proteins: affinity purification and a model for coat assembly. J. Cell Biol. 105:1989-1998

Keen JH, Chestnut MH, Beck KA (1987) The clathrin coat assembly polypeptide complex. J Biol Chem 262:3864-3871

Pearse BMF (1983) Isolation of coated vesicles. Methods Enzymol. 98:320-326

Pearse, BMF (1988) Receptors compete for adaptors found in plasma membrane coated pits. EMBO J 7:3331-3336

Tanchak MA, Griffing LR, Mersey BG, Fowke LC (1984) Endocytosis of catonized ferritin by coated vesicles of soybean protoplasts. Planta 162:481-486

# Uncoating of Plant Clathrin Coated Vesicles by Uncoating ATPase from Peas

T. Kirsch and L. Beevers
Department of Botany and Microbiology
University of Oklahoma
Norman, Oklahoma 73019
USA

In plants, clathrin coated vesicles are involved in the transport of proteins to the vacuole/protein body (Harley, 1989). This transport system requires a mechanism for uncoating the clathrin from the vesicle prior to or concomitant with fusion of the vesicle with the target membrane and deposition of the cargo proteins. In yeast and mammals, a cytosolic ATPase which is a member of the HSP 70 family has been shown to uncoat clathrin coated vesicles in vitro (Chappel, 1986; Gao, 1991). The present study shows that there is also a cytosolic uncoating ATPase in plants. When a postmicrosomal supernatant obtained from peas was chromatographed on ATP agarose under conditions previously described for the uncoating ATPase from yeast (Gao, 1991), the protein fraction that was eluted from this column with 1 mM ATP (Fraction 2) showed two conspicuous bands in the molecular range of 70 KD (Figure 1, lane 2) on SDS gels. Fraction 2 was applied to Phenyl Sepharose in 30% ammonium sulfate and adsorbed proteins were eluted stepwise with a buffer containing 10% ammonium sulfate, 0% ammonium sulfate or 80% ethylene glycol. The 70 KD proteins were markedly enriched in the fraction that eluted in 0% ammonium sulfate (Figure 1, lane 4). This fraction (fraction 4) was incubated with clathrin coated vesicles from peas or bovine brain in the presence of ATP and an ATP regenerating system. After separation of vesicles from soluble proteins by high speed centrifugation, some of the clathrin was contained in the fraction of soluble proteins as shown by SDS Page (Figure 1, lanes 8 and 9). When fraction 4 was omitted from the incubation mixture, much less clathrin was solubilized (Figure 1, lanes 6 and 7). This shows, that fraction 4 contains an uncoating activity that releases clathrin from both pea and bovine brain coated vesicles. This uncoating activity was ATP dependent as in the absence of ATP, no net uncoating was observed (Figure 1, lanes 10 and 11). The fractions eluted from Phenyl Sepharose with 10% ammonium sulfate or 80% ethylene glycol which contained less 70 KD proteins than fraction 4 (Figure 1, lanes 3 and 4) dissociated only little clathrin from

NATO ASI Series, Vol. H 74
Molecular Mechanisms of Membrane Traffic
Edited by D. J. Morré, K. E. Howell, and J. J. M. Bergeron
© Springer-Verlag Berlin Heidelberg 1993

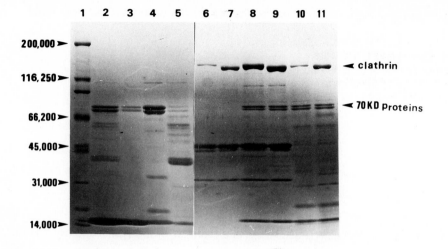

Figure 1. Uncoating of clathrin coated vesicles by uncoating activity enriched on Phenyl Sepharose. The protein fraction eluted from ATP agarose with 1 mM ATP (lane 2) was applied to Phenyl Sepharose in 30% ammonium sulfate and absorbed proteins eluted in 10% ammonium sulfate (lane 3), 0% ammonium sulfate (lane 4) or 80% ethylene glycol (lane 5). Lanes 6-11, high speed supernatants after incubation of clathrin coated vesicles in the absence (lanes 6 and 7) or presence (lanes 8-11) of protein fraction eluted from Phenyl Sepharose with 0% ammonium sulfate. Incubation was in the presence (lanes 6-9) or absence (lanes 10 and 11) of ATP and an ATP regenerating system. Lane 1, molecular standards in daltons.

coated vesicles (data not shown). This suggests, that the uncoating activity is associated with the 70 KD proteins. When fraction 4 was analyzed by Western blotting using antibodies against HSP 70 from pea leaves[*], mainly the upper of the two 70 KD bands was recognized by the antibody (data not shown).

Literature References

Chappell T G, Welch W J, Schlossman D M, Palter K B, Schlesinger M J, Rothman E J (1986) Uncoating ATPase is a member of the 70 kilodalton family of stress proteins. Cell 45: 3-13
Gao B, Biosca J, Craig E A, Greene L E, Eisenberg E (1991) Uncoating of coated vesicles by yeast hsp 70 proteins. J Biol Chem 266: 19565-19571
Harley S M, Beevers L (1989) Coated vesicles are involved in the transport of storage proteins during seed development in C. Pisum sativum. Plant Physiol 91: 674-678

[*]We thank Drs. E. Vierling and Amy De Rocher for the Antibodies against HSP 70 proteins. Supported by NSF Grant DCB 8916621.

# SELECTIVE DEGRADATION OF CYTOSOLIC PROTEINS BY LYSOSOMES

J. Fred Dice
Department of Physiology
Tufts University School of Medicine
136 Harrison Avenue
Boston, MA 02111

Both cytosolic and lysosomal pathways of proteolysis operate in most eukaryotic cells (Dice, 1987). In well-nourished cells lysosomes appear to be able to internalize proteins by a process called microautophagy in which the lysosomal membrane invaginates at multiple locations (Dice, 1987; Olson et al., 1992). Microautophagy appears to be nonselective in that several proteins and inert particles are internalized at similar rates (Ahlberg et al., 1982).

When cultured cells reach confluence an additional lysosomal pathway of proteolysis called macroautophagy is stimulated (Cockle and Dean, 1982; Knecht et al., 1984. Macroautophagy involves formation of autophagic vacuoles which sequester areas of cytoplasm (Pfeifer, 1987). These vacuoles then acquire lysosomal hydrolases to form autophagosomes. Macroautophagy also appears to be nonselective in that many different organelles and proteins are sequestered at approximately the same rates (Kopitz et al., 1990).

We have studied an additional pathway of lysosomal proteolysis that is activated in confluent cell monolayers in response to serum withdrawal. This pathway is restricted to cytosolic proteins that contain peptide sequences biochemically related to Lys-Phe-Glu-Arg-Gln (KFERQ) (Dice, 1990). The mechanism by which proteins with KFERQ-like peptide regions are targeted to lysosomes for degradation is similar in certain respects to the direct import of newly synthesized proteins into mitochondria or into the lumen of the endoplasmic reticulum (Figure 1). Therefore, proteins with KFERQ-like peptide regions may enter lysosomes by directly crossing a membrane bilayer.

NATO ASI Series, Vol. H 74
Molecular Mechanisms of Membrane Traffic
Edited by D. J. Morré, K. E. Howell, and J. J. M. Bergeron
© Springer-Verlag Berlin Heidelberg 1993

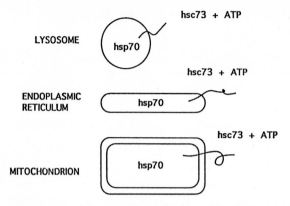

Figure 1: Similarities between selective lysosomal protein degradation and protein transport into endoplasmic reticulum and mitochondria. The common features include stimulation by cytosolic hsc73 and ATP and the presence of an hsp70 within the organelle. The role of the hsp70 within lysosomes has not yet been established.

To try to elucidate the mechanism of degradation of KFERQ motif-containing proteins, we developed an in vitro assay using lysosomes isolated from IMR-90 fibroblasts over two consecutive discontinuous density gradients (Chiang et al., 1990; S. R. Terlecky and J. F. Dice, unpublished). Maximal degradation of [³H]RNase S-peptide, a KFERQ-containing model peptide, by isolated lysosomes requires both ATP and a heat shock protein of 70 kDa (hsp70). This degradation is specific because proteins which do not contain a KFERQ motif are degraded little, if at all, under the same conditions.

Degradation of [³H]RNase S-peptide can be inhibited by reducing the temperature, and degradation appears to occur within acidified organelles because it is inhibited by ammonium chloride. Additional experiments show that the lysosomal uptake of [³H]RNase S-peptide is saturable. Furthermore, at 0-4° C. specific binding of [³H]RNase S-peptide to a lysosomal membrane protein occurs. Presumably, this binding component is a receptor or a peptide transport channel.

The hsp70 that stimulates lysosomal uptake and degradation of KFERQ motif-containing proteins has been identified as the constitutively expressed heat shock cognate protein of 73 kilodaltons (hsc73) (Terlecky et al., 1992). A yeast analog of hsc73 is also required for transport of proteins into mitochondria and the endoplasmic reticulum (Deshaies et al., 1988; Chirico et al., 1988).

The mechanisms by which hsc73 promotes lysosomal degradation of proteins containing KFERQ-like peptide regions are not known. In response to serum withdrawal hsc73 may shift its subcellular distribution to the cytosol where the proteins containing the KFERQ-like regions reside (Chiang et al., 1989). Activation of the pathway could also involve changes in the substrate proteins that expose the KFERQ-like regions. Finally, activation of the pathway may be due to regulation of a component distal to hsc73 binding to the protein substrate. For example, the activity of the putative receptor or peptide transporter on the lysosome surface may be regulated.

Another similarity between the selective lysosomal degradation pathway and protein transport into endoplasmic reticulum and mitochondria is the presence of an hsp70 within the organelle (Figure 1). In the case of mitochondria and endoplasmic reticulum, this lumenal hsp70 appears to be required as the "pulling force" for protein import (Vogel et al., 1990; Kang et al., 1990).

Research in the author's laboratory is supported by NIH grants AG06116 and AG07472. I also thank former and present members of the lab who have been responsible for much of the work presented.

References

Ahlberg J, Marzella L, Glaumann H (1982) Uptake and degradation of proteins by isolated rat liver lysosomes. Suggestion of a microautophagic pathway of proteolysis. Lab Invest 47:523-532

Chiang H-L, Terlecky SR, Plant CP, Dice JF (1989) A role for a 70-kilodalton heat shock protein in lysosomal proteolysis of intracellular proteins. Science 246:282-285

Chirico WJ, Waters MG, Blobel G (1988) 70K heat shock related proteins stimulate protein translocation into microsomes. Nature 332:805-810

Cockle SM, Dean RT (1982) The regulation of proteolysis in normal fibroblasts as they approach confluence. Evidence for participation of the lysosomal system. Biochem J 208:243-249

Deshaies RJ, Koch BD, Werner-Washburne M, Craig EA, Schekman R (1988) 70 kD stress protein homologues facilitate translocation of secretory and mitochondrial precursor polypeptides. Nature 332:800-805

Dice JF (1987) Molecular determinants of protein half-lives in eukaryotic cells. FASEB J 1:349-357

Dice JF (1990) Peptide sequences that target cytosolic proteins for lysosomal proteolysis. Trends Biochem Sci 15:305-309

Kang P-J, Ostermann J, Shilling J, Neupert W, Craig EA, Pfanner N (1990) Hsp70 in the mitochondrial matrix is required for translocation and folding of precursor proteins. Nature 348:137-143

Knecht E, Hernandez-Yago, J, Grisolia S (1984) Regulation of lysosomal autophagy in transformed and non-transformed mouse fibroblasts under several growth conditions. Exp Cell Res 154:224-232

Kopitz J, Kisen GO, Gordon PB, Bohley P, Seglen PO (1990) Non-selective autophagy of cytosolic enzymes in isolated rat hepatocytes. J Cell Biol 111:941-954

Olson TS, Terlecky SR, Dice JF (1992) Pathways of intracellular protein degradation in eukaryotic cells. In: Ahern TJ and Manning MC (eds) Stability of Protein Pharmaceuticals: In Vivo Pathways of Degradation and Stratagies for Protein Stabilization. Plenum Publishing Corporation, New York, in press

Pfeifer, U 1987 Functional morphology of the lysosomal apparatus. In: Glaumann H, Ballard FJ (eds) Lysosomes: Their Role in Protein Breakdown. Academic Press, New York p 3-59

Terlecky SR, Chiang H-L, Olson TS, Dice JF (1992) Protein and peptide binding and stimulation of in vitro lysosomal proteolysis by the 73-kDa heat shock cognate protein. J Biol Chem 267:9202-9209

Vogel JP, Misra LM, Rose MD (1990) Loss of Bip/GRP78 function blocks translocation of secretory proteins in yeast. J Cell Biol 110:1885-1896

# ROLE OF CALCIUM, PROTEIN PHOSPHORYLATION AND THE CYTOSKELETON IN HEPATOCYTIC AUTOPHAGY

P.O. Seglen, I. Holen and P.B. Gordon
Department of Tissue Culture
Institute for Cancer Research
The Norwegian Radium Hospital
Montebello, 0310 Oslo 3, Norway

Hepatocytic autophagy, measured as the sequestration of electroinjected, cytosolic [$^3$H]raffinose into sedimentable autophagic vacuoles, is markedly inhibited by epinephrine through an $\alpha_1$-adrenergic, receptor-mediated mechanism involving release of $Ca^{2+}$ from the endoplasmic reticulum. More effective $Ca^{2+}$-releasing agents like thapsigargin and $t$BHQ suppress autophagy completely. Autophagy is also inhibited by protein phosphatase inhibitors like okadaic acid, the potency of which suggests the involvement of a type 2A protein phosphatase (PP2A) in autophagy control. The effect of okadaic acid can be reversed by a calmodulin antagonist (W-7) and by inhibitors (KN-62, K-252a, KT-5926) of $Ca^{2+}$/calmodulin-dependent protein kinase II (CaMK-II), indicating that this protein kinase is responsible for autophagy-inhibitory protein phosphorylations. Okadaic acid, furthermore, causes an extensive disruption of the hepatocytic cytoskeleton which becomes evident as structural disassembly of the cell corpses following electrodisruption of the plasma membrane. The cytoskeletal disruption can be specifically prevented by the autophagy-protective protein kinase inhibitors. One explanation of these observations would be that autophagic activity is dependent upon the integrity or formation of cytoskeletal elements that assemble or disassemble according to a dynamic phosphorylation balance governed by CaMK-II and PP2A.

Autophagy is a non-selective process by which cells sequester and degrade portions of their cytoplasm in response to amino acid deprivation or growth-inhibitory signals (Seglen and Bohley, 1992). The process can be measured biochemically as the sequestration of endogenous cytosolic enzymes (Kopitz et al., 1990) or electroinjected, radiolabelled sugar probes (Gordon et al., 1985; Seglen et al., 1986; Høyvik et al., 1986) into autophagic vacuoles that sediment with the cell corpses after electrodisruption of the plasma membrane (Gordon and Seglen, 1982). In rat hepatocytes autophagy is modulated by $Ca^{2+}$-releasing and adenylate cyclase-activating hormones as well as by cyclic nucleotides (Seglen et al., 1991), suggesting regulatory roles for $Ca^{2+}$ and protein phosphorylation. In the present study we have used a variety of $Ca^{2+}$-releasing agents, protein phosphatase inhibitors and protein kinase inhibitors to further elucidate the roles of $Ca^{2+}$ and protein phosphorylation in the control of hepatocytic autophagy.

---

**Abbreviations**: CaMK-II, $Ca^{2+}$/calmodulin-dependent protein kinase II; OA, okadaic acid; PKC, protein kinase C; PP2A, protein phosphatase type 2A; $t$BHQ, 2,5-di-($tert$-butyl)-1,4-benzohydroquinone.

NATO ASI Series, Vol. H 74
Molecular Mechanisms of Membrane Traffic
Edited by D. J. Morré, K. E. Howell, and J. J. M. Bergeron
© Springer-Verlag Berlin Heidelberg 1993

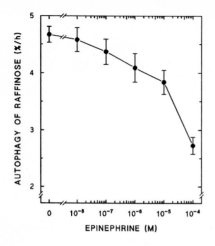

**Fig. 1. Inhibition of hepatocytic autophagy by epinephrine.** Rat hepatocytes were isolated by collagenase perfusion (Seglen, 1976), electroloaded with [³H]raffinose (Seglen et al., 1986) and incubated for 3h at 37°C in a buffered salt solution containing 15 mM pyruvate as an energy substrate (Holen et al., 1992) and epinephrine at the concentration indicated. The net sequestration of [³H]raffinose during the incubation was measured, and the autophagic rate expressed as % (of the total cellular radioactivity) sequestered per hour (Seglen et al., 1986).

### Hormonal regulation of autophagy

Autophagic sequestration of [³H]raffinose in hepatocytes is subject to regulation by various hormones; for example, insulin inhibits sequestration by synergizing with the more basic amino acid control, while glucagon antagonizes the amino acid inhibition (Seglen et al., 1991). Adrenergic hormones were reported to accelerate hepatic autophagy *in vivo* (Pfeifer, 1984), but in isolated hepatocytes epinephrine surprisingly inhibited autophagic sequestration (Fig. 1).

**Table 1. Effects of adrenergic agonists and antagonists on hepatocytic autophagy.** Hepatocytes electroloaded with [³H]raffinose were incubated for 3h with the various adrenergic agonists and antagonists indicated, and the net autophagic sequestration of raffinose during this period was recorded. Effects of the added agents have been expressed as % inhibition *vs.* control cells (no additions), a + sign indicating stimulation rather than inhibition. Each value is the mean ± S.E. of the number of experiments given in parentheses.

| Adrenergic agonist (100 µM) | Inhibition of [³H]raffinose autophagy (%) | | |
| --- | --- | --- | --- |
| | No antagonist | Alprenolol (ß) (10 µM) | Prazosin ($\alpha_1$) (1 µM) |
| None | 0 | 13.0 ± 5.0 (2) | 4.8 ± 1.6 (5) |
| Epinephrine ($\alpha_1,\alpha_2,\beta_1,\beta_2$) | 41.5 ± 3.2 (11) | 44.0 ± 5.0 (2) | 11.0 ± 5.1 (3) |
| Norepinephrine ($\alpha_1,\alpha_2,\beta_1$) | 31.0 ± 3.2 (4) | – | – |
| Salbutamol ($\beta_2$) | +8.0 ± 5.5 (3) | – | – |
| Phenylephrine ($\alpha_1$) | 35.0 ± 2.4 (10) | 50.5 ± 0.5 (2) | 12.3 ± 5.4 (3) |
| Clonidine ($\alpha_2$) | 10     (1) | – | – |

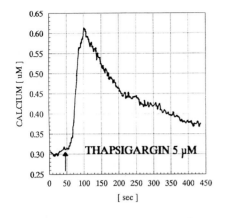 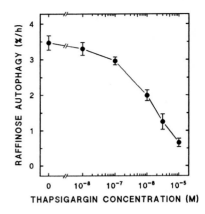

**Fig. 2. Stimulation of intracellular Ca$^{2+}$ release and inhibition of autophagy by thapsigargin.** (Left) Intracellular free Ca$^{2+}$ in a single hepatocyte preloaded with Fura-2 (Monck et al., 1988) and treated with 5 μM thapsigargin (added at arrow); (right) rate of autophagic [$^3$H]raffinose sequestration in hepatocytes during a 3-h incubation at 37°C in the presence of various concentrations of thapsigargin. Each value is the mean ± S.E. of 3-6 experiments.

Since hepatocytes possess both $\alpha_1$ and $\beta_2$ adrenergic receptors, various receptor-specific adrenergic agonists and antagonists were examined. As shown in Tab. 1, the specific $\alpha_1$ agonist phenylephrine suppressed autophagy as efficiently as did epinephrine, while the $\beta_2$ agonist salbutamol and the $\alpha_2$ agonist clonidine were ineffective. The effects of epinephrine and phenylephrine were suppressed by the $\alpha_1$ antagonist prazosin but not by the $\beta$-blocker alprenolol, suggesting that adrenergic inhibition of autophagy is mediated by $\alpha_1$ receptors.

The hepatocytic $\alpha_1$ receptor is known to generate a variety of secondary signals, including an inositol triphosphate-dependent release of Ca$^{2+}$ from intracellular stores (Blackmore et al., 1982; Williamson et al., 1985). The effect of other Ca$^{2+}$-releasing agents on hepatocytic autophagy was therefore investigated.

## Inhibition of autophagy by Ca$^{2+}$-releasing agents

Thapsigargin is a Ca$^{2+}$ pump (ATPase) inhibitor capable of eliciting Ca$^{2+}$ release from intracellular stores (Thastrup et al., 1990). Thapsigargin at 5 μM rapidly doubled the level of free Ca$^{2+}$ in isolated rat hepatocytes, and suppressed autophagy strongly during a 3-h incubation (Fig. 2). Similar results were obtained with *tert*-butylbenzohydroquinone (*t*BHQ), another effective Ca$^{2+}$ release agent in hepatocytes (Duddy et al., 1989), and by low concentrations (10 μM) of the Ca$^{2+}$ ionophore A23187 (results not shown). Release of Ca$^{2+}$ from some unidentified store is thus capable of inducing autophagic arrest.

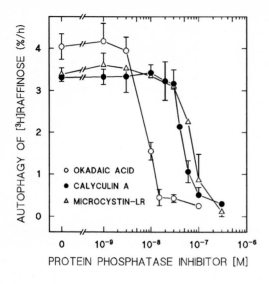

Fig. 3. Suppression of hepatocytic autophagy by protein phosphatase inhibitors. Rat hepatocytes electroloaded with [³H]raffinose were incubated at 37°C for 3h in the presence of okadaic acid (o), calyculin A (●) or microcystin LR (△) at the concentration indicated. Autophagic sequestration of raffinose was measured and expressed as %/h. Each point is the mean ± S.E./range of 2-7 experiments.

## Protein phosphorylation in the regulation of autophagy: identification of an autophagy-promoting protein phosphatase

Inhibition of autophagy by $Ca^{2+}$-releasing agents could be due to $Ca^{2+}$ depletion (Kuznetsov et al., 1992) of organelles involved in the autophagic process, or to structural alterations (Booth and Koch, 1989) induced by such depletion. The inhibition could also be a result of the increase in cytosolic free $Ca^{2+}$ levels, with consequent activation of protein-phosphorylating enzymes like protein kinase C (PKC) (Pittner and Fain, 1991) or the $Ca^{2+}$/calmodulin-dependent protein kinases (Connelly et al., 1987). The PKC activator TPA (tetradecanoyl phorbol acetate), known to stimulate hepatocytic PKC maximally at 1 µM (García-Sáinz et al., 1985), had no significant effect on autophagy at concentrations up to 10 µM (results not shown). PKC activation would thus seem unlikely to be involved in the autophagy-inhibitory effect of intracellular $Ca^{2+}$ release.

To more systematically investigate the possible role of protein phosphorylation in autophagy control, we made use of specific protein phosphatase inhibitors. Microcystin-LR, calyculin A and okadaic acid (OA) are all potent inhibitors of the major serine/threonine protein phosphatases (Bialojan and Takai, 1988; MacKintosh et al., 1990; Ishihara et al., 1989), the latter inducing a generalized hyperphosphorylation of hepatocytic phosphoproteins (Haystead et al., 1989). All three inhibitors suppressed autophagy strongly, okadaic acid being an order of magnitude more potent than the other two (Fig. 3). Both the extremely low concentrations of okadaic acid needed to achieve half-maximal inhibition (<10 nM) and the selective effect of okadaic acid vs. the other two (Eriksson et al., 1990; Ishihara et al., 1989) would tend to suggest inhibition of a protein phosphatase of type 2A (PP2A) as the mechanism of action involved.

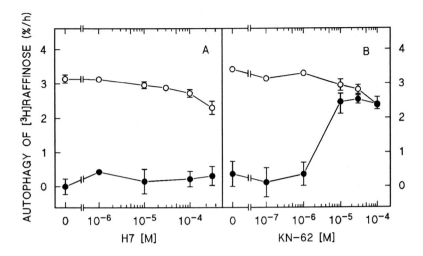

**Fig 4. Okadaic acid-antagonistic effects of protein kinase inhibitors.** Hepatocytes electroloaded with [³H]raffinose were incubated for 3h at 37°C in the absence (o) or presence (●) of 15 nM OA and various concentrations of the protein kinase inhibitors H-7 (A) or KN-62 (B). The rate of autophagic raffinose sequestration during the incubation period was measured and expressed as %/h. Each value is the mean ± range/S.E. of 2-3 experiments.

Okadaic acid had little or no effect on hepatocytic protein synthesis and ATP levels at concentrations (10-15 nM) that inhibited autophagy virtually completely, and no morphological alterations or other signs of general toxicity were observed at these concentrations.

**Identification of an autophagy-suppressive protein kinase**

Since all protein phosphorylations are dynamic equilibrium processes involving both protein kinase and protein phosphatase activity, the suppression of autophagy by OA must depend on the activity of some protein kinase. In an attempt to identify the type of protein kinase involved we have applied a panel of protein kinase inhibitors, assuming that inhibitors of the autophagy-suppressive protein kinase would be able to antagonize the effect of OA. General protein kinase inhibitors like the isoquinoline-sulfonamide H-7 (Hidaka et al., 1984) (Fig. 4A) or staurosporine (not shown) did not detectably antagonize the OA effect, nor did the specific PKC inhibitor calphostin C (Kobayashi et al., 1989) or H-89, an inhibitor of the cyclic nucleotide-dependent protein kinases (Chijiwa et al., 1990) (Tab. 2). None of these major protein kinases would thus seem to be responsible for the OA-induced suppression of autophagy.

**Tab. 2. Ability of various protein kinase inhibitors to antagonize the autophagy-inhibitory effect of okadaic acid.** Hepatocytes electroloaded with [³H]raffinose were incubated for 3h at 37°C in the absence or presence of 15 nM OA and the protein kinase inhibitor indicated. The rate of autophagic raffinose sequestration during the incubation period was measured and expressed as %/h relative to the total cellular radioactivity. Each value is the mean ± S.E/range of the number of experiments indicated in parentheses. [a]P<0.05; [b]P<0.005; [c]P<0.001 for significance vs. OA alone according to the t-test.

| Protein kinase inhibitor | Conc. (µM) | Autophagic sequestration of [³H]raffinose (%/h) | |
|---|---|---|---|
| | | − Okadaic acid | + Okadaic acid |
| None | | 3.16 ± 0.14 (12) | 0.45 ± 0.19 (12) |
| H-7 | 100 | 2.57 ± 0.08 (4) | 0.53 ± 0.38 (4) |
| Calphostin C | 100 | 3.48 (1) | 0.61 (1) |
| H-89 | 20 | 3.65 (1) | 0.82 (1) |
| KN-62 | 30 | 2.70 ± 0.06 (2) | 2.43 ± 0.11 (2)[b] |
| K-252a | 10 | 2.60 ± 0.20 (3) | 1.81 ± 0.03 (3)[b] |
| KT-5926 | 60 | 2.54 ± 0.21 (3) | 2.38 ± 0.08 (3)[c] |
| W-7 | 100 | 2.40 ± 0.24 (4) | 1.43 ± 0.35 (4)[a] |

A variety of other specific protein kinase inhibitors, including inhibitors of tyrosine protein kinases, have likewise been found to be unable to reverse the OA effect (Holen et al., 1992). On the other hand, KN-62, a highly specific inhibitor of $Ca^{2+}$/calmodulin-dependent protein kinase II (CaMK-II) (Tokumitsu et al., 1990), reversed the effect of OA completely (Fig. 4B). Other, less specific inhibitors of CaMK-II, like K-252a and KT-5926 (Hashimoto et al., 1991), and the calmodulin antagonist W-7 (Inagaki et al., 1986) were also able to antagonize OA (Tab. 2), suggesting an involvement of CaMK-II in autophagy control.

**Regulation of cytoskeletal integrity by dynamic phosphorylation**

While hepatocytes treated with low concentrations of OA (10-15 nM) maintain a normal morphology during 3h of incubation, we found that they more or less disintegrated to a homogenate upon electrodisruption, in contrast to untreated cells which formed coherent cell corpses (Gordon and Seglen, 1986). OA would thus appear to induce a general disassembly of

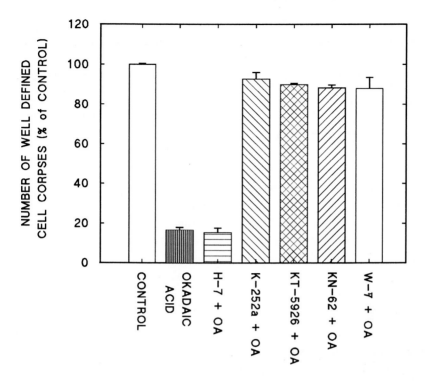

**Fig. 5. Disruption of the hepatocytic cytoskeleton by okadaic acid: antagonistic effects of protein kinase inhibitors.** Hepatocytes were incubated for 3h at 37°C with or without 15 nM OA in combination with protein kinase inhibitors KN-62 (10 μM), K-252a (100 μM), KT-5926 (100 μM), H-7 (300 μM) or the calmodulin antagonist W-7 (150 μM). Following electrodisruption the number of structurally recognizable cell corpses (Gordon and Seglen, 1982) per unit volume of cell suspension was counted in the microscope and expressed as % of the control value. Each value is the mean ± S.E. of 3-6 experiments.

the cytoskeleton, quantifiable as the number of structurally recognizable cell corpses present per unit volume after electrodisruption. As shown in Fig. 5, the CaMK-II inhibitors K-252a, KT-5926 and KN-62 as well as the calmodulin antagonist W-7 effectively prevented the cytoskeleton-disruptive effect of OA. H-7, which does not inhibit CaMK-II, and which failed to antagonize the autophagy-inhibitory effect of OA, was also unable to reverse the cytoskeleton-disruptive effect of the latter. These observations suggest that CaMK-II exerts a disruptive effect on the hepatocytic cytoskeleton parallelling its inhibitory effect on autophagy. CaMK-II has been shown to be capable of phosphorylating isolated cytoskeletal proteins like vimentin and myosin light chains as well as the microtubule-associated proteins MAP-2, tau and synapsin and Fujisawa, 1982; Yamamoto et al., 1983; Vallano et al., 1986; Colbran et al., 1989), and it is likely that its cytoskeleton-disruptive effect is due to the phosphorylation of such proteins. The ultrastructural alterations previously observed in hepatocytes treated with high concentrations of OA and related protein phosphatase inhibitors, changes which include reorganization of the microfilament and intermediate filament networks, organelle segregation and reduced attachment to the culture substratum (Miura et al., 1989; Eriksson et al., 1989; Bøe et al., 1991; Falconer and Yeung, 1992), are thus probably mediated by CaMK-II.

### Possible involvement of the cytoskeleton in autophagy

The apparent ability of CaMK-II to interfere with cytoskeletal integrity as well as with autophagy could be taken to indicate an involvement of cytoskeletal proteins in the autophagic process. It is not unreasonable to assume that the initial spreading and folding of the sequestering organelle, the phagophore (Seglen, 1987), may require the formation of a filamentous scaffold, and/or anchorage to the cytoskeleton. Neither the microtubule poison vinblastine (Seglen, 1987) nor the microfilament inhibitor cytochalasin B (Grinde, 1985) have much effect on autophagic-lysosomal protein degradation, leaving intermediate filaments as the most likely cytoskeletal element to be involved in autophagy.

Cytoskeletal rearrangements and low autophagic activity are characteristics of cancer cells and other rapidly proliferating cells, and it should be pointed out that OA and microcystin are potent tumour promoters in many tissues, including the liver (Suganuma et al., 1988; Nishiwaki-Matsushima et al., 1991). Several growth factors have been shown to induce activation of CaMK-II (Ohta et al., 1988; MacNicol et al., 1990), and the enzyme has been implicated in early mitotic events (disassembly of the nuclear envelope) (Baitinger et al., 1990). Abnormalities in the organization of cytoskeletal proteins are, furthermore, observed in a variety of human diseases, including the aggregation of hyperphosphorylated cytoskeletal proteins in the amyloid plaques of Alzheimer's disease (Colbran et al., 1989). If CaMK-II plays a role in malignant cell growth or in other pathological conditions, a possible therapeutic potential of CaMK-II inhibitors would be indicated.

### Acknowledgments

We wish to thank Mona Birkeland and Margrete Fosse for expert technical assistance. This work has been generously supported by The Norwegian Cancer Society.

# References

Baitinger C, Alderton J, Poenie M, Schulman H, Steinhardt RA (1990) Multifunctional $Ca^{2+}$/calmodulin-dependent protein kinase is necessary for nuclear envelope breakdown. J Cell Biol 111:1763-1773

Bialojan C, Takai A (1988) Inhibitory effect of a marine-sponge toxin, okadaic acid, on protein phosphatases. Specificity and kinetics. Biochem J 256:283-290

Blackmore PF, Hughes BP, Shuman EA, Exton JH (1982) $\alpha$-Adrenergic activation of phosphorylase in liver cells involves mobilization of intracellular calcium without influx of extracellular calcium. J Biol Chem 257:190-197

Booth C, Koch GLE (1989) Perturbation of cellular calcium induces secretion of luminal ER proteins. Cell 59:729-737

Bøe R, Gjertsen BT, Vintermyr OK, Houge G, Lanotte M, Døskeland SO (1991) The protein phosphatase inhibitor okadaic acid induces morphological changes typical of apoptosis in mammalian cells. Exp Cell Res 195:237-246

Chijiwa T, Mishima A, Hagiwara M, Sano M, Hayashi K, Inoue T, Naito K, Toshioka T, Hidaka H (1990) Inhibition of forskolin-induced neurite outgrowth and protein phosphorylation by a newly synthesized selective inhibitor of cyclic AMP-dependent protein kinase, $N$-[2-($p$-bromocinnamylamino)ethyl]-5-isoquinolinesulfonamide (H-89), of PC12D pheochromocytoma cells. J Biol Chem 265:5267-5272

Colbran RJ, Schworer CM, Hashimoto Y, Fong Y-L, Rich DP, Smith MK, Soderling TR (1989) Calcium/calmodulin-dependent protein kinase II. Biochem J 258:313-325

Connelly PA, Sisk RB, Schulman H, Garrison JC (1987) Evidence for the activation of the multifunctional $Ca^{2+}$/calmodulin-dependent protein kinase in response to hormones that increase intracellular $Ca^{2+}$. J Biol Chem 262:10154-10163

Duddy SK, Kass GEN, Orrenius S (1989) $Ca^{2+}$-mobilizing hormones stimulate $Ca^{2+}$ efflux from hepatocytes. J Biol Chem 264:20863-20866

Eriksson JE, Paatero GIL, Meriluoto JAO, Codd GA, Kass GEN, Nicotera P, Orrenius S (1989) Rapid microfilament reorganization induced in isolated rat hepatocytes by microcystin-LR, a cyclic peptide toxin. Exp Cell Res 185:86-100

Eriksson JE, Toivola D, Meriluoto JAO, Karaki H, Han Y-G, Hartshorne D (1990) Hepatocyte deformation induced by cyanobacterial toxins reflects inhibition of protein phosphatases. Biochem Biophys Res Commun 173:1347-1353

Falconer IR, Yeung DSK (1992) Cytoskeletal changes in hepatocytes induced by *Microcystis* toxins and their relation to hyperphosphorylation of cell proteins. Chem-Biol Interact 81:181-196

García-Sáinz JA, Mendlovic F, Martínez-Olmedo MA (1985) Effects of phorbol esters on $\alpha_1$-adrenergic-mediated and glucagon-mediated actions in isolated rat hepatocytes. Biochem J 228:277-280

Gordon PB, Tolleshaug H, Seglen PO (1985) Use of digitonin extraction to distinguish between autophagic-lysosomal sequestration and mitochondrial uptake of [$^{14}$C]sucrose in hepatocytes. Biochem J 232:773-780

Gordon PB, Seglen PO (1982) Autophagic sequestration of [$^{14}$C]sucrose, introduced into isolated rat hepatocytes by electropermeabilization. Exp Cell Res 142:1-14

Gordon PB, Seglen PO (1986) Use of electrical methods in the study of hepatocytic autophagy. Biomed Biochim Acta 45:1635-1645

Grinde B (1985) Autophagy and lysosomal proteolysis in the liver. Experientia 41:1090-1095

Hashimoto Y, Nakayama T, Teramoto T, Kato H, Watanabe T, Kinoshita M, Tsukamoto K, Tokunaga K, Kurokawa K, Nakanishi S, Matsuda Y, Nonomura Y (1991) Potent and preferential inhibition of $Ca^{2+}$/calmodulin-dependent protein kinase II by K252a and its derivative, KT5926. Biochem Biophys Res Commun 181:423-429

Haystead TAJ, Sim ATR, Carling D, Honnor RC, Tsukitani Y, Cohen P, Hardie DG (1989) Effects of the tumour promoter okadaic acid on intracellular protein phosphorylation and metabolism. Nature 337:78-81

Hidaka H, Inagaki M, Kawamoto S, Sasaki Y (1984) Isoquinolinesulfonamides, novel and potent inhibitors of cyclic nucleotide dependent protein kinase and protein kinase C. Biochemistry 23:5036-5041

Holen I, Gordon PB, Seglen PO (1992) Protein kinase-dependent effects of okadaic acid on hepatocytic autophagy and cytoskeletal integrity. Biochem J 284:633-636

Høyvik H, Gordon PB, Seglen PO (1986) Use of a hydrolysable probe, [$^{14}$C]lactose, to distinguish between pre-lysosomal and lysosomal steps in the autophagic pathway. Exp Cell Res 166:1-14

Inagaki M, Kawamoto S, Itoh H, Saitoh M, Hagiwara M, Takahashi J, Hidaka H (1986) Naphtalenesulfonamides as calmodulin antagonists and protein kinase inhibitors. Mol Pharmacol 29:577-581

Ishihara H, Martin BL, Brautigan DL, Karaki H, Ozaki H, Kato Y, Fusetani N, Watabe S, Hashimoto K, Uemura D, Hartshorne DJ (1989) Calyculin A and okadaic acid: Inhibitors of protein phosphatase activity. Biochem Biophys Res Commun 159:871-877

Kobayashi E, Nakano H, Morimoto M, Tamaoki T (1989) Calphostin C (UCN-1028C), a novel microbial compound, is a highly potent and specific inhibitor of protein kinase C. Biochem Biophys Res Commun 159:548-553

Kopitz J, Kisen GØ, Gordon PB, Bohley P, Seglen PO (1990) Non-selective autophagy of cytosolic enzymes in isolated rat hepatocytes. J Cell Biol 111:941-953

Kuznetsov G, Brostrom MA, Brostrom CO (1992) Demonstration of a calcium requirement for secretory protein processing and export. Differential effects of calcium and dithiothreitol. J Biol Chem 267:3932-3939

MacKintosh C, Beattie KA, Klumpp S, Cohen P, Codd GA (1990) Cyanobacterial microcystin-LR is a potent and specific inhibitor of protein phosphatases 1 and 2A from both mammals and higher plants. FEBS Lett 264:187-192

MacNicol M, Jefferson AB, Schulman H (1990) $Ca^{2+}$/calmodulin kinase is activated by the phosphatidylinositol signaling pathway and becomes $Ca^{2+}$-independent in PC12 cells. J Biol Chem 265:18055-18058

Miura GA, Robinson NA, Geisbert TW, Bostian KA, White JD, Pace JG (1989) Comparison of in vivo and in vitro toxic effects of microcystin-LR in fasted rats. Toxicon 27:1229-1240

Monck JR, Reynolds EE, Thomas AP, Williamson JR (1988) Novel kinetics of single cell $Ca^{2+}$ transients in stimulated hepatocytes and A10 cells measured using fura-2 and fluorescent videomicroscopy. J Biol Chem 263:4569-4575

Nishiwaki-Matsushima R, Nishiwaki S, Ohta T, Yoshizawa S, Suganuma M, Harada K, Watanabe MF, Fujiki H (1991) Structure-function relationships of microcystins, liver tumor promoters, in interaction with protein phosphatase. Jpn J Cancer Res 82:993-996

Ohta Y, Ohba T, Fukunaga K, Miyamoto E (1988) Serum and growth factors rapidly elicit phosphorylation of the $Ca^{2+}$/calmodulin-dependent protein kinase II in intact quiescent rat 3Y1 cells. J Biol Chem 263:11540-11547

Pfeifer U (1984) Application of test substances to the surface of rat liver in situ: opposite effects of insulin and isoproterenol on cellular autophagy. Lab Invest 50:348-354

Pittner RA, Fain JN (1991) Activation of membrane protein kinase C by glucagon and $Ca^{2+}$-mobilizing hormones in cultured rat hepatocytes. Role of phosphatidylinositol and phosphatidylcholine hydrolysis. Biochem J 277:371-378

Seglen PO (1976) Preparation of isolated rat liver cells. Meth Cell Biol 13:29-83

Seglen PO, Gordon PB, Tolleshaug H, Høyvik H (1986) Use of [$^3$H]raffinose as a specific probe of autophagic sequestration. Exp Cell Res 162:273-277

Seglen PO (1987) Regulation of autophagic protein degradation in isolated liver cells. In: Glaumann H, Ballard FJ (eds)Lysosomes: Their Role in Protein Breakdown. Academic Press, London, pp 369-414

Seglen PO, Gordon PB, Holen I, Høyvik H (1991) Hepatocytic autophagy. Biomed Biochim Acta 50:373-381

Seglen PO, Bohley P (1992) Autophagy and other vacuolar protein degradation mechanisms. Experientia 48:158-172

Suganuma M, Fujiki H, Suguri H, Yoshizawa S, Hirota M, Nakayasu M, Ojika M, Wakamatsu K, Yamada K, Sugimura T (1988) Okadaic acid: An additional non-phorbol-12-tetradecanoate-13-acetate-type tumor promoter. Proc Natl Acad Sci USA 85:1768-1771

Thastrup O, Cullen PJ, Drobak BK, Hanley MR, Dawson AP (1990) Thapsigargin, a tumor promoter, discharges intracellular $Ca^{2+}$ stores by specific inhibition of the endoplasmic reticulum $Ca^{2+}$-ATPase. Proc Natl Acad Sci USA 87:2466-2470

Tokumitsu H, Chijiwa T, Hagiwara M, Mizutani A, Terasawa M, Hidaka H (1990) KN-62, 1-[N,O-Bis-(5-isoquinolinesulfonyl)-N-methyl-L-tyrosyl]-4-phenylpiperazine, a specific inhibitor of $Ca^{2+}$/calmodulin-dependent protein kinase II. J Biol Chem 265:4315-4320

Vallano ML, Goldenring JR, Lasher RS, DeLorenzo RJ (1986) Association of calcium/calmodulin-dependent kinase with cytoskeletal preparations: phosphorylation of tubulin, neurofilament, and microtubule-associated proteins. Ann N Y Acad Sci 466:357-374

Williamson JR, Cooper RH, Joseph SK, Thomas AP (1985) Inositol triphosphate and diacylglycerol as intracellular second messengers in liver. Am J Physiol 248:C203-C216

Yamamoto H, Fukunaga K, Tanaka E, Miyamoto E (1983) $Ca^{2+}$- and calmodulin-dependent phosphorylation of microtubule-associated protein 2 and tau factor, and inhibition of microtubule assembly. J Neurochem 41:1119-1125

Yamauchi T, Fujisawa H (1982) Phosphorylation of microtubule-associated protein 2 by calmodulin-dependent protein kinase (kinase II) which occurs only in the brain tissues. Biochem Biophys Res Commun 109:975-981

# SIGNALS FOR TRANSPORT FROM ENDOSOMES TO LYSOSOMES

J. Paul Luzio, Tomomi Kuwana and Barbara M. Mullock

Department of Clinical Biochemistry, University of Cambridge,

Addenbrooke's Hospital, Hills Road, Cambridge   CB2 2QR,   U.K.

It has been proposed either that endocytosed ligands are delivered to lysosomes by a process of endosome maturation (Murphy, 1991) or by vesicular transport (Griffiths and Gruenberg, 1991). The latter mechanism implies the possibility of generating cell-free systems in which endosome membrane vesicles fuse with lysosomes and which can be used to identify factors required for targeting and fusion. In our studies we have investigated the interactions between rat liver endosome and lysosome fractions previously shown to be involved in the endocytic pathway for asialoglycoproteins in hepatocytes.

There is an extensive literature showing that asialoglycoproteins are taken up into hepatocytes by receptor mediated endocytosis and delivered to the lysosomes for digestion via endosomal compartments (Geuze et al., 1986; Mueller and Hubbard, 1986; Mullock et al., 1989). Using isopycnic centrifugation on shallow Ficoll gradients we have shown that $^{125}I$-asialofetuin (ASF) taken up by rat liver *in vivo* appears sequentially in sinusoidal plasma membrane and three separated endosomal fractions (Branch et al., 1987). It was found that ASF appeared first in light endosomes (LE), then in dense endosomes (DE) and then in very dense endosomes (VDE) and lysosomes. ASF reaching the lysosomes was rapidly digested so that little label was seen in this fraction unless the protease inhibitor leupeptin was administered.

NATO ASI Series, Vol. H 74
Molecular Mechanisms of Membrane Traffic
Edited by D. J. Morré, K. E. Howell, and J. J. M. Bergeron
© Springer-Verlag Berlin Heidelberg 1993

On the basis of this work we have developed cell-free systems from rat liver to examine the later stages of the movement of ASF from endosomes to lysosomes.

The DE and VDE components of liver post-mitochondrial supernatants were labelled by injecting rats with [125]I-labelled ASF at an appropriate time before killing. When the post-mitochondrial supernatants were incubated *in vitro* under suitable conditions, the ASF became specifically associated with lysosomes (Mullock et al., 1989). LE did not participate in the association. This cell-free system, however, contained many irrelevant components. We subsequently used gradient centrifugation techniques (Branch et al., 1987; Mullock et al., 1983; 1987) to purify a DE fraction, which reacted with purified lysosomes *in vitro* under suitable conditions (Luzio and Mullock, 1992).

Use of purified DE vesicle fractions has enabled reactions to be examined not only by centrifugal techniques, but also by direct measurement of membrane fusion using a fluorescence dequenching method (Hoekstra et al., 1984; Hoekstra, 1990; Mullock and Luzio, 1992). The development of a simple fluorescence dequenching assay has allowed us to monitor the purification from lysosomes of a protein factor involved in mediating endosome-lysosome membrane fusion. We have also developed an assay to measure content mixing after cell-free fusion of endosomes and lysosomes.

**Preparation of endosome, lysosome and cytosol fractions.**
Endosomes (DE), lysosomes, mitochondria and filtered cytosol were prepared as previously described (Mullock and Luzio, 1992). DE were purified >6 fold relative to the endoplasmic reticulum marker glucose 6-phosphatase and >35 fold relative to the lysosomal marker N-acetyl-β-glucosaminidase. For centrifugation experiments DE fractions were used without freezing; for fluorescent dequenching experiments aliquots were stored in liquid nitrogen. DE were labelled with self quenching concentrations of octadecylrhodamine (R18) as previously described (Mullock and Luzio, 1992). In the

lysosome preparation ~90% of the N-acetyl-β-glucosaminadase was latent. Other organelles were present at <2% homogenate concentration. Crude mitochondria were obtained as a side fraction in the course of lysosome preparation. Cytosol was prepared by centrifuging liver post-mitochondrial supernatant at 288,000g for 45 min. Small molecules were removed by gel filtration through Bio-Gel P6 (Biorad Laboratories) and aliquots were then stored in liquid nitrogen. The filtered cytosol still contained 0.25M sucrose, 10mM N-tris (hydroxymethyl) methyl-2-aminoethane-sulphonic acid (TES), pH 7.4 and 1mM $Mg^{2+}$.

Endosome-lysosome incubations, analysis of interactions by analytical density gradient centrifugation and analysis of membrane fusion by fluorescence dequenching were as previously described (Mullock et al, 1989; Mullock and Luzio, 1992).

## Characteristics of endosome-lysosome interaction analysed by centrifugation.

After incubation of DE with lysosomes in the presence of cytosol and an ATP regenerating system at 37°C for 30 min, 28 ± 4% (8) of $^{125}$I-ASF originally in DE was detected at the lysosomal position following centrifugation in analytical Nycodenz density gradients. In contrast <4% of the radioactivity was observed at the lysosomal position on the Nycodenz gradients after incubation at 37°C in the absence of lysosomes or after incubating the complete system at 4°C. Only 14±3% (6) of radioactivity was observed at the lysosomal position after incubations were carried out at 37°C in the absence of cytosol. These data show that the DE interaction with lysosomes is maximal in the presence of energy and cytosol but that some DE appear to be primed for interaction with lysosomes in the absence of cytosol. No evidence for DE maturation into lysosomes was obtained.

Interaction of DE and lysosomes analysed on Nycodenz gradients was inhibited ~40% by 1.5mM N-ethylmaleimide, 45% by 1mM GDP-βS and ~40% by 1mM vanadate.

**Characterisation of endosome-lysosome interaction analysed by fluorescence dequenching.**

Fluorescence dequenching experiments were carried out to show membrane fusion rather than association between DE and lysosomes. They were performed in the absence of cytosol since it contains phospholipid exchange proteins. Fluorescence dequenching of R18-labelled DE membranes when incubated with lysosomes at 37°C was time dependent and also dependent on the concentration of lysosomes (Mullock and Luzio, 1992). Mitochondrial fractions could not substitute for lysosomes. Dequenching after 1h incubation was greater in the presence of 2mM ATP (~30%) than in the presence of apyrase without ATP (~10%). Increased fluorescence always remained associated with pelletable membranes. Vanadate (1mM) inhibited fluorescence dequenching ~40% but N-ethylmaleimide (5mM) and GDP-βS were without effect.

Extraction of lysosomes with 10mM TES buffer, pH 7.4 greatly reduced the ability of lysosome membranes to function in the fusion assay with R18-labelled DE. Recombination of extract with extracted lysosomes produced a response at least double that produced by either alone. Proteolytic digestion showed that the fusogenic factor(s) in the extract was protein. This fusogenic factor has been partially purified >200 fold from the extract using successive ammonium sulphate precipitation, ion exchange chromatography and gel filtration steps. It appears to be a single polypeptide chain of $M_r$ 22,000 by SDS-PAGE.

**Characterisation of endosome-lysosome interaction analysed by content mixing.**

The destructive nature of the interior of the lysosome *in vivo* has made it difficult to load lysosomes with reagents for measuring content mixing. In previous experiments using perfused liver we have shown that whereas much of a dose of [125]I-labelled pIgA entering the liver is transcytosed across hepatocytes and appears in bile, some of the remainder enters the lysosomal pathway (Perez et al., 1988). Radiolabelled pIgA can be immunoprecipitated from liver lysosomes purified from rats which have received an intravenous injection of $^{125}I$-pIgA 20-50 min before killing. Such immunoprecipitated $^{125}I$-pIgA appears to be relatively intact judged by SDS-PAGE. The ability to prepare lysosomes containing intact endocytosed ligand has enabled the establishment of a content mixing assay to assess endosome-lysosome fusion.

A donor endosome fraction was prepared from a rat liver loaded for 4 min with avidin conjugated ASF (this incubation time should load LE and DE). Acceptor endosomes or lysosomes were prepared from a rat liver loaded for 20 min with biotinylated $^{125}I$-pIgA (to load all endosome compartments and lysosomes). Donor and acceptor fractions were then incubated for 30 min in the presence of filtered cytosol, 50mM KCl, 1.5mM $MgCl_2$, 1mM dithiothreitol and biotin-labelled insulin to mop up any avidin leakage . After incubation, more biotin-labelled insulin and protease inhibitors were added, followed by detergents. The mixture was incubated with rabbit anti-avidin antibodies. The avidin-anti-avidin complex was immunoprecipitated using an anti-rabbit Ig immunoadsorbent and counted. Total immunoprecipitable counts were measured in the absence of biotin-labelled insulin.

Using this system endosome-lysosome fusion was demonstrated after incubation in the presence of an energy regenerating system at 37ºC (but not at 0ºC) with ~20% efficiency. Apyrase inhibited fusion. Similarly endosome-lysosome fusion was demonstrated with ~50% efficiency, again with inhibition in the presence of apyrase. Vanadate was an effective inhibitor of endosome-lysosome fusion assessed by content mixing with >50% inhibition at 5mM vanadate.

## Discussion

The three assays presented here demonstrate that it is possible to prepare endosome fractions that are capable of interaction and fusion with lysosomes in cell-free systems. Whilst this does not rule out a role for maturation in the passage of ligands through the endosomal system it indicates that membrane fusion events also occur even at late points in the pathway. It also suggests a mechanism for delivery of material from endosomes to pre-existing lysosomes.

The different assays reported measure different aspects of endosome-lysosome interaction and it is therefore unsurprising that inhibitors have differential effects in each assay. To date only vanadate has been shown to inhibit all three assays. The effects of N-ethylmalemide and GDP-βS on the interaction assay assessed by centrifugation suggest it will be of interest to know whether the specific NSF needed for early endosome fusions (Wessling-Resnick and Braell, 1990) and rab 7, a specific low $M_r$ GTP-binding protein localised to late endosomes in MDCK cells (Chavrier et al, 1990), are required at some steps in endosome-lysosome interaction. The partial purification of a protein required for endosome-lysosome membrane fusion from lysosomal extracts indicates a role for additional protein factors.

## Acknowledgment

This work was supported by the Medical Research Council.

## References

Branch, W.J., B.M. Mullock and J.P. Luzio. 1987. Rapid subcellular fractionation of the rat liver endocytic compartments involved in transcytosis of polymeric immunoglobulin A and endocytosis of asialofetuin. Biochem. J. 244: 311-315.

Chavrier, P., R.G. Parton, H.P. Hauri, K. Simons and M. Zerial. 1990. Localisation of low molecular weight GTP binding proteins to exocytic and endocytic compartments. Cell 62: 317-329.

Geuze, H.J., H.A. van der Donk, C.F. Simmons, J.W. Slot, G.J. Strous and A.L. Schwarz. 1986. Receptor-mediated endocytosis in liver parenchymal cells. Int Rev Exp Pathol 29: 113-117.

Griffiths, G. and J. Gruenberg, J. 1991. The arguments for pre-existing early and late endosomes. Trends Cell Biol 1: 5-9.

Hoekstra, D., T. de Boer, K. Klappe and J. Wilschut. 1984. Fluorescence method for measuring the kinetics of fusion between biological membranes. Biochemistry 23: 5675-5681.

Hoekstra, D. 1990. Fluorescence assays to monitor membrane fusion: potential application in biliary lipid secretion and vesicle interactions. Hepatology 12: 615-665.

Luzio, J.P. and B.M. Mullock. 1992. The interaction of late endosomes with lysosomes in a cell-free system. In: Courtoy, P. J. (ed.) Endocytosis: From Cell Biology to Health, Disease and Therapy. Springer-Verlag, Berlin, Heidelberg, New York (NATO ASI Series H: Cell Biology Vol. 62), pp 123-129.

Mueller, S.C. and A.L. Hubbard. 1986. Receptor-mediated endocytosis of asialoglyco-proteins by rat hepatocytes: receptor-positive and receptor-negative endosomes. J Cell Biol 102: 932-942.

Mullock, B.M. and J.P. Luzio. 1992. Reconstitution of rat liver endosome-lysosome fusion in vitro. Methods Enzymol 219: 52-60.

Mullock, B.M., J.P. Luzio and R.H. Hinton. 1983. Preparation of a low-density species of endocytic vesicle containing immunoglobulin A. Biochem J 214: 823-827.

Mullock, B.M., R.H. Hinton, J.V. Peppard, J.W. Slot and J.P. Luzio. 1987. The preparative isolation of endosome fractions: a review. Cell Biochem Function 5: 235-243.

Mullock, B.M., W.J. Branch, M. van Schaik, L.K. Gilbert and J.P. Luzio. 1989. Reconstitution of an endosome-lysosome interaction in a cell-free system. J Cell Biol 108: 2093-2099.

Murphy, R.F. 1991. Maturation models for endosome and lysosome biogenesis. Trends Cell Biol 1: 77-82.

Perez, J.H., W.J. Branch, L. Smith, B.M. Mullock and J.P. Luzio. 1988. Investigation of endosomal compartments involved in endocytosis and transcytosis of polymeric immunoglobulin A by subcellular fractionation of perfused isolated rat liver. Biochem J 251: 763-770.

Wessling-Resnick, M. and W.A. Braell. 1990. Characterisation of the mechanism of endocytic vesicle fusion *in vitro*. J Biol Chem 265: 7827-7831.

# A NEW RECEPTOR FOR LYSOSOMAL PROENZYMES

Ann H. Erickson, Gail F. McIntyre, Gene D. Godbold, and Richard L. Chapman
Department of Biochemistry and Biophysics
University of North Carolina
Chapel Hill, North Carolina 27599-7260

Lysosomal enzymes are segregated from secretory proteins in the trans Golgi network by virtue of their ability to bind to mannose 6-phosphate receptors (MPRs) which carry the lysosomal enzymes to late endosomes or prelysosomes (Review, Kornfeld and Mellman, 1989). In these acidic vesicles, the enzymes are released from the MPRs, which recycle back to the Golgi. It is not clear how the lysosomal enzymes, now soluble in the lumen of these sorting vesicles, are targeted on to lysosomes. We have identified a new receptor which binds lysosomal proenzymes at pH 5 (McIntyre and Erickson, 1991). The pH-dependence of the binding reaction suggests that the receptors might be localized in late endosomes and therefore could participate in this sorting process.

The proform of cathepsin L, a lysosomal cysteine protease, interacts specifically with microsomal membranes at pH 5 by an MPR-independent mechanism. When microsomal membranes were lysed by repeated freezing and thawing in a pH 5 buffer containing mannose 6-phosphate (10 mM) and saponin (0.05%), the inactive proenzyme remained with the membranes (Fig. 1). Washing the membranes with pH 7 buffer solubilized most of the proenzyme and a pH 10 wash, which strips the membranes of peripheral membrane proteins, released the remaining procathepsin L. Consistent with these results, the proform but not the mature active forms of the enzyme can be rebound at pH 5 but not at pH 7 to membranes stripped of peripheral membrane proteins (McIntyre and Erickson, 1991). Similar results were obtained for procathepsin B (data not shown), a second lysosomal cysteine protease, and for procathepsin D (McIntyre and Erickson, 1991), a lysosomal aspartic protease. In each case the proform of the lysosomal protease behaves like a peripheral membrane protein at pH 5, the pH of late endosomes.

NATO ASI Series, Vol. H 74
Molecular Mechanisms of Membrane Traffic
Edited by D. J. Morré, K. E. Howell, and J. J. M. Bergeron
© Springer-Verlag Berlin Heidelberg 1993

360

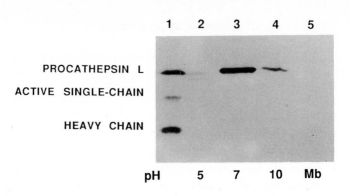

PROCATHEPSIN L

ACTIVE SINGLE-CHAIN

HEAVY CHAIN

pH     5     7     10     Mb

Fig. 1. **Procathepsin L is bound to microsomal membranes at pH 5.** Cathepsin L was immunoprecipitated from KNIH cells (*lane 1*), the soluble fractions generated when microsomal membranes prepared from KNIH cells were lysed at pH 5 (*lane 2*), pH 7 (*lane 3*), or pH 10 (*lane 4*) or from the membranes after the pH 10 lysis (*lane 5*). Forms of cathepsin L were resolved by polyacrylamide gel electrophoresis and visualized on a Western blot using enhanced chemiluminesence (Amersham).

Binding of procathepsin L to microsomal membranes could be due to a nonspecific interaction with random proteins or lipids in the membranes, or it could result from an interaction with a specific membrane protein or receptor. Two hallmarks of a specific, receptor-mediated interaction are saturation and competition. Binding of procathepsin L secreted into KNIH tissue culture medium to microsomal membranes stripped of peripherial proteins was clearly saturable at pH 5 (Fig. 2). Furthermore, this binding reaction could be inhibited by synthetic peptides based on the N-terminal sequence of the cathepsin L propeptide. Peptide A, consisting of the first 24 amino acid residues (Thr18-Gly41) of mouse procathepsin L (which lacks the signal peptide but contains the 96-residue N-terminal propeptide) plus alanyl-cysteine amide, inhibited the binding of procathepsin L to stripped microsomal membranes with a $K_I$ of 7.3 $\mu$M. Peptide B, comprised of the C-terminal 9 amino acids of peptide A (Lys33-Gly41) plus alanyl-cysteine amide, inhibited binding with a $K_I$ of 165 $\mu$M, while Peptide C, a control 12-residue peptide based on the N-terminus of mature cathepsin L (Lle114-Gly124, plus tyrosine amide), had little or no affect on the binding of procathepsin L to membranes at pH 5. These data indicate that a specific protein sequence, consisting in part of Peptide B, mediates the association of procathepsin L with microsomal membranes at pH 5.

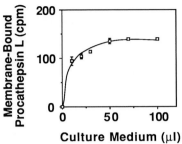

Fig. 2. **Binding of procathepsin L to microsomal membranes is saturable.** KNIH microsomal membranes stripped of peripheral proteins were incubated for 2 hr at 4 °C with increasing amounts of conditioned KNIH culture medium containing [$^{35}$S]methionine-labeled secreted procathepsin L. Cathepsin L was immunoprecipitated from the pelleted membranes, resolved by polyacrylamide gel electrophoresis, and quantitated by scintillation counting of gel slices.

A 6-residue sequence (Ser34-Leu39) from mouse procathepsin L present in both peptides A and B resembles sequences shown to be critical for correct sorting of yeast carboxypeptidase Y (Johnson et al., 1987; Valls et al., 1987) and proteinase A (Klionsky et al., 1988) to vacuoles, the functional equivalents of mammalian lysosomes (Fig. 3). In addition, the 6-residue sequence is present in the propeptides of procathepsins B and D, which have also been shown to bind to membranes at pH 5 (Fig. 3). Thus this consensus sequence defined by phenotypically similar types of amino acids (positive, hydrophobe) may constitute a binding site which is recognized by a receptor or a family of receptor proteins.

The similarity of the procathepsin L binding site to the yeast vacuolar sorting sequences suggests that pH 5-dependent membrane association may be critical for correct sorting of the proenzyme to lysosomes. After release from an MPR in a late endosome, procathepsin L may bind to a lysosomal proenzyme receptor (LPR) for transport to lysosomes. On reaching the lysosome, cleavage of the propeptide from the LPR-bound proenzyme would release the enzymatically active mature protein within the lysosome. Thus mammalian cells may have evolved the MPR-mediated sorting pathway to function in series with an LPR-mediated pathway, which resembles the vacuolar-sorting pathway used by the more primitive unicellular eukaryotes.

```
                              +     + + +
          Peptide B           K S T H R R L Y G
                              +     + + +
    Mouse Procathepsin L      . . . K S T H R R L Y G . . .
                                    + +
    Mouse Procathepsin B      . . . I S Y L K K L C G . . .
                                    + +
    Mouse Procathepsin D      . . . T S I R R T M T E . . .
                                    +
  Yeast Carboxypeptidase Y    . . . I S L Q R P L G I . . .
                                    + - +
     Yeast Proteinase A       . . . F S R E H P F F T . . .
                                    |     |   |
                                    1 2 3 4 5 6
       Consensus Sequence         S . . + . Hydrophobe
```

Fig. 3. **Alignment of peptide B from procathepsin L with similar sequences from mouse procathepsins and vacuolar sorting sequences from yeast proenzymes.** Sequences shown are mouse procathepsin L-(33-41) (Portnoy et al., 1986), mouse procathepsin B-(52-60) (Chan et al., 1986), mouse procathepsin D-(30-38) (Grusby et al., 1990), yeast carboxypeptidase Y-(21-29) (Johnson et al., 1987; Valls et al., 1990), and yeast proteinase A-(67-75) (Klionsky et al., 1988).

## REFERENCES

Chan, SJ, Segundo, BS, McCormick, MB, Steiner, DH (1986) Nucleotide and predicted amino acid sequences of cloned human and mouse preprocathepsin B cDNAs. Proc Natl Acad Sci USA 83:7721-7725

Grusby, MJ, Mitchell, SC, Glimcher, LH (1990) Molecular cloning of mouse cathepsin D. Nucleic Acids Res 18:4008

Johnson, LM, Bankaitis, VA, Emr, SD (1987) Distinct sequence determinants direct intracellular sorting and modification of a yeast vacuolar protease. Cell 48:875-885

Klionsky, DJ, Banta, LM, Emr, SD (1988) Intracellular sorting and processing of a yeast vacuolar hydrolase: Proteinase A propeptide contains vacuolar targeting information. Mol Cell Biol 8:2105-2116

Kornfeld, S and Mellman, I (1989) The biogenesis of lysosomes. Ann Rev Cell Biol 5:483-525

McIntyre, GF, Erickson, AH (1991) Procathepsins L and D are membrane-bound in acidic microsomal vesicles. J Biol Chem 266:15438-15445

Portnoy, DA, Erickson, AH, Kochan, J, Ravetch, JV, Unkeless, JC (1986) Cloning and characterization of a mouse cysteine proteinase. J Biol Chem 261:14697-14703

Valls, LA, Hunter, CP, Rothman, JH, Stevens, TH (1987) Protein sorting in yeast: The localization determinant of yeast vacuolar carboxypeptidase Y resides in the propeptide. Cell 48:887-897

# A Protein Kinase/Lipid Kinase Complex Required for Yeast Vacuolar Protein Sorting

Jeffrey H. Stack, Paul K. Herman* and Scott D. Emr
Division of Cellular and Molecular Medicine
Howard Hughes Medical Institute
UC San Diego School of Medicine
La Jolla, CA 92093-0668

The delivery of proteins to the yeast vacuole serves as an excellent paradigm for intracellular protein sorting pathways. Genetic selections for mutants defective in this process have identified a large number of *vps* (vacuolar protein sorting defective) mutants that define more than 40 complementation groups (reviewed in Klionsky et al., 1991). This high degree of genetic complexity presumably reflects the biochemical complexity of the multi-step protein sorting reaction. Specific cellular components must recognize vacuolar proteins, package them into specific transport vesicles and then mediate the delivery and fusion of these vesicles with the vacuolar membrane. The list of potential activities and structures required for the sorting and transport of vacuolar proteins can easily accommodate the large number of gene products implicated by the genetic studies. Two of the *VPS* gene products encode proteins of known biochemical function and each has provided significant new insights into the regulation of the sorting reaction. *VPS15* encodes a 1,455 amino acid serine/threonine protein kinase homolog and *VPS34* encodes a 875 amino acid phosphatidylinositol kinase homolog (Herman and Emr, 1990; Herman et al., 1991a; Hiles et al., 1992).

## Vps15 Protein Kinase

The N-terminus of the *VPS15* gene product exhibits substantial sequence similarity to the catalytic domain of protein kinases. An extensive mutational analysis of the Vps15 protein kinase has shown that alterations in residues in

---

* Department of Molecular and Cellular Biology, University of California, Berkeley, Berkeley, CA 94720

NATO ASI Series, Vol. H 74
Molecular Mechanisms of Membrane Traffic
Edited by D. J. Morré, K. E. Howell, and J. J. M. Bergeron
© Springer-Verlag Berlin Heidelberg 1993

Vps15p that are highly conserved among protein kinases result in biological inactivation of the protein (Herman et al., 1991a,b). The soluble vacuolar hydrolase, carboxypeptidase Y (CPY), was completely missorted and secreted as a Golgi-modified precursor in severe *vps15* kinase domain mutants. In addition to a *ts* growth defect, *vps15* kinase mutants also are found to be defective for the *in vivo* phosphorylation of Vps15p itself, most likely reflecting diminished kinase catalytic activity required for an autophosphorylation event (Herman et al., 1991a).

Analysis of a temperature-sensitive *vps15* allele has indicated that Vps15p plays a direct role in the sorting of soluble vacuolar hydrolases. Deletion of as little as 30 amino acids from the C-terminus of Vps15p results in a temperature-conditional vacuolar protein sorting defect (Herman et al., 1991b). Vacuolar proteins are sorted and processed normally at the permissive temperature but a complete block in CPY processing is imposed immediately after shifting the mutant cells to the non-permissive temperature. The rapid onset and efficient reversibility of the block suggest that the primary role of Vps15p in yeast cells is to mediate the delivery of soluble vacuolar proteins.

## Vps34 Lipid Kinase

The *VPS34* gene encodes a protein with significant similarity to the catalytic subunit of mammalian phosphatidylinositol 3-kinase (PI 3-kinase) (Hiles et al., 1992). PI 3-kinase phosphorylates membrane phosphatidylinositol and in mammalian cells is found to associate with many signal transducing receptor tyrosine kinases (reviewed in Cantley et al., 1991). While PI 3-kinase activity is readily detected in extracts from wild-type yeast cells (Auger et al., 1989), extracts from yeast strains deleted for the *VPS34* gene, or containing *vps34* point mutations, exhibit extremely low levels of PI 3-kinase activity (P.Schu, K.Takegawa, J.Stack and S.Emr, unpublished observations). In addition, strains overproducing Vps34p show elevated levels of PI 3-kinase. Altogether, these data indicate that *VPS34* encodes a PI 3-kinase in yeast and further suggest that PI 3-kinase activity is involved in regulating intracellular protein sorting within the eukaryotic secretory pathway.

## Vps15p/Vps34p Protein Complex

The Vps15 protein kinase and Vps34 lipid kinase have been found to functionally and physically interact in yeast cells. Overproduction of Vps34p from a *VPS34* multicopy plasmid can suppress the growth and vacuolar protein sorting defects associated with *vps15* kinase domain mutants (J.Stack, P. Herman and S.Emr, unpublished observations). In contrast, neither the growth nor sorting defects of a strain deleted for the *VPS15* gene are suppressed by the overproduction of the Vps34 lipid kinase. This lack of suppression indicates that the overproduction of Vps34p cannot bypass the requirement for Vps15p in vacuolar protein sorting. Non-denaturing immunoprecipitations and chemical cross-linking experiments have demonstrated that Vps15p and Vps34p physically interact in yeast. Vps15p and Vps34p can be co-immunoprecipitated from yeast cell extracts by antisera specific for either protein. Subcellular fractionation and sucrose gradient centrifugation have shown that a substantial proportion of Vps15p and Vps34p colocalize to a specific membrane fraction in yeast cell extracts. The combined genetic and biochemical data therefore demonstrate that Vps15p and Vps34p act together within a membrane-associated hetero-oligomeric protein complex to facilitate yeast vacuolar protein delivery.

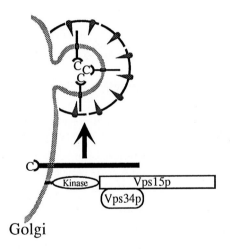

Figure 1. Possible Role for Vps15p and Vps34p in Vacuolar Protein Sorting
The Vps15p/Vps34p complex is shown associating with a CPY-specific transmembrane receptor present in a late Golgi compartment. The action of Vps15p/Vps34p facilitates vacuolar protein sorting possibly through the packaging of ligand-bound receptors into transport vesicles or by catalyzing the formation and/or stabilization of the vesicles themselves. The transport vesicle containing CPY-receptor complexes is shown with associated adaptors and coat proteins.

Model for Vacuolar Protein Sorting

Vps34p-mediated phosphorylation of membrane phosphatidylinositol may serve as a signal that triggers the specific interaction, or stabilization, of other proteins, such as coat proteins, required for transport to the vacuole (Fig.1). It is also possible that the Vps15 protein kinase may directly regulate Vps34 lipid kinase activity through a specific protein phosphorylation reaction. Vps15p kinase activity may itself be regulated by direct interaction with particular transmembrane receptors, such as a p2CPY-specific receptor. Ligand binding by the receptor could result in the activation of both Vps15p and Vps34p. In such a model, Vps15p and Vps34p effectively act as components of a signal transduction complex. Ligand activation of specific transmembrane receptors leads to production of a second messenger (PI 3-phosphate) that may trigger the action of as yet unknown effector proteins to mediate vacuolar protein sorting.

References

Auger KR, Carpenter CL, Cantley LC, Varticovski L (1989) Phosphatidylinositol 3-kinase and its novel product, phosphatidylinositol 3-phosphate, are present in *Saccharomyces cerevisiae*. J Biol Chem 264:20181-20184

Cantley LC, Auger KR, Carpenter C, Duckworth B, Graziani A, Kapeller R, Soltoff S (1991) Oncogenes and signal transduction. Cell 64:281-302

Herman PK, Emr SD (1990) Characterization of *VPS34*, a gene required for vacuolar protein sorting and vacuole segregation in *Saccharomyces cerevisiae*. Mol Cell Biol 10:6742-6754

Herman PK, Stack JH, DeModena JA, Emr SD (1991) A novel protein kinase essential for protein sorting to the yeast lysosome-like vacuole. Cell 64:425-437

Herman PK, Stack JH, Emr SD (1991) A genetic and structural analysis of the yeast Vps15 protein kinase: evidence for a direct role of Vps15p in vacuolar protein delivery. EMBO J 10:4049-4060

Hiles ID, Otsu M, Volinia S, Fry MJ, Gout I, Dhand R, Panayotou G, Ruiz-Larrea F, Thompson A, Totty NF, Hsuan JJ, Courtneidge SA, Parker PJ, Waterfield MD (1992) Phosphatidylinositol 3-kinase: structure and expression of the 110 kd catalytic subunit. Cell (in press)

Klionsky DJ, Herman PK, Emr SD (1990) The fungal vacuole: composition, function and biogenesis. Microbiol Rev 54:266-292

# THE MORPHOLOGY BUT NOT THE FUNCTION OF ENDOSOMES AND LYSOSOMES IS AFFECTED BY BREFELDIN A

Salli A. Wood and William J. Brown
357 Biotechnology Building
Section of Biochemistry, Molecular and Cell Biology
Cornell University
Ithaca, New York 14853

The organelles of both the biosynthetic and endocytic systems interact via highly regulated, generally vesicle-mediated, transport pathways that are disrupted in a very specific and characteristic manner by the fungal metabolite, brefeldin A (BFA) (Klausner et al., 1992). For example, by preventing the association of the coatomer complex with nascent buds in the Golgi complex, BFA blocks vesicle formation, inducing the formation of tubular processes (Donaldson et al., 1990; Orci et al., 1991; Cluett and Brown, 1991). These membranous tubules extend from the Golgi complex and fuse with the endoplasmic reticulum (ER), forming a hybrid organelle from which transport is blocked (Lippincott-Schwartz et al., 1989, 1990).

In an analogous manner, BFA prevents the association of clathrin adaptins with the trans Golgi network (TGN), which blocks the formation of clathrin coated vesicles at the TGN (Robinson and Kreis, 1992; Wong and Brodsky, 1992), and induces the formation of tubules from the TGN (Wood et al., 1991; Lippincott-Schwartz et al., 1991). These tubules can be labeled by indirect immunofluorescence using antibodies directed against the mannose-6-phosphate receptor (M6PR). Although M6PRs are normally concentrated in the TGN and pre-lysosomal compartment (PLC), they are rapidly redistributed to the TGN in the presence of BFA, so that the M6PR enriched tubules are derived from the TGN and not the PLC. These TGN-derived tubules extend along microtubules and fuse with early endosomes forming an extensive network (Wood et al., 1991).

To demonstrate that these tubules form first from the TGN and then fuse with early endosomes, cells with early endosomes labeled by uptake of lucifer yellow were treated with BFA for 2 min, fixed, and then the TGN was labeled with anti-M6PR antibodies. Under these conditions, the TGN tubulates but the early endosomes remain as distinct peripheral vesicles. It is only after at least 10 min of BFA treatment that fusion of the TGN with early endosomes can be

NATO ASI Series, Vol. H 74
Molecular Mechanisms of Membrane Traffic
Edited by D. J. Morré, K. E. Howell, and J. J. M. Bergeron
© Springer-Verlag Berlin Heidelberg 1993

demonstrated by the colocalization of M6PR and lucifer yellow, endocytosed for 5 min, in an extensive tubular network. With increased incubation time in BFA (>1 h), the fused TGN/early endosomal tubular network collapses to a cluster of tubulo-vesicular elements in the juxtanuclear region. Nevertheless, endocytosed material is still delivered to the fused compartment within 5 min of uptake.

Remarkably, despite these dramatic alterations in early endosomal morphology, endocytosed material delivered to the hybrid TGN/early endosome can be chased into the centrally located, morphologically unaltered PLC. Thus a 5 min pulse and a 10 min chase with lucifer yellow is sufficient to label the central vesicles of the PLC in untreated cells and in cells treated with BFA for a total of either 15 min or 2 h.

Transport from the TGN/early endosome to the PLC remains selective in BFA, so that recycling receptors return from the fused TGN/early endosome to the cell surface with normal kinetics (Damke et al., 1991). However, the M6PR, which normally cycles between the TGN and endosomes, with only some transport to the cell surface, is dramatically redistributed by BFA. Rapidly upon addition of BFA, the PLC is depleted of M6PR, while levels of M6PR in the TGN increase. After the BFA-induced fusion of the TGN and early endosomes, levels of M6PR on the cell surface increase 5-fold over control cells (Wood et al. 1991). Even with long-term BFA treatment, M6PRs are excluded from the PLC. Thus, unlike endocytosed material, the M6PR is not transported to the PLC from the fused TGN/early endosome, but instead is preferentially transported to the cell surface.

The morphology of the PLC is not significantly altered by BFA, and it appears that transport of the M6PR from the PLC to the TGN continues, suggesting the functioning of the PLC is unaffected by BFA. If this is correct, transport to lysosomes should continue. In fact, in cells treated with BFA for 4 h, lucifer yellow pulsed into the PLC could be chased into lysosomes, which were labeled with antibodies to cathepsin D (CD), a resident lysosomal enzyme. The colocalization of CD, visualized at the EM level by immunoperoxidase staining, and colloidal gold particles, endocytosed and chased into lysosomes during a 4 h incubation with BFA, confirms these results. These data demonstrate that transport throughout the entire endocytic pathway continues in the presence of BFA. These results support findings by Misumi et al. (1986) that lysosomal degradation of endocytosed material continues in BFA at up to 80% of normal levels. However, these results differ from those of Lippincott-Schwartz et al. (1991), who found a transport block in chick embryo fibroblasts at the PLC, using rhodamine-ovalbumin as a tracer. This discrepancy is probably due to cell-specific differences in either the BFA-sensitivity of the pathway or transport mechanisms.

BFA is thought to act by preventing the association of coat proteins with nascent vesicles, blocking specific transport events in the Golgi complex and TGN (Klausner et al., 1992). The continuation of transport through the endocytic pathway in the presence of BFA suggests that transport between compartments of the endosome-lysosome system does not use vesicles but instead occurs through maturation (Murphy, 1991). Alternatively, transport through the endocytic pathway may occur via coated vesicles that are resistant to BFA. Clathrin coated vesicle formation at the plasma membrane is unaffected by BFA (Robinson and Kreis, 1992), indicating that BFA's effect on vesicle formation is highly selective.

Surprisingly, in the presence of BFA, lysosomes, defined both kinetically by the delivery of lucifer yellow, and immunocytochemically, by the presence of CD, are not large spherical structures loosely clustered in the perinuclear region as in untreated cells. Instead, with BFA treatment, lysosomes first scatter, and then gradually elongate, until after 4 h, they have formed long tubules which extend into the cell periphery. As with other BFA-induced morphological changes, the tubulation of lysosomes is rapidly reversible. Upon removal of the drug lysosomes regain their normal morphology and positioning within 5 min.

As with the formation of BFA-induced tubules from the Golgi complex and TGN (Lippincott-Schwartz et al., 1990; Cluett and Brown, 1991), lysosomal tubulation is energy dependent. Depletion of cellular ATP by pretreatment of cells with 50mM deoxy-glucose and 0.05% sodium azide, inhibits the formation of lysosomal tubules during a 4 h incubation in BFA. However, once lysosomal tubules have formed in BFA, ATP is not required for their maintenance. Interestingly, when ATP is depleted during recovery from 4 h of BFA exposure, lysosomal tubules are maintained, indicating that recovery is an energy dependent process.

Lysosomes are maintained in a loose perinuclear position by associations with microtubules. Treatment of cells with nocodazole, causes lysosomes to scatter (Matteoni and Kreis, 1987). Double-label immunofluorescence, demonstrates that BFA-induced lysosomal tubules, labeled with antibodies to CD, closely associate with microtubules visualized with anti-tubulin antibodies, suggesting that the formation of lysosomal tubules is microtubule-dependent. To confirm this hypothesis, cells were pretreated with nocodazole to disrupt microtubules and then incubated with BFA. As expected, these conditions prevent lysosomal tubulation; however, lysosomes form clusters which are not seen with nocodazole treatment alone. These results suggest that BFA has effects on lysosomes beyond those requiring microtubules.

The formation of tubules from lysosomes in the presence of BFA raises several intriguing questions: If, as has been proposed, tubulation results from a block in vesiculation, then what transport from lysosomes has been disrupted by BFA? Do lysosomes fuse with a target organelle? If so, what is the target? Transport from lysosomes to the PLC or the plasma membrane may occur via vesicle mediated transport. If so these pathways may be targets for BFA.

# References

Cluett EB, Brown WJ (1991) Golgi membranes have a propensity to form tubules in the absence of vesicle associated coat proteins. J Cell Biol 115: 411a

Damke H, Klumperman J, von Figura K, Braulke T. (1991). Effects of brefeldin A on the endocytic route. Redistribution of mannose 6-phosphate /insulin-like growth factor II receptors to the cell surface. J Biol Chem 266: 24829-24833.

Donaldson JG, Lippincott-Schwartz J, Bloom GS, Kreis TE, Klausner RD.(1990). Dissociation of a 110-kD peripheral membrane protein from the Golgi apparatus is an early event in Brefeldin A action. J Cell Biol 111: 2295-2306.

Klausner RD, Donaldson JG, Lippincott-Schwartz J (1992) Brefeldin A: insights into the control of membrane traffic and organelle structure. J Cell Biol 116: 1071-1080.

Lippincott-Schwartz J, Yuan LC, Bonifacino JS,Klausner RD (1989). Rapid redistribution of Golgi proteins into the ER in cells treated with brefeldin A: evidence for membrane recycling from the Golgi to ER. Cell 56: 801-813.

Lippincott-Schwartz J, Donaldson JG, Schweizer A,Berger EG, Hauri HP, Yuan HP, Klausner RD (1990) Microtubule-dependent retrograde transport of proteins into the ER in the presence of brefeldin A suggests an ER recycling pathway. Cell 60: 821-836.

Lippincott-Schwartz J, Yuan LC, Tipper C, Amherdt M, Orci L,Klausner RD (1991) Brefeldin A's effects on endosomes, lysosomes, and the TGN suggest a general mechanism for regulating organelle structure and membrane traffic. Cell 67: 601-617.

Matteoni R, Kreis TE (1987) Translocation and clustering of endosomes and lysosomes depends on microtubules. J Cell Bio 105: 1253-1265.

Misumi Y, Miki K, Takatsuki A, Tamura G, Ikehara Y. (1986) Novel blockade by brefeldin A of intracellular transport of secretory proteins in cultured rat hepatocytes. J Biol Chem 261: 11398-11403.

Murphy R (1991) Maturation models for endosome and lysosome biogenesis. Trends Cell Biol 1:77-82.

Orci L, Tagaya M, Amherdt M, Perrelet A, Donaldson JG, Lippincott-Schwartz JG, Klausner RD, Rothman JE (1991) Brefeldin A, a drug that blocks secretion, prevents the assembly of non-clathrin-coated buds on Golgi cisternae. Cell 64:1183-1195.

Robinson MS, Kreis T. (1992) Recruitment of coat proteins onto Golgi membranes in intact and permeabilized cells: effects of brefeldin A and G-proteins. Cell 69:129-138.

Wood SA, Park JE, Brown WJ. (1991) Brefeldin A causes a microtubule-mediated fusion of the trans Golgi network and early endosomes. Cell 67: 591-600.

# PURIFICATION OF THE N-ACETYLGLUCOSAMINE-1-PHOSPHODIESTER α-N-ACETYLGLUCOSAMINDASE FROM HUMAN LYMPHOBLASTS

Ke-Wei Zhao and Arnold L. Miller
Department of Neurosciences, 0624
University of California, San Diege
La Jolla, CA 92093 U.S.A.

The GlcNAc-1-phosphodiester α-N-acetylglucosaminidase ("uncovering enzyme") plays a key role in the intracellular targeting of lysosomal enzymes by catalyzing the removal of N-acetylglucosamine exposing mannose-6-phosphate residues on high mannose and/or hybrid oligosaccharide chains of lysosomal enzymes. The subcellular localization of the uncovering enzyme has been proposed to be associated with the *cis* Golgi cisternae, but no definitive evidence exists as to either its exact location or structure.

The uncovering enzyme was first extracted from cultured human lymphoblast cells (25 gm wet weight) with Tergitol NP-10, partially purified by lentil lectin Sepharose 4B, DEAE-Sephacel and Affigel 501 column chromatography (Table 1). The partially purified enzyme preparation was free of GlcNAc phosphotransferase activity. Further purification was achieved by isoelectric focusing (pH 3-10) and preparative SDS-PAGE. Analytical SDS-PAGE of the final preparation revealed two closely migrating protein bands ($M_r$ = 141 kD and 125 kD, respectively) coincident with the uncovering enzyme activity (Fig. 1).

The present study is the first report on the purification of the uncovering enzyme to near homogeneity from human lymphoblasts. A purified uncovering enzyme will lead to a better understanding of its structure, function and location. Current studies are focused on optimizing the

**Table 1.** *Purification of the GlcNAc-1-phosphodiester α–N-acetylglucosaminidase*

| Step | Total Protein (mg) | Total Activity (pmol/h) | Specific Activity (pmol/h/mg) | Purfication Fold | Yield (%) |
|------|------|------|------|------|------|
| Cell Homogenate | 1817 | 47,840 | 26 | 1 | 100 |
| Washed Membranes | 1410 | 42,336 | 30 | 1.1 | 88 |
| Solubilized Enzyme | 136 | 34,440 | 252 | 9.7 | 72 |
| Lentil Lectin-Sepharose 4B | 8.7 | 19,157 | 2192 | 84 | 40 |
| DEAE-Sephacel/Affigel 501 | 5.3 | 19,040 | 3584 | 137 | 40 |

NATO ASI Series, Vol. H 74
Molecular Mechanisms of Membrane Traffic
Edited by D. J. Morré, K. E. Howell, and J. J. M. Bergeron
© Springer-Verlag Berlin Heidelberg 1993

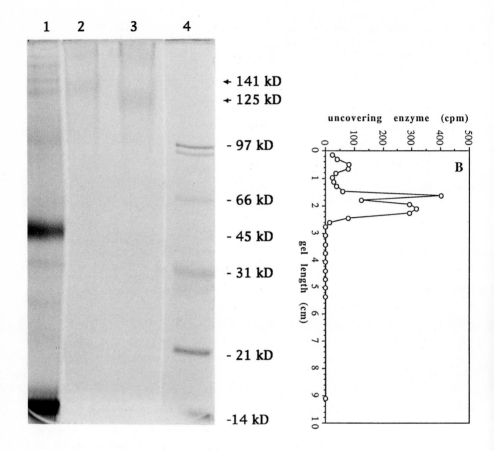

**Fig. 1.** SDS-PAGE of purified uncovering enzyme. Panel A:Coomassie blue stained gel depicting protein patterns in the enzyme preparation after isoelectric focusing (lane 1) and after preparative SDS-PAGE purification (lanes 2 and 3). Lane 4 contains low M✱ standard proteins. Panel B: uncovering enzyme activity monitored in gel slices of a corresponding unstained lane which contains SDS-PAGE purified enzyme.

purification procedure to obtain apparently homogeneous enzyme that can be used for the subsequent production of polyclonal antibodies, examination of the enzyme's subunit structure, and amino acid sequencing. (Supported by USPHS grant NS12138.)

References

Varki A, Kornfeld S (1981) Purification and characterization of rat liver α-N-acetylglucosaminyl phosphodiesterase. J Biol Chem 256: 9937-9943
Waheed A, Hassilik A, Von Figura K (1981) Processing of the phosphorylated recognition marker in lysosomal enzymes. J Biol Chem 256: 5717-5721

# ALTERED REGULATION OF PROTEIN DEGRADATION IN TRANSFORMED HUMAN BRONCHIAL EPITHELIAL CELLS

[1]Hsiang-Kuang Lee, [2]Roy A.M., Myers, and [1,2]Louis Marzella
[1]Department of Pathology, School of Medicine, and
[2]Maryland Institute for Emergency Medical Services Systems
University of Maryland, Baltimore, Maryland 21201, USA

In many transformed cells the post-translational processing and/or translocation of lysosomal proteases and integral membrane proteins is altered. This results in increased expression of lysosomal proteins on the cell surface (Mane et al., 1989; Sloane et al., 1990). In addition, lysosomal protease inhibitors are deregulated. These alterations increase the capacity of transformed cells for degradation of extracellular matrix constituents and are thought to contribute to tumor invasion (Rochefort et al., 1990; Lee et al., 1993). We have tested the hypothesis that intracellular protein degradation is also deregulated in transformed cells. This deregulation may contribute to tumorigenesis by conferring a growth advantage to transformed cells, particularly in conditions of nutritional stress (Lee et al., 1989 and 1992).

To this end, we have compared protein turnover and the induction of lysosomal protein degradation in three transformed bronchial epithelial cells in culture [BEAS-2B, a SV40 transformed clone of normal human bronchial cells (NHBE); BZR, a ras transfected tumorigenic clone of BEAS-2B; Calu-1, an epidermoid carcinoma cell line] with NHBE. Protein degradation and synthesis were measured as the release of $^{14}$C-valine from cells and incorporation of $^{14}$C-valine into proteins. Autophagic degradation of cytoplasm was quantified by ultrastructural morphometry.

We found that basal proteolysis was decreased (by 27 %) in two of the transformed cells (BEAS-2B and BZR). However, the decrease in basal protein degradation did not appear to be related to the induction of the malignant phenotype. In the transformed bronchial cells, basal protein degradation was unaffected by cell density and did not appear to be regulated in parallel with protein synthesis and cell proliferation. Prolonged nutritional deprivation (70-day culture) up-regulated (1.5-fold) basal proteolysis in Calu-1 cells; this change was associated with the development of a more differentiated cellular phenotype and with an increase in autophagy. A relatively long period of nutritional stress was required for this specific augmentation of autophagy reduction (30 day post-confluent culture did not alter basal protein degradation). The application of lysosomal inhibitors, 3-methyladenine (block of sequestration) and chloroquine (acidotropic agent), to Calu-1 cells in serum and nutrient-fully supplemented

NATO ASI Series, Vol. H 74
Molecular Mechanisms of Membrane Traffic
Edited by D. J. Morré, K. E. Howell, and J. J. M. Bergeron
© Springer-Verlag Berlin Heidelberg 1993

media had little effect upon protein degradation, suggesting that in these transformed cells, lysosomes play a minor role in proteolysis in basal conditions.

To examine the induction of the autophagic-lysosomal pathway in transformed cells, cultures were maintained in amino acid-free media for 4 hrs. Amino acid deprivation stimulated autophagic protein degradation equally well (2-fold) in transformed and normal cells. Serum growth factors did not appear to regulate protein degradation in either normal or transformed cells. However, chronic nutritional stress markedly up-regulated autophagic protein degradation (4-fold) and also enhanced (16 %) protein degradation in response to deprivation of growth factors in Calu-1 cells.

The impact of cell differentiation upon protein degradation was also determined. Culture of normal bronchial cells in high $Ca^{2+}$ (1 mM) induced phenotypic changes and further increased (30 %) the protein degradation in nutrient-free media. On the other hand, in transformed cells (Calu-1) $Ca^{2+}$ caused only phenotypic changes.

We conclude that in some bronchial epithelial cells, transformation decreases basal degradation of cytoplasmic proteins by the non-lysosomal pathway. Down-regulation of protein degradation occurs independently of protein synthesis. In the transformed cells the lysosomal protein degradation is not down-regulated, is augmented by nutritional stress, but is not further enhanced by inducing cell differentiation. These alterations in the regulation of proteolysis in the transformed cells may contribute to tumorigenesis.

## REFERENCES

Lee H-K, Jones RT, Myers RAM, Marzella L (1992) Regulation of protein degradation in normal and transformed human bronchial epithelial cells in culture. Arch Biochem Biophys:in press

Lee H-K, Marzella L (1993) Regulation of intracellular protein degradation with special reference to lysosomes: role in cell physiology and pathology. Int Rev Exp Pathol:in press

Lee H-K, Myers RAM, Marzella L (1989) Stimulation of protein degradation by nutrient deprivation in a differentiated murine teratocarcinoma (F9 12-1a) cell line. Exp Mol Pathol 50:134-146

Mane SM, Marzella L, Bainton DF, Holt VK, Cha Y, Hildreth JEK, August JT (1989) Purification and characterization of human lysosomal membrane glycoproteins. Arch Biochem Biophys 268:360-378

Rochefort H, Capony F, Garcia M (1990) Cathepsin D: a protease involved in breast cancer metastasis. Cancer Metastasis Rev 9:321-331

Sloane BF, Moin K, Krepela E, Rozhin J (1990) Cathepsin B and its endogenous inhibitors: the role in tumor malignancy. Cancer Metastasis Rev 9:333-352

# AN EXAMINATION OF THE STRUCTURE AND FUNCTIONS OF THE BOVINE VACUOLAR ATPASE USING ANTISENSE OLIGONUCLEOTIDES

Jane E. Strasser, Ying-xian Pan[1], Randal Morris[2], and Gary E. Dean
Departments of Molecular Genetics, Biochemistry, and Microbiology; [1] Physiology and Cell Biophysics; and [2] Anatomy and Cell Biology
University of Cincinnati College of Medicine
Cincinnati, Ohio 45267-0524

The V-ATPase (V-ATPase) is a multimeric proton pump responsible for the acidification of endosomes, the trans-Golgi network (TGN), lysosomes, and secretory granules (Forgac, 1989). This acidification is widely believed to be necessary for accurate intracellular trafficking. Disruption of several of the V-ATPase subunit genes in *Saccharomyces* is already known to result in conditional lethality (Nelson and Nelson, 1990) and causes constitutive secretion of normally soluble vacuolar proteins (Kane *et al*, 1989). These proteins are believed to be sorted away from the secretory pathway at the TGN. We were curious to know what effects alteration of V-ATPase levels would have in mammalian cells. To this end, we have used antisense oligonucleotides specific to subunit B of the V-ATPase in Madin Darby Bovine Kidney (MDBK) cells. A cDNA encoding subunit B of the bovine V-ATPase was cloned (Y.-X. Pan and G. E. Dean, unpublished) and, using the derived DNA sequence, 18-mer phosphorothioate oligonucleotides were designed in both sense and anti-sense directions at the 5'-end. These oligomers were then added to culture dish wells into which cells had been plated.

To ascertain the effects of the oligonucleotides on cell viability, MDBK cells were incubated with 2.5 uM of the antisense oligonucleotide (test oligo), sense oligomer (control #1), a mixture containing 2.5 uM of both sense and antisense that had been allowed to hybridize to each other prior to exposure to cells (control #2), or cells grown under identical conditions to which no oligomer had been added (control #3). Viable cells were then counted daily. These data indicated that the antisense oligonucleotide specifically diminished cell viability, while the controls showed little or no difference.

To determine the effects of the antisense oligonucleotide on V-ATPase subunit mRNA levels, quantitative Northern analysis was performed with RNA isolated from antisense-treated and control cells (1 - 3 as above) after four days of treatment. The RNA was blotted and the relative levels of mRNA expression determined for subunit B and for two other V-ATPase subunits, A (Pan *et al*, 1991) and c (Mandel *et al*, 1988). RNA load and transfer efficiency were adjusted by means of a standardization probe. Antisense oligonucleotide treatment increased the levels of subunit B mRNA 12-fold

NATO ASI Series, Vol. H 74
Molecular Mechanisms of Membrane Traffic
Edited by D. J. Morré, K. E. Howell, and J. J. M. Bergeron
© Springer-Verlag Berlin Heidelberg 1993

relative to controls, while expression of subunits A and c was unaffected, indicating that the treatment resulted in a specific effect.

To determine whether the antisense oligonucleotide affected intracellular acidification, cells treated as above were examined for their ability to sequester the weak bases 3-(2,4-dintroanilino)-3'-amino-N-methyldipropylamine (DAMP) and acridine orange. Within two days of incubation with antisense oligomer, DAMP sequestration (as judged by indirect immunocytochemistry) was radically diminished relative to controls, indicating that antisense-treated cells are much less effective at properly acidifying their internal compartments relative to controls. Similar results were seen when assayed by acridine orange accumulation.

To examine the effects of antisense treatment on sorting of lysosomal enzymes, antisense-treated and control cells were assayed for their ability to correctly localize acid phosphatase and cathepsin D. On the second day of treatment with antisense oligonucleotide, these lysosomal enzymes appeared to be distributed in the same manner as control cells. On days 3 through 5, however, there was progressively less intracellular accumulation and more secretion of these lysosomal enzymes, implying that intracellular sorting was perturbed.

Electron microscopic examination of these cells reveals that antisense treatment specifically induces bodies similar to those seen in endosomal acidification mutants (Park *et al*, 1991) and in lysosomal storage diseases.

Using antisense oligonucleotide technology, we have shown that subunit B is involved in intra-organellar acidification, that exposure to oligonucleotides antisense to subunit B specifically induces the occurrence of intracellular bodies similar to those seen in endosomal acidification mutants and in lysosomal storage diseases, and that by perturbing the V-ATPase we can disrupt targeting of lysosomal hydrolases. We now have established a system permitting examination of the biogenesis of the V-ATPase, its structure-function relationships, and its role in cellular physiology.

Literature References:

Forgac M. (1989) Structure and function of vacuolar class of ATP-driven proton pumps. *Physiol. Rev.* **69**, 765-796.

Kane P. M. , Yamashiro C. T. , Rothman J. H. and Stevens T. H. (1989) Protein sorting in yeast: the role of the vacuolar proton- translocating ATPase. *J. Cell Sci. Suppl.* **11**, 161-178.

Mandel M. , Moriyama Y. , Hulmes J. D. , Pan Y. -C. E. , Nelson H. and Nelson N. (1988) cDNA sequence encoding the 16-kDa proteolipid of chromaffin granules implies gene duplication in the evolution of $H^+$-ATPases. *Proc. Natl. Acad. Sci. USA* **85**, 5521-5524.

Nelson H. and Nelson N. (1990) Disruption of genes encoding subunits of yeast vacuolar $H^+$- ATPase causes conditional lethality. *Proc. Natl. Acad. Sci. USA* **87**, 3503-3507.

Pan Y. X. , Xu J. , Strasser J. E. , Howell M. and Dean G. E. (1991) Structure and Expression of Subunit-A from the Bovine Chromaffin Cell Vacuolar ATPase. *FEBS Lett.* **293**, 89-92.

Park J. E. , Draper R. K. and Brown W. J. (1991) Biosynthesis of lysosomal enzymes in cells of the End3 complementation group conditionally defective in endosomal acidification. *Somatic Cell Mol. Genet.* **17**, 137-150.

# REGULATION OF ENDOCYTOSIS BY THE SMALL GTP-ASE RAB5

R G Parton, C Bucci, B Hoflack and M Zerial
European Molecular Biology Laboratory EMBL
Meyerhofstrasse 1
6900 Heidelberg
Germany

Summary

Over the last few years, work in yeast and mammalian cells has led to the discovery that small GTPases of the Ypt1/Sec4/Rab family function as specific regulators of intracellular transport. We have carried out functional studies on rab5, which is associated with the plasma membrane and early endosomes. The data obtained indicate that this protein is a rate-limiting GTPase that regulates the kinetics of both fusion of plasma membrane-derived endocytic vesicles with early endosomes and lateral fusion of early endosomes.

Introduction.

GTP-binding proteins are involved in the regulation of a wide spectrum of biological functions. The best known examples are the heterotrimeric G proteins which play a role in signal transduction processes, and the elongation factor EF-Tu which functions in protein synthesis (for a review see Bourne et al., 1991). During the last few years, there has been growing interest in a large family of small GTP-binding proteins structurally related to p21[ras], the protooncogene product. These proteins share structural similarities but are functionally heterogeneous. While members of the ras subfamily are involved in control of cell growth (Bourne et al., 1990), rho proteins are implicated in the control of cytoskeletal organization in mammalian cells (Paterson et al, 1990) or in bud formation in yeast (Johnson and Pringle, 1990). The two other subfamilies of small GTP-binding proteins, Ypt1/Sec4/rab and Arf/Sar, are involved in the regulation of vesicular traffic (Bourne, 1988; Pfeffer, 1992).

NATO ASI Series, Vol. H 74
Molecular Mechanisms of Membrane Traffic
Edited by D. J. Morré, K. E. Howell, and J. J. M. Bergeron
© Springer-Verlag Berlin Heidelberg 1993

| Proteins | Organelles | References |
|---|---|---|
| Ypt1 | ER-transport vesicles-Golgi apparatus | Segev et al. (1988) Rexach and Schekman (1991) |
| Sec4 | secretory vesicles- plasma membrane | Salminen and Novick (1987) Goud et al. (1988) |
| Rab1b | ER-Golgi | Plutner et al. (1991) |
| Rab2 | CGN (also called intermediate or salvage compartment) | Chavrier et al. (1990a) |
| Rab3a | synaptic vesicles Chromaffin granules | Fischer v. Mollard et al. (1990) Darchen et al. (1990) Mizoguchi et al. (1990) |
| Rab4 | early endosomes | van der Sluijs et al. (1991) |
| Rab5 | plasma membrane, clathrin- coated vesicles, early endosomes | Chavrier et al. (1990a) Bucci et al. (1992) |
| Rab6 | middle + trans cisternae of Golgi apparatus | Goud et al. (1990) |
| Rab7 | late endosomes | Chavrier et al. (1990a) |
| Rab8 | post-Golgi secretory vesicles | Chavrier et al. (1990b) Huber et al. (submitted) |
| Rab9 | late endosomes TGN | Lombardi et al. (submitted) |
| Rab10 | post-Golgi secretory vesicles? | Chavrier et al. (1990b) |

Table 1. Summary of the localization of Ypt1, Sec4 and rab proteins in yeast and mammalian cells. The predicted location of rab8 and rab10 is post-Golgi secretory vesicles and plasma membrane because of the high structural similarity with Sec4p (Chavrier et al., 1990b).

Studies performed on yeast *Saccharomyces cerevisiae* first showed that YPT1 and SEC4 genes encode two 23kD GTPases regulating secretion. These two proteins are required in two separate transport events. Ypt1p controls ER to Golgi transport (Segev et al., 1988; Baker et al., 1990; Rexach and Shekman, 1991) and Sec4p the fusion of post-Golgi vesicles with the plasma membrane (Salminen and Novick, 1987; Goud et al., 1988). It has been proposed by analogy to EF-Tu, that the conformational change accompanying GTP-hydrolysis might be utilized to direct the energy-dependent unidirectional delivery of vesicles to their target organelles (Bourne, 1988; Walworth et al., 1989).

Ypt1p, Sec4p and the other ras-related proteins contain four conserved regions which fold forming the GTP-binding site. The high aminoacid conservation in these regions has greatly facilitated the identification of Ypt1p/Sec4p-related proteins expressed in mammalian cells, termed rab proteins (Touchot et al., 1987). To date, almost 30 different rab proteins have been identified using different strategies: screening cDNA libraries using oligonucleotides corresponding to the highly conserved sequences (Touchot et al., 1987; Chavrier et al., 1990b), using rab cDNA probes at low stringency hybridization conditions (Zahraoui et al., 1989; Elferink et al., 1992), or using a polymerase chain reaction (PCR) approach (Ngsee et al., 1991; Chavrier et al., 1992).

Morphological and biochemical studies have shown that rab proteins are localized to specific sub-compartments of exocytic and endocytic organelles (Table 1), thereby providing further support to the notion that they may control different steps of membrane traffic.

Correct targeting of rab proteins to their site of function requires a determinant contained in the C-terminal hypervariable region. We have defined this region by transiently expressing in BHK cells a series of hybrid proteins between rab5, located on early endosomes, and rab7, which is associated with late endosomes. The C-terminal 34 aminoacid residues of rab7 transplanted onto rab5 directed the hybrid protein to late endosomes (Chavrier et al., 1991). However, the C-terminal cysteines, which undergo isoprenylation (Peter et al.,1992), are required for membrane association but do not play any role in the targeting process (Chavrier et al., 1991).

Mutational analysis on Ypt1 and Sec4 proteins in yeast suggests that these proteins are required at specific locations of the secretory pathway. However, the precise function of rab proteins in mammalian cells remains largely unknown.

We have focused our studies on the function of rab5. This protein is associated with the plasma membrane and early endosomes (Chavrier et al., 1990). This localization suggests that rab5 is involved in the machinery regulating transport along the early endocytic pathway. Therefore, we investigated the involvement of rab5 in the first steps of endocytosis and in the dynamics of early endosome fusion.

Rab5 is required in the *in vitro* lateral fusion of early endosomes.

The lateral fusion of early endosomes *in vitro* is thought to reflect the dynamic state of this organelle *in vivo*. Since rab5 is associated with early endosomes and the fusion reaction is inhibited by GTPγS, we tested the possible involvement of rab5 in the early endosome fusion process using a cell-free assay (Gorvel et al., 1991). Addition of anti-rab5 antibodies strongly inhibited fusion whereas control antibodies were not effective. Cytosols containing either wild type rab5 or mutant proteins expressed at high level using the vaccinia T7 RNA polymerase system (Fuerst et al., 1986) were tested in the assay measuring fusion between early endosomes prepared from untransfected cells. Overexpression of rab5 increased the efficiency of fusion and rescued the inhibitory effect caused by the anti-rab5 antibody. In addition, cytosol containing overexpressed rab5 lacking the cysteines required for membrane association (Chavrier et al., 1990a) had no effect. In contrast, cytosol containing rab5ile133, a mutant unable to bind GTP as measured by ligand blotting, could inhibit fusion at high concentration. Alltogether, these data indicate that a functional rab5 protein is required in the process of early endosome fusion *in vitro*.

Rab5 functions as a regulatory factor *in vivo*.

The results described above prompted us to investigate the function of rab5 *in vivo* (Bucci et al.,1992). Our strategy was to interfere with the function of endogenous rab5 by expressing in BHK cells either a dominant negative mutant, the GTP-binding defective rab5ile133 protein, or wild type rab5. We used the vaccinia virus-based transient expression system in order to circumvent potential problems of lethality accompanying the isolation of stable transfected cell lines. We worked out conditions to express the two proteins 15-fold higher than endogenous rab5 over a short period of time (30 min).

We observed that expression of the mutant rab5ile133 led to a 50% decrease in the rate of uptake of a fluid phase endocytic marker, horseradish peroxidase (HRP), compared to control cells. Internalization of [125]I-transferrin, a marker of receptor-mediated endocytosis, was similarly inhibited, while the kinetics of recycling were not dramatically affected.

Confocal immunofluorescence and electron microscopy analysis revealed drastic alterations in the morphology of early endosomes. In cells overexpressing rab5ile133, the early endocytic structures were smaller than in control cells. In non-transfected cells, HRP labelled typical early endosomes within 5 min of internalization, whereas in the rab5 mutant expressing cells the marker labelled small tubules and coated vesicles. HRP reached more typical early endosomes only after 10-20 min of internalization (Fig. 1). These results agree with the biochemical studies, showing that expression of rab5ile133 protein leads to a reduction in the rate of endocytosis. We interpret our morphological results as indicating that inhibition of rab5 function results in accumulation of transport intermediates and in fragmentation of early endosomes.

Overexpression of the wild type protein led to the opposite effect. Both the rates of HRP and [125]I-transferrin uptake were increased in cells overexpressing rab5. Morphological studies confirmed these observations and further indicated an increase in the size of early endosomes in these cells.

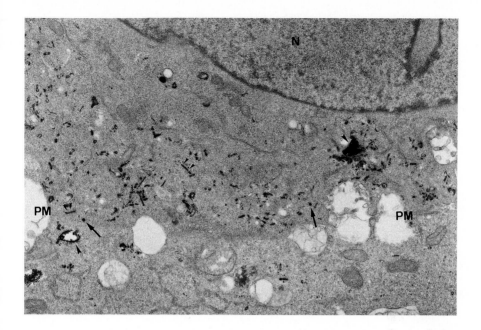

Fig. 1. Early endocytic structures in cells overexpresing rab5ile33 protein. Cells were incubated with HRP for 20 min at 37°C before fixation and processing for Epon embedding. In the semi-thick section shown here HRP reaction product is evident in small tubular and vesicular structures in the cell periphery. PM, plasma membrane, N, nucleus. Large arrowheads indicate typical HRP-labelled early endosomes, small arrowheads show tubular endosomal elements. Arrows indicate possible coated vesicles.

Based on these data, we favour a model in which rab5 regulates the kinetics of fusion of endocytic vesicles derived from the plasma membrane with early endosomes (Fig. 2 A-C). Rab5 present on the plasma membrane moves with the endocytic vesicles to early endosomes. Fusion of these vesicles with early endosomes is perturbed by expression of the mutant rab5ile133 protein thus causing their accumulation (Fig. 2B). Conversely, an increase in the expression of wild type rab5 increases the efficiency of the fusion process (Fig. 1C).

This model is consistent with our calculations on the life time of coated pits. The faster rate of endocytosis in cells overexpressing wild type rab5 protein correlated with a decreased lifetime of surface coated pits, from ~1min in control cells to ~13.5sec. In contrast, overexpression of the rab5 mutant protein prolonged the lifetime of a coated pit to ~2min. These results might indicate that rab5 controls coated pit/coated vesicles formation. However, the accumulation of

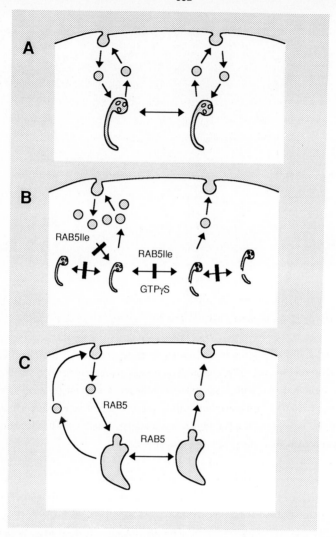

Fig. 2. Hypothetical scheme of rab5 function *in vivo*. (A) In untransfected cells transport between plasma mambrane and early endosomes is achieved by endocytic and recycling vesicles which bud and fuse with normal kinetics. (B) Overexpression of rab5ile133 slows the kinetics of fusion of endocytic vesicles with early endosomes and lateral fusion between early endosomes. This effect leads to accumulation of transport vesicles and to fragmentation of early endosomes in the small tubules viewed in Fig. 1. (C) In cells overexpressing wild type rab5 the kinetics of transport are accelerated. Endocytic markers are internalized more rapidly, lateral fusion between early endosomes is stimulated and abnormally large early endosomes appear.

vesicles caused by the mutant protein argues that rab5 is mainly involved in vesicle fusion with early endosomes.

Our model takes into account our previous results which indicated that rab5 is involved in the fusion process of early endosomes *in vitro*. In cells expressing the rab5ile133 protein, we observed tubular structures and small vesicles accumulating in the cell periphery. This pattern is consistent with fragmentation of the early endosomes (Fig. 2B). In contrast, overexpression of the wild type protein induces an increase in the size of early endosomes (Fig. 2C). These data suggest that the fusion of early endosomes is stimulated by rab5 and inhibited by rab5ile133 protein *in vivo* thus providing strong support for the dynamic nature of the early endosome compartment.

Conclusion and perspectives.

The functional studies performed on rab5 indicate that this is a key regulatory protein involved in controlling the kinetics of both early endosome fusion *in vitro* and *in vivo* and in the fusion of plasma membrane derived endocytic vesicles with early endosomes. This provides the first piece of evidence that the endocytic pathway as a whole can be regulated by small GTPases. Other rab proteins are likely to play a regulatory role in exocytosis and endocytosis. An important implication of these findings is that rab proteins could modulate the kinetics of transport in different cell types from various organs, both in the adult and during development, depending on their level of expression.

Owing to the recent identification of novel members, the subfamily of rab proteins is increasing in complexity. If all rab proteins are involved in vesicular trafficking such variety might account for a fine and accurate tuning of homotypic and heterotypic transport events.

Much effort will be necessary to unravel the functional mechanisms of these small GTPases. How many factors are necessary to regulate the GTPase and nucleotide exchange activity of rab proteins? What receptor machinery recognizes the targeting signal of rab proteins and mediates their specific membrane association? Which factors belong to the docking machinery on the early endosomes? We are currently addressing these questions. The identification of these molecules will hopefully provide important insights into the machinery regulating membrane traffic.

References

Baker D, Wuestehube L, Schekman R, Botstein D, Segev N (1990) GTP-binding Ypt1 protein and Ca$^{2+}$ function independently in a cell-free protein transport reaction. Proc Natl Acad Sci USA 87:355-359.

Bourne HR (1988) Do GTPases direct membrane traffic in secretion? Cell 53:669-671.

Bourne HR, Sanders DA, McCormick F (1990) The GTPase superfamily: conserved structure and molecular mechanism. Nature 349:117-127.

Bucci C, Parton, RG, Mather IH, Stunneberg H, Simons K., Hoflack B and Zerial M (1992) The small GTPase rab 5 functions as a regulatory factor in the early endocytic pathway. Cell 70: September 4 issue.

Chavrier P, Parton RG, Hauri HP, Simons K, Zerial M (1990a) Localization of low molecular weight GTP binding proteins to exocytic and endocytic compartments. Cell 62:317-329.

Chavrier, P., Vingron, M., Sander, C., Simons, K. and Zerial, M. (1990b). Molecular cloning of YPT1/SEC4-related cDNAs from an epithelial cell line. Mol Cell Biol 10: 6578-6585

Chavrier P, Gorvel JP, Stelzer E, Simons K, Gruenberg J, Zerial M (1991) Hypervariable C-terminal domain of rab proteins acts as a targeting signal. Nature 353:769-772.

Chavrier P, Simons K, Zerial M (1992) The complexity of the Rab and Rho GTP binding protein subfamilies revealed by a PCR cloning approach. Gene 112: 261-264.

Darchen F, Zahraoui A, Hammel F, Monteils, M-P, Tavitian A, Scherman D (1990) Association of the GTP-binding protein Rab3A with bovine adrenal chromaffin granules. Proc Natl Acad Sci USA 87:5692-5696.

Elferink LA, Anzai K, Scheller RH (1992) Rab15, a novel low molecular weight GTP-binding protein specifically expressed in rat brain. J Biol Chem 267:5768-5775.

Fisher v. Mollard G, Mignery GA, Baumert M, Perin MS, Hanson TJ, Burger PM, Jahn R, Sudhof TC (1990) Rab3 is a small GTP-binding protein exclusively localized to synaptic vesicles. Proc Natl Acad Sci USA 87:1988-1992.

Fuerst TR, Niles EG, Studier FW, Moss B (1986) Eukaryotic transient-expression system based on recombinant vaccinia virus that synthesizes bacteriophage T7 RNA polymerase. Proc Natl Acad Sci USA 83:8122-8126.

Gorvel JP, Chavrier P, Zerial M, Gruenberg J (1991) Rab5 controls early endosome fusion *in vitro*. Cell 64:915-925.

Goud B, Salminen A, Walworth NC, Novick PJ (1988) A GTP-binding protein required for secretion rapidly associates with secretory vesicles and the plasma membrane in yeast. Cell 53:753-768.

Goud B, Zahraoui A, Tavitian A, Saraste J (1990) Small GTP-binding protein associated with Golgi cisternae. Nature, 345:553-556.

Johnson DI and Pringle JR (1990) Molecular characterization of CDC42, a Saccharomyces cerevisiae gene involved in the development of cell polarity. J Cell Biol 111: 143-152.

Mizoguchi A, Kim S, Ueda T, Kikuchi A, Yorifuji H, Hirokawa N, Takai, Y (1990) Localization and subcellular distribution of smg p25A, a ras p21-like GTP-binding protein, in rat brain. J Biol Chem 265:11872-11879.

Ngsee JK, Elferink LA, and Scheller RH (1991) A family of ras-like GTP-binding proteins expressed in electromotor neurons. J Biol Chem 266: 2675-2680.

Paterson, HF, Self AJ, Garrett, MD, Just, I, Aktories, K, and Hall A (1990) Microinjection of recombinant p21rho induces rapid changes in cell morphology. J Cell Biol 111: 1001-1007.

Peter M, Chavrier P, Nigg EA, and Zerial M (1992) Isoprenylation of rab proteins on structurally distinct cysteine motifs. J Cell Sci 102: August issue.

Pfeffer S (1992) GTP binding proteins in intracellular transport. Trends in Cell Biol 2: 41-46.

Plutner H, Cox AD, Pind S, Khosravi-Far R, Bourne JR, Schwaninger R, Der CJ, and Balch B (1991) Rab1b regulates vesicular transport between the endoplasmic reticulum and successive Golgi compartments. J Cell Biol 115: 31-43.

Rexach MF, Schekman RW (1991) Distinct biochemical requirements for the budding, targeting, and fusion of ER-derived transport vesicles. J Cell Biol 114:219-229.

Salminen A, Novick PJ (1987) A *ras*-like protein is required for a post-Golgi event in yeast scretion. Cell 49:527-538.

Segev N, Mulholland J, Botstein, D (1988) The yeast GTP-binding YPT1 protein and a mammalian counterpart are associated with the secretion machinery. Cell 52:915-924.

Touchot N, Chardin P, and Tavitian A (1987) Four additional members of the ras gene superfamily isolated by an oligonucleotide strategy: molecular cloning of YPTrelated cDNAs from a rat brain library. Proc. Natl Acad Sci USA 84: 8210-8214.

Valencia A, Chardin P, Wittinghofer A, Sander C (1991) The *ras* protein family: evolutionary tree and role of conserved amino acids. Biochemistry 30:4637-4648.

van der Sluijs P, Hull M, Zahraoui A, Tavitian A, Goud B, Mellman I (1991) The small GTP-binding protein rab4 is associated with early endosomes. Proc Natl Acad Sci USA 88:6313-6317.

Walworth NC, Goud B, Kastan Kabcenell A, Novick PJ (1989) Mutational analysis of *SEC4* suggests a cyclical mechanism for the regulation of vesicular traffic. EMBO J 8:1685-1693.

Zahraoui A, Touchot N, Chardin P, and Tavitian A (1989). The human *Rab* genes encode a family of GTP-binding proteins related to yeast YPT1 and SEC4 products involved in secretion. J Biol Chem 264: 12394-12401

# THE NUCLEOTIDE CYCLE OF *SEC4* IS IMPORTANT FOR ITS FUNCTION IN VESICULAR TRANSPORT

Peter Novick, Patrick Brennwald, Michelle D. Garrett, Mary Moya, Denise Roberts and Robert Bowser.
Department of Cell Biology,
Yale University School of Medicine,
333 Cedar Street
New Haven, CT 06510

Sec4 is a *ras*-like, GTP-binding protein required for exocytosis in yeast (Novick et al., 1980; Salminen and Novick, 1987). The purified protein displays a high affinity for GTP and a very low intrinsic GTPase activity (Kabcenell et al., 1990). We have found that the addition of yeast lysate greatly stimulates the GTPase activity of the protein, reflecting the presence of a Sec4-GTPase activating protein (GAP) (Walworth et al., 1992). Excess Sec4 bound to GTPγS will inhibit the interaction, but excess Ypt1-GTPγS will not, indicating that the GAP is specific. A mutation ($^{79}$Q->L) in a site of Sec4p which is predicted to interact with the phosphoryl group of GTP, lowers the intrinsic hydrolysis rate to unmeasurable levels. When purified from yeast, the mutant protein is predominantly GTP bound (Walworth et al., 1992), while the wild-type protein is predominantly GDP bound (Kabcenell et al., 1990). This mutant protein can be stimulated by the yeast Sec4-GAP activity, but only to 30% of the GAP-stimulated rate seen with wild type Sec4p. The *sec4-L79* allele can function as the only copy of *sec4* in yeast (Walworth et al., 1992). However, the mutation causes recessive cold sensitivity, slowing of invertase secretion, accumulation of secretory vesicles and displays synthetic lethality with a subset of other secretory mutants, indicating a partial loss of function. This is in contrast to the equivalent mutation in *ras* which is completely insensitive to stimulation by ras-GAP and results in a gain of ras function seen ultimately as oncogenic transformation. Taken together this argues that, unlike the case of *ras*, where the absolute level of the GTP-bound form determines activity, it is the ability of Sec4 to cycle between its GTP- and GDP-bound forms that is critical for Sec4p function

NATO ASI Series, Vol. H 74
Molecular Mechanisms of Membrane Traffic
Edited by D. J. Morré, K. E. Howell, and J. J. M. Bergeron
© Springer-Verlag Berlin Heidelberg 1993

in vesicle transport.

The cycle of nucleotide binding and hydrolysis may be coupled to a cycle of Sec4 localization within the cell. Sec4 binds to secretory vesicles which then fuse with the plasma membrane by exocytosis (Goud et al., 1988). Sec4 can recycle from the plasma membrane through a soluble pool to rebind onto a new round of vesicles. We have found a soluble yeast activity which can release Sec4 from membranes, but only when Sec4 is in its GDP-bound form. Pre-binding GTPγS to Sec4 inhibits release. This activity is comparable to that of the GDI protein isolated from bovine brain described in Araki et al. (1990). Association of Sec4 with secretory vesicles requires prior prenylation by Bet2, a component of a geranyl-geranyl transferase (Rossi et al., 1991). Correct targeting of Sec4 may involve the carboxy terminal hypervariable region of Sec4 as well as upstream sequence information since chimeras of Sec4 with the carboxy terminus of the related protein, Ypt1, result in partial mislocalization to the Ypt1 compartment.

*DSS4-1*, a dominant suppressor of the *sec4-8* temperature-sensitive mutation, encodes an exchange protein (Moya et al., submitted). Lysates derived from yeast strains overproducing the suppressor protein or purified Dss4 protein produced in *E. coli* stimulate the dissociation of both GDP and GTP from Sec4. The wild-type Dss4 protein is predominantly insoluble in yeast, while the suppressing allele encodes a partially soluble protein. Disruption of *dss4* is not lethal, yet it can mimic and as well as aggravates a partial loss of *SEC4*, suggesting that Dss4 normally serves to facilitate Sec4 function.

The cycle of Sec4 may function to allow the assembly and subsequent disassembly of a set of proteins necessary for exocytosis. Candidates for members of this set of proteins are encoded by *sec* genes which show strong genetic interactions with *sec4-8*. Two of these, *SEC8* and *SEC15* encode large proteins which are in a 20 *S* complex and which can peripherally associate with the plasma membrane (Salminen et al., 1989; Bowser and Novick, 1991; Bowser et al., 1992). The predicted amino acid sequence of Sec8 includes a 200 amino acid domain that shares 25% sequence identity with the region of yeast adenylate cyclase that is proposed to interact with Ras. By analogy with the Ras/cyclase interaction, the Sec8/Sec15 complex may respond to activated Sec4 to facilitate exocytosis.

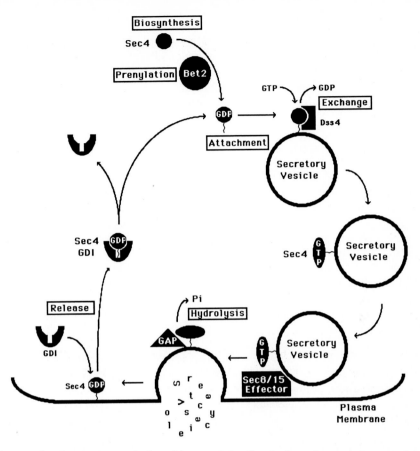

*Figure 1.* As presented in this model, Sec4 functions in a cycle of nucleotide binding and hydrolysis that is coupled to a cycle of subcellular localization. We will consider the start of the cycle to be Sec4 in its GDP-bound state. Since the GDP-bound form can be solubilized by interaction with GDI, the cycle begins with Sec4 as a complex with GDI in the cytosol. Interaction with an exchange protein removes the GDP allowing a GTP to bind. This causes dissociation of GDI allowing Sec4 membrane attachment. The site of membrane association may be dictated by the location of the exchange protein. We propose that membrane association occurs in the trans Golgi at the site of secretory vesicle budding. Sec4 remains on the secretory vesicle in its GTP-bound state until the vesicle interacts with an effector complex, containing Sec8 and Sec15, on the plasma membrane. This interaction triggers a series of events leading to exocytosis. Sec4-GAP also may be a component of the effector complex causing Sec4 to hydrolyse the bound GTP. This leaves Sec4 in its GDP-bound state which allows solubilization by GDI and initiation of a new cycle.

*References*

Araki,S., A. Kikuchi, Y. Hata, M. Isomura, and Y. Takai. 1990. Regulation of reversible binding of *smg* p25A, a *ras* p21-like GTP-binding protein, to synaptic plasma membranes and vesicles by its specific regulatory protein, GDP dissociation inhibitor. J Biol Chem 265:13007-13015.

Bowser, R. and P. Novick. 1991. Sec15 protein, an essential component of the exocytotic apparatus, is associated with the plasma membrane and with a soluble 19.5*S* particle. J Cell Biol 112: 1117-1131.

Bowser, R., H. Müller, B. Govindan and P. Novick. 1992. Sec8p and Sec15p are components of a 19.5*S* particle that interacts with Sec4p to control exocytosis. (In press) J Cell Biol

Kabcenell, A. K., B. Goud, J. K. Northup and P. Novick. 1990. Binding and hydrolysis of guanine nucleotides by Sec4p, a yeast protein involved in the regulation of vesicular traffic. J Biol Chem 265: 9366-9372.

Goud, B., A. Salminen, N.C. Walworth and P. Novick, 1988. A GTP-binding protein required for secretion rapidly associates with secretory vesicles and the plasma membrane in yeast. Cell 53: 753-768.

Moya, M., D. Roberts and P. Novick. *DSS4-1*, a dominant suppressor of *sec4-8* encodes a nucleotide exchange protein that facilitates Sec4p function. (Submitted).

Novick, P., C. Field and R. Schekman, 1980. Identification of 23 complementation groups required for post-translational events in the yeast secretory pathway. Cell 21: 205.

Rossi, G., Y. Jiang, A. Newman and S. Ferro-Novick. 1991. Dependence of Ypt1 and Sec4 membrane attachment on Bet2. Nature 351: 158-161.

Salminen, A. and P. Novick, 1987. A ras-like protein is required for a post-Golgi event in yeast secretion. Cell 49: 527-538.

Salminen, A. and P. Novick, 1989. The Sec15 protein responds to the function of the GTP binding protein, Sec4, to control vesicular traffic. J Cell Biol 109: 1023-1036.

Sasaki ,T., K. Kaibuchi, A.K. Kabcenell, P. Novick, and Y. Takai. 1991. A mammalian inhibitory GDP/GTP exchange protein (GDI) for *smg*p25A is active on the yeast *SEC4* protein. Mol Cell Biol 11: 2909-2912.

Walworth, N. C., P. Brennwald, A. K. Kabcenell, M. D. Garrett and P. Novick. 1992. Hydrolysis of GTP by Sec4 protein plays an important role in vesicular transport and is stimulated by a GTPase activating protein in yeast. Mol Cell Biol 12: 2017-2028.

ACTIVATORS OF TRIMERIC G-PROTEINS STIMULATE AND INHIBIT INTERCOMPARTMENTAL
GOLGI TRANSPORT *IN VITRO*

P.J. Weidman
Department of Biochemistry and Molecular Biology
St. Louis University Medical School
1402 S. Grand Blvd.
St. Louis, Missouri 63004

Small GTP-binding proteins of the RAS-superfamily play a
major role in regulating vesicular transport within the secretory
and endocytic pathways (reviewed by Pfeffer, 1992). New evidence
suggests that transport might also be regulated by trimeric GTP-
binding proteins (G-proteins, reviewed by Burgoyne, 1992). To
investigate the potential role of trimeric G-proteins in
vesicular transport, the effects of mastoparan (MAS) and
benzalkonium Cl (BAC) on *in vitro* intercompartmental Golgi
transport were analyzed. MAS and BAC are cationic, amphiphilic
molecules that trigger activation of G-protein $\alpha$ subunits by
simulating ligand-activated receptors (Higashijima et al., 1990).

This analysis employed a cell-free intercompartmental Golgi
transport system that reconstitutes transport of vesicular
stomatitis virus (VSV[1]) glycoprotein from a donor Golgi
population lacking n-acetylglucosamine (GlcNAc) transferase I, to
an acceptor Golgi population containing the active enzyme (Balch
et al., 1984). Transport is measured by the incorporation of [3]H-
GlcNAc into VSV glycoprotein upon transfer to acceptor Golgi.
Figure 1 shows that both MAS (open circles) and BAC (closed
circles) are potent inhibitors of intercompartmental Golgi
transport *in vitro*. Inhibition by both compounds is
irreversible, requires a 37°C incubation, and is partially
relieved by high levels of cytosolic protein. Pretreatment of
either donor or acceptor Golgi results in loss of transport
function, suggesting that processes associated with both vesicle
formation and fusion are affected. MAS is distinctive in that

---

[1]Abbreviations used: BAC, benzalkoniumchloride; GlcNAc,
n-acetylglucosamine; MAS, mastoparan; VSV, vesicular stomatitis
virus

NATO ASI Series, Vol. H 74
Molecular Mechanisms of Membrane Traffic
Edited by D. J. Morré, K. E. Howell, and J. J. M. Bergeron
© Springer-Verlag Berlin Heidelberg 1993

low concentrations stimulate transport (Figure 1, open circles) with as much as 150% stimulation observed when cytosol is limiting in the reaction (Figure 1, triangles). Both stimulation and inhibition are specific for MAS-like peptides that activate trimeric G-proteins, consistent with the possibility that both effects are mediated by trimeric G-proteins.

To determine whether BAC and MAS inhibit the same step(s) in transport, the rate at which transport of VSV-glycoprotein becomes resistant to inhibition by each compound was determined. Figure 2 (closed circles) shows that transport becomes resistant to BAC 30 minutes after initiating a 60 minute incubation. In contrast, transport remains sensitive to inhibition by MAS until very late in the incubation (open circles), suggesting that MAS and BAC inhibit different transport steps. Additional studies reveal that BAC inhibits previously described steps in transport that require palmitoyl coenzyme A as cofactor (Pfanner et al., 1989, 1990). BAC is thus most likely acting as a mild detergent and not a specific activator of trimeric G-proteins in this system.

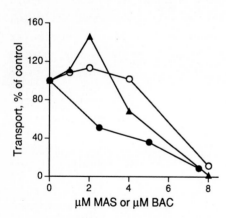

Figure 1. MAS and BAC inhibit *in vitro* Golgi transport. Transport reactions were performed as described by Balch et al., 1984. Assays contained MAS (O) or BAC (●) with 0.8 mg/ml cytosol; MAS (▲) with 0.4 mg/ml cytosol.

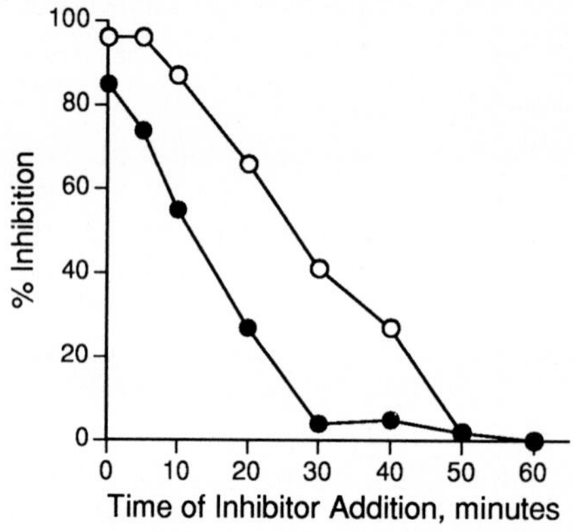

Figure 2. MAS and BAC inhibit different transport steps. MAS (10 μM, O) or BAC (8 μM, ●) were added to transport reactions at the indicated times during a 60 minute incubation.

The late steps in transport have recently been shown to correspond to steps in VSV-glycoprotein glycosylation (Hiebsch and Wattenburg, 1992). Figure 3 shows that VSV-glycoprotein accumulated in the acceptor Golgi during a transport reaction can then be efficiently glycosylated in a subsequent incubation with UDP-$^3$H-GlcNAc (open circles). If the Golgi are treated with MAS <u>after</u> VSV-glycoprotein is accumulated in the acceptor compartment, subsequent glycosylation is markedly diminished (Figure 3, closed circles). A thorough analysis reveals, however, that mastoparan does not inhibit

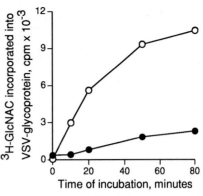

Figure 3. MAS inhibits glycosylation of VSV-glycoprotein. VSV-glycoprotein was accumulated in the acceptor Golgi in a standard transport incubation without UDP-$^3$H-GlcNAc. The mebranes were then treated with 10 μM MAS, reisolated by centrifugation, and incubated for the indicated times at 37°C in buffer containing UDP-$^3$H-GlcNAc and UMP-Kinase. Untreated Golgi(O), MAS-treated Golgi(●).

uptake of UDP-$^3$H-GlcNAc, does not permeabilize the Golgi membranes, and does not directly or indirectly inhibit the activity of GlcNAc transferase. Sedimentation analyses indicate that mastoparan treatment induces changes in Golgi size/shape, suggesting that partitioning of MAS into the lipid bilayer may be disrupting Golgi structure.

Scanning electron microscopy was used to examine changes in Golgi morphology after treatment with MAS. When Golgi are treated with stimulatory concentrations of MAS, normal Golgi morphology is observed with a slight increase in the number of vesicles per Golgi complex relative to untreated Golgi. This suggests that MAS may stimulate transport by stimulating vesicle production. Treatment of Golgi with inhibitory levels of MAS resulted in swelling and distortion of the Golgi and/or loss of recognizable Golgi morphology. This suggests that the inhibitory effects of MAS are due to disruption of membrane structure, and not to specific activation of an inhibitory trimeric G-protein.

In summary, the evidence suggests that BAC and MAS inhibit *in vitro* intercompartmental Golgi transport by mechanisms unrelated to the activation of trimeric G-proteins. In contrast, low concentrations of MAS stimulate transport, possibly by stimulating vesicle production. It has recently been reported that MAS stimulates binding of B-COP, a constituent of Golgi transport vesicles, to Golgi membranes via a pertussis toxin sensitive G-protein (Ktistakis et al., 1992). These results raise the possibility that vesicular transport is positively regulated by a trimeric G-protein.

This research was supported by American Cancer Society Grant BE-87. The author gratefully acknowledges the assistance of Dr. John Heuser, Robyn Roth, and Elana Ignatova in the electron microscopic portion of this study.

Balch WE, Dunphy WG, Braell WA, and Rothman JE (1984) Reconstitution of the transport of protein between successive compartments of the Golgi measured by the coupled incorporation of n-acetylglucosamine. Cell 39:405-416

Burgoyne RD (1992) Trimeric G-proteins in Golgi transport. Trends in Biochem 17:87-88

Hiebsch RR and Wattenberg BW (1992) A re-assessment of proposed intermediates in protein transport through the Golgi apparatus *in vitro*. Molecular cloning reveals a required component of the assay (POP) enhances the glycosylation used to mark protein transport. Biochem in press

Higashijima T, Burnier J, and Ross EM (1990) Regulation of $G_i$ and $G_o$ by mastoparan, related amphiphilic peptides, and hydrophobic amines. (1990) J Biol Chem 265:14176-14186

Ktistakis N, Linder ME, and Roth MG (1992) Action of brefeldin A blocked by activation of a pertussis-toxin-sensitive G protein. Nature 356:344-336

Pfanner N, Orci L, Glick BS, Amherdt M, Arden SR, Malhotra V, and Rothman JE (1989) Fatty acyl-coenzyme A is required for budding of transport vesicles from Golgi cisternae. Cell 59:95-102

PFanner N, Glick BS, Arden SR, and Rothman JE (1990) Fatty acylation promotes fusion of transport vesicles with Golgi cisternae. J Cell Biol 110:955-961

Pfeffer SR (1992) GTP-binding proteins in intracellular transport. Trends in Cell Biol 2:41-45

# EVIDENCE OF A ROLE FOR HETEROTRIMERIC GTP-BINDING PROTEINS AND ARF IN ENDOSOME FUSION.

L. Mayorga, M.I. Colombo, J.M. Lenhard, and P. Stahl.
Department of Cell Biology and Physiology
Washington University School of Medicine
St. Louis, MO 63110, U.S.A.

Several steps in the transport of macromolecules along the endocytic and exocytic pathways are mediated by carrier vesicles. The mechanism by which vesicles recognize and fuse with the target compartment is poorly understood.

A growing body of evidence indicates a role for several families of GTP-binding proteins on vesicular transport. Small GTP-binding proteins (smgs) are required for secretion in yeast and mammalian cells. Smgs have also been implicated in endocytosis (Goud and McCaffrey, 1991). Lately, evidence coming from several laboratories indicates that heterotrimeric GTP-binding proteins (G proteins) may also regulate secretion (Barr et al., 1992).

Using an assay that reconstitutes fusion between early endosomes we present evidence that indicates that both ARF, a member of the smg family, and G proteins are involved in endosome fusion. The assay is based in two molecules that are internalized by receptor-mediated endocytosis in J774-E clone cells, a murine macrophage cell line that express the mannose receptor. The two probes are dinitrophenol-derivatized $\beta$-glucuronidase and mannosylated anti-dinitrophenol IgG. Endosomes are loaded with either probe by a 5 min incubation at $37^\circ C$. After homogenization, endosome-enriched fractions are isolated from both preparations and incubated together as previously described (Colombo et al., 1992). Fusion is assessed by the formation of immune complexes between the enzyme and the antibody. The complexes are immunoprecipitated and quantitated measuring $\beta$-glucuronidase activity.

NATO ASI Series, Vol. H 74
Molecular Mechanisms of Membrane Traffic
Edited by D. J. Morré, K. E. Howell, and J. J. M. Bergeron
© Springer-Verlag Berlin Heidelberg 1993

## Results and Discussion

The first evidence of a role for GTP-binding proteins in endocytosis came from the observation that GTPγS activates endosome fusion at suboptimal concentrations of cytosol and inhibits at saturating concentrations of cytosol (Fig. 1A). GTPγS is able to activate both smgs and G proteins. To assess whether G proteins were involved in endosomal fusion, several compounds that affect this class of proteins were tested in the in vitro fusion assay. Normally, G proteins are activated by interacting with a ligand-activated receptor. Mastoparan is a peptide from wasp venom that increases nucleotide exchange in some G proteins.  This peptide forms an amphiphilic alpha-helix that mimics ligand-activated receptors.  Mastoparan reversed the effect of GTPγS on endosome fusion (Fig. 1A).  Mas 17, an analog that does not promote nucleotide exchange, was ineffective (Fig. 1A). Other compounds, known to stimulate nucleotide exchange in certain G

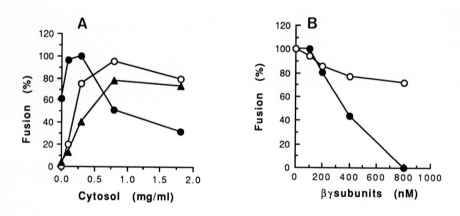

Figure 1.  Effect of GTPγS, mastoparan and Gβγ subunits on in vitro endosome fusion. **A**. Endocytic vesicles were mixed with increasing concentrations of cytosol in the presence of 20 μM GTPγS (●), 20 μM GTPγS and 10 μM mastoparan (▲) or without additions (○). **B**.  Vesicles were incubated at 37°C in the presence of 0.2 mg/ml cytosol with increasing concentrations of Gβγ subunits .  After 10 min, 20 μM GTPγS was added and the mixture was incubated for additional 45 min (●).  As a control, Gβγ subunits were added after a 10 min incubation in the presence of GTPγS (○).

proteins, such as cationic peptides (melittin, HR1) and hydrophobic amines (benzalkonium chloride, methyl benzetonium and compound 48-80) also inhibited GTPγS stimulated fusion (data not shown).

To specifically show that G proteins are involved in fusion, we tested the effect of purified Gβγ subunits in the in vitro assay. Addition of these subunits inhibited endosome fusion (Fig 1B). This effect was abrogated by heat-denaturation of Gβγ subunits (data not shown). Moreover, incubations of vesicles and cytosol with GTPγS before the addition of Gβγ subunits, significantly reduced the inhibitory effect (Fig. 1B). This result is consistent with the low affinity of Gβγ subunits for the GTP-bound form of the Gα subunit.

There is good evidence showing that ARF is involved in secretion. However, there was no indication of a role for ARF in endocytosis. Recently, it has been shown that the N-terminus of ARF regulates several important functions of this protein. The N-terminal peptide interacts with membranes by forming an amphiphilic alpha-helix. We tested the effect of a peptide corresponding to the amino terminal 16 amino acids of human ARF1. Similar to mastoparan, the ARF1 peptide reversed the effect of GTPγS on

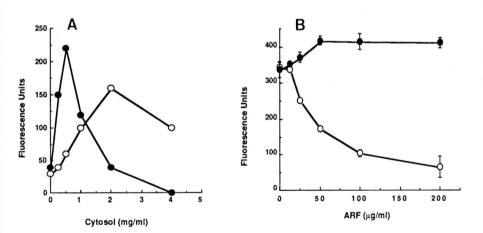

Figure 2. Effect of the N-terminal peptide of ARF and ARF protein on endosome fusion. **A.** Endosomes were incubated with 20 μM GTPγS and increasing concentrations of cytosol in the presence (O) or absence (●) of 50 μM ARF1 (2-17) peptide. **B.** Endosomes were incubated with 1 mg/ml cytosol and 20 μM GTPγS in the presence of increasing concentrations of myrARF (O) or non-myrARF (●).

endosomal fusion (Fig. 2A). A control peptide corresponding to amino acids 68-88 of ARF1 was ineffective (data not shown). The observation that the N-terminus of ARF antagonized the effect of GTPγS on fusion indicates that ARF is involved in the process. Unlike other smgs, ARF is myristoylated at the amino terminus. Myristoylation is required for biological activity. Addition of myristoylated ARF (myrARF) inhibited endosome fusion in the presence of GTPγS (Fig. 2B). Inhibition was not observed by the addition of non-myr ARF. Moreover, myrARF did not affect fusion in the absence of GTPγS (data not shown). We also observed that ARF peptide reversed the inhibitory effect of myrARF (data not shown).

These results indicate that both ARF and G proteins are involved in endosomal fusion. Several pieces of evidence suggest that these proteins interact with each other. First, ARF is required for cholera toxin-catalyzed ADP-ribosylation of Gαs subunits. Second, Gβγ subunits prevent the association of ARF with Golgi membranes (Barr et al., 1992). Third, mastoparan and ARF peptide have identical effects on endosome fusion. Several hypothesis may be proposed to explain our present data. One working model would be that the cytoplasmic tail of receptors activate a G protein in the endosomal membrane. An intermediate stage is formed with ARF interacting with the Gα subunit. Dissociation of ARF from the Gα subunit may allow other components of the fusion machinery to be assembled on the membrane. Thus, GTPγS would promote dissociation of Gβγ from Gα and favor the binding of ARF to Gα. In the presence of excess activated ARF, dissociation of this protein would be prevented, and fusion would not proceed. Mastoparan and ARF peptide would compete with the association of ARF to Gα. More experimental data are necessary to understand the complex interaction between these proteins and other components of the fusion machinery (e.g., N-ethylmaleimide sensitive factor, rab5, rab4).

## Literature references

Barr FA, Leyte A , Huttner WB (1992) Trimeric G proteins and vesicle formation. Trends Cell Biol 2: 91-94

Colombo MI, Mayorga LS, Casey PJ, Stahl PD (1992) Evidence of a role for heterotrimeric GTP-binding proteins in endosome fusion. Science 255: 1695-1697

Goud B, McCaffey M (1991) Small GTP-binding proteins and their role in transport. Curr Opinion Cell Biol 3: 626-633

Lenhard JM, Kahn RA, Stahl PD (1992) Evidence for ADP-ribosylation factor ARF) as a regulator of in vitro endosome-endosome fusion. J Biol Chem 267:13047-13052

# THE SMALL GTPase RAB4 CONTROLS AN EARLY ENDOCYTIC SORTING EVENT

Peter van der Sluijs, Michael Hull and Ira Mellman
Department of Cell Biology
Yale University School of Medicine
333 Cedar Street, New Haven, Connecticut 06510, USA

Small GTP binding proteins of the rab family have been implicated in the control of vesicular traffic in eukaryotes. It is thought that rab proteins ensure the specificity or directionality of a vesicle budding or fusion event by coupling the completion of the reaction to the hydrolysis of bound GTP (Bourne, 1988). Rab4 and rab5, two distinct rab proteins, have now been localized to early endosomes (Chavrier et al., 1990; van der Sluijs et al., 1991). Although the functions of the endosome-associated rabs are not clear, in vitro evidence indicates that rab5 is involved in lateral fusion between early endosomes (Gorvel et al., 1991). Although an activity for rab4 has yet to be established, it seems likely that it will play a role in the pathway of receptor recycling since it co-localizes with the transferrin receptor (Tfn-R) (van der Sluijs et al., 1991).

The fact that at least two rab proteins are associated with early endosomes is consistent with the fact that these organelles engage in multiple transport and sorting events, each possibly requiring distinct GTP-binding proteins. Consequently, it is of importance to determine the role of rab4 on the endocytic pathway. To this end, we chose to search for specific alterations in the endocytic pathway in cells stably overexpressing wild type and mutant rab4. In the first mutant, the cdc2 kinase site at serine 196 (van der Sluijs et al, 1992a) was changed to a glutamine (rab4-$S^{196}Q$). The second mutant had an asparagine to isoleucine change in rab4's GTP-binding domain (rab4-$N^{121}I$) and was found unable to bind GTP *in vitro*. The resulting cell lines overproduced human rab4 relative to the endogenous protein in CHO cells by ~80-fold (wild type and $S^{196}Q$) and ~15-fold (rab4-$N^{121}I$ mutant).

*Accumulation of fluid phase markers is inhibited by overexpression of wild type rab4.*

To evaluate whether the sorting functions of early endosomes were affected in the rab4 transfectants, we monitored the accumulation of horseradish peroxidase (HRP), a marker of fluid phase endocytosis. After internalization, fluid phase markers are delivered to early endosomes and then either sequestered in late endosomes-lysosomes or recycled back to the plasma membrane and released into the medium (Besterman et al., 1981). Thus the accumulation of HRP also reflects the efficiency of intracellular retention.

NATO ASI Series, Vol. H 74
Molecular Mechanisms of Membrane Traffic
Edited by D. J. Morré, K. E. Howell, and J. J. M. Bergeron
© Springer-Verlag Berlin Heidelberg 1993

Control cells and cell expressing the rab4 mutants exhibited comparable rates of HRP accumulation for at least 1 hr. After 2 hr, HRP uptake by cells expressing rab4-S$^{196}$Q was, if anything, slightly increased relative to controls. In contrast, HRP accumulation in CHO cells overexpressing wild type rab4 was 2-3-fold lower than that observed in any of the other cell lines tested. Thus, overexpression of wild type rab4 exerted a significant reduction in fluid phase endocytosis. This was likely to reflect a decreased ability of the cells to retain internalized HRP as opposed to a decrease in the rate of endocytosis *per se*, because the rate of Semliki forest virus (SFV) internalization was unchanged by rab4 expression; SFV and HRP have previously been shown to be internalized via the same endocytic coated vesicles in fibroblasts (Marsh and Helenius, 1980).

*Overexpression of rab4 alters the transferrin receptor cycle*

To more directly analyze the effect of rab4 overexpression on early endosome function, we determined whether any aspect of Tfn-R recycling was altered by rab4 expression. We first measured the distribution of human Tfn-R in cells that were or were not transfected with wild type or mutant rab4. In cells not transfected with rab4 cDNA, almost 80% of the total Tfn-R was found intracellularly while only 20-25% was present on the plasma membrane. In cells overexpressing wild type rab4, however, this situation was reversed: almost 80% of the total cell Tfn-R was now found on the cell surface, as opposed to only 20% found intracellularly. The distribution of Tfn-R in cells overexpressing the rab4-S$^{196}$Q mutant was indistinguishable from the non-transfected controls, while cells expressing the GTP-binding-deficient rab4-N$^{121}$I mutant exhibited a partial alteration (~50% of the receptors found on the plasma membrane). Thus, overexpression of wild type rab4, and to a lesser extent the rab4-N$^{121}$I mutant, greatly decreased the proportion of Tfn-R was found in the intracellular pool.

*Rab4 overexpression prevents transferrin from reaching acidic early endosomes.*

Since recycled $^{125}$I-Tfn was slow to be released from the transfectants in the absence of excess unlabeled Tfn (van der Sluijs et al., 1992), it may have returned to the cell surface still in its diferric holo form (which dissociates slowly from surface receptors). Accordingly, we reasoned that incoming $^{125}$I-Tfn might not reach endosomes of sufficiently low pH to facilitate the dissociation of Tfn-bound iron. We therefore investigated the acidification properties of the endocytic compartments containing internalized Tfn. Endosome-enriched fractions were prepared from cells allowed to internalize FITC-Tfn (Fuchs et al., 1989). The capacity of these vesicles for ATP-driven H$^{+}$-transport was then determined *in vitro* by monitoring the pH-dependent decrease in FITC fluorescence spectrofluorometrically. As expected, endosomes from control cells, or from cells expressing the rab4-S$^{196}$Q mutant, exhibited significant acidification activity upon addition of ATP. These pH gradients were immediately reversed upon addition of the carboxylic ionophore nigericin, demonstrating that acidification occured intravesicularly. In contrast, FITC-Tfn-containing endosomes from cells overexpressing wild type rab4 or the rab4-N$^{121}$I mutant exhibited a marked reduction in both the rate and extent of acidification activity. The small amount of ATP-driven H$^+$ transport observed, however, was reversed by nigericin.

The acidification "defect" was specific for early Tfn-containing compartments. When cells were allowed to take up FITC-dextran for 1 hr to label late endosomes and lysosomes (Schmid et al., 1988), all cell lines exhibited equivalent amounts of *in vitro* acidification activity. Thus, it is apparent that overexpression of wild type rab4 (and the rab4-N$^{121}$I mutant) reduced the ability of Tfn to reach or accumulate in acidic early endosomes without causing a generalized defect in acidification. It was less likely that rab4 expression inhibited acidification directly because CHO cells with genetic defects in early endosome acidification do not exhibit the alterations in the Tfn cycle observed in the transfectants (Klausner et al., 1984).

In conclusion, we here described an alternative approach to analyze the function of the early endosome-associated protein rab4. By overexpressing wild type or mutant rab4, we have found that this protein played a critical role in the sorting functions of endosomes. While we have not identified the precise step controlled by rab4, we have directly demonstrated in intact cells that an endosome-associated rab protein does plays an important role that affects the sorting or transit of Tfn-R through early endosomes.

## REFERENCES

Besterman, J M, Airhart, J A, Woodworth, R C, and Low, R B (1981) Exocytosis of pinocytosed fluid in cultured cells: kinetic evidence for rapid turnover and compartmentation. J Cell Biol 99, 716-727.

Bourne, H R (1988) Do GTPases direct membrane traffic in secretion? Cell 53, 669-671.

Chavrier, P, Parton, R G, Hauri, H P, Simons, K, and Zerial, M (1990) Localization of low molecular weight GTP binding proteins to exocytic and endocytic compartments. Cell 62, 317-329.

Fuchs, R, Male, P, and Mellman, I (1989) Acidification and ion permeabilities of highly purified rat liver endosomes. J Biol Chem 264, 2212-2220.

Gorvel, J P, Chavrier, P, Zerial, M, and Gruenberg, J (1991) rab5 controls early endosome fusion in vitro. Cell 64, 915 925.

Klausner, R D, van Rensvoude, J, Kempf, C, Rao, K, Bateman, J L, and Robbins, A R (1984) Failure to release iron from transferrin in a CHO cell mutant pleiotropically defective in endocytosis. J Cell Biol 98, 1098-1101.

Marsh, M, and Helenius, A (1980) Adsorptive endocytosis of Semliki Forest Virus. J Mol Biol 142, 439-454.

Schmid, S L, Fuchs, R, Male, P, and Mellman, I (1988) Two distinct subpopulations of endosomes involved in membrane recycling and transport to lysosomes. Cell 52, 73-83.

van der Sluijs, P , Hull, M . Webster, P , Goud, B , and Mellman, I (1992) The small GTP binding protein rab4 contrc s an early sorting event on the endocytic pathway. Cell in press,

van der Sluijs, P , Hull, M , Zahraoui, A , Tavitian, A , Goud, B , and Mellman, I (1991) The small GTP binding protein rab4 is associated with early endosomes. Proc Natl Acad Sci USA. 88, 6313-6317.

van der Sluijs, P , Hull, Л , Male, P , Goud, B , лnd Mellman, I (1992a) Reversible phosphorylation-dephosphorylation determines the localization of rab4 during the cell cycle. EMBO J in press.

# PROTEIN MODIFICATIONS AND THEIR SIGNIFICANCE IN rab5 FUNCTION

J.C. Sanford and M. Wessling-Resnick
Department of Nutrition
Harvard School of Public Health
Boston, MA 02115

Rab5 belongs to a subset of proteins that are post-translationally modified at their carboxytermini via geranylgeranylation and carboxymethylation. We are studying hydrophobic modifications of cell-free synthesized rab5 and how these structural additions contribute to rab5 function. Prenylation of nascent peptide can be identified by *in vitro* translation in the presence of the appropriate radiolabelled precursor. We find that cell-free synthesized rab5 exists in at least two isoforms, with the addition of mevalonate (100 $\mu$M) shifting the peptide to a greater mobility on urea-gradient gels (Fig. 1). The isoform of greater mobility incorporates radiolabel upon addition of [3H]mevalonate to the *in vitro* system, clearly demonstrating that this form of the peptide is polyisoprenylated. Cell-free synthesized rab5 will also incorporate radiolabel in the presence of [3H]S-adenosyl methionine, however, we have found this reaction to be substoichiometric and non-specific. Additional factors appear to be required for the methylesterification of rab5 *in vitro*. In contrast, expression of rab5 in the cell-free reaction strongly enhances the stable methylation of an endogenous protein that runs with a broad mobility of 17-19kD (Fig. 2). The novel 17/19kD species is also methylated in response to the presence of proteins implicated to participate in vesicle traffic, including $\beta\gamma$ subunits of trimeric G proteins. While the alpha subunit of Gi$_3$ also supports modification of p17/19kD, related factors (e.g. alpha subunits of Gi$_1$ and Gs) do not. The pattern of methylation in response to various GTP-binding proteins indicates that elements specifically involved in vesicle traffic interact with p17/19 to influence its methylesterification. What role this may play, for example, in rab5 endocytic function is still unclear. To identify domains of interaction, a series of truncated rab5 molecules have been constructed and expressed *in vitro*. Our analysis of the functional ability to support methylation of p17/19 indicates that while rab5 molecules lacking C-terminal cysteines, and thus non-prenylated forms, support methylesterification of p17/19,

NATO ASI Series, Vol. H 74
Molecular Mechanisms of Membrane Traffic
Edited by D. J. Morré, K. E. Howell, and J. J. M. Bergeron
© Springer-Verlag Berlin Heidelberg 1993

deletion of 16 upstream amino acids produces a form of rab5 that is apparently unable to influence the methylation of p17/19. Thus, rab5 C-terminal structural elements, exclusive of the isoprenylated cysteine residues, are required for interaction with p17/19. We are continuing studies to define the relative contributions of hydrophobic modifications of both rab5 and p17/19 in endocytic vesicle traffic.

Fig. 1.

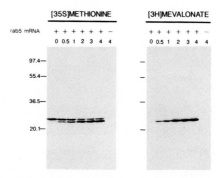

**Time Course of Prenylation.** The conversion of rab5 to a greater mobility isoform via prenylation is detected via acrylamide/urea-gradient gel electrophoresis. 80% conversion is attained after 3h of incubation.

Fig. 2.

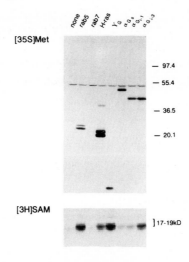

**Vesicle Traffic Proteins Support Methylation of p17/19.** Various proteins were analyzed for their potential to enhance the methylation of p17/19 via translation in the presence of [3H]S-adenosyl methionine.

# LOW MOLECULAR WEIGHT GTP-BINDING PROTEINS IN ROUGH ENDOPLASMIC RETICULUM MEMBRANES FROM RAT LIVER AND RAT HEPATOCELLULAR CARCINOMAS

Joël Lanoix and Jacques Paiement
Département d'anatomie
Université de Montréal
C.P. 6128, Succ. A
Montréal, Québec
Canada   H3C 3J7

Physiological concentrations of GTP have been shown to stimulate a variety of biochemical and morphological events in rough endoplasmic reticulum (RER) membranes. This includes $Ca^{++}$ release, protein translocation, core glycosylation, membrane permeability changes, phospholipid metabolism and membrane fusion. In effort to identify GTP-binding proteins involved in specific RER functions we have looked for similar proteins in a rat liver membrane preparation enriched in derivatives of RER. Furthermore, we have studied the effects of a variety of post-translational modifications of GTP-binding proteins on GTP-dependent membrane fusion.

## Detection of low molecular weight GTP-binding proteins in rough microsomal membranes

$[\alpha\text{-}^{32}P]$GTP blot overlay technique was used to probe for the presence of low molecular weight GTP-binding proteins after separation by one- and two-dimensional (1-D and 2-D) gel electrophoresis and transfer onto nitrocellulose sheets. Radioautograms of 1-D Western blots revealed 24, 23 and 22.5 kDa $[\alpha\text{-}^{32}P]$GTP-binding proteins. Radioautograms of 2-D Western blots revealed that the 24, 23 and 22.5 kDa proteins were composed of numerous charge isomers (Lanoix et al., 1989). Cation requirement and specificity of nucleotide binding to the low molecular weight GTP-binding proteins were determined as well as their cytosolic membrane orientation as demonstrated by mild trypsinization (Lanoix et al., 1989).

## Analysis of ras and ras-related antigenicity in rough microsomal membranes

Anti-ras antibodies and both immunochemistry and immunocytochemistry were used to study ras antigenicity in rough microsomal membranes. By immunoprecipitation a 23.5 kDa $[\alpha\text{-}^{32}P]$GTP-binding protein was detected (Dominguez et al., 1991). Using electron microscopic immunocytochemistry ras antigenicity was observed associated with contaminating smooth membranes (Dominguez et al., 1991). It was concluded that ras antigenicity was not associated with microsomal derivatives of the RER.

## Post-translational modifications of low molecular weight GTP-binding proteins

a)  Phosphorylation

When rough microsomes were incubated in the presence of the catalytic subunit of cyclic AMP-dependent protein kinase, phosphorylation was observed to two proteins which exhibited similar electrophoretic mobilities to known $[\alpha\text{-}^{32}P]$GTP-binding proteins (Lanoix and Paiement, 1991).

NATO ASI Series, Vol. H 74
Molecular Mechanisms of Membrane Traffic
Edited by D. J. Morré, K. E. Howell, and J. J. M. Bergeron
© Springer-Verlag Berlin Heidelberg 1993

# LOW MOLECULAR WEIGHT GTP-BINDING PROTEINS ASSOCIATED WITH THE MEMBRANES INVOLVED IN POST-GOLGI TRANSPORT OF RHODOPSIN

Dusanka Deretic and David S. Papermaster

Department of Pathology

University of Texas Health Science Center

7703 Floyd Curl Drive

San Antonio, TX 78284-7750

Newly synthesized rhodopsin is sorted to the rod outer segment (ROS) of frog rod photoreceptor cells by a population of post-Golgi vesicles which have a relatively simple protein content (Deretic and Papermaster, 1991). In addition to rhodopsin, several low molecular weight (21-25 kD) GTP-binding proteins can be identified on these membranes by $\alpha$ $^{32}$P GTP overlays. We have compared the GTP-binding protein content of the retinal post-Golgi vesicle fraction to the content of the Golgi enriched and ROS fractions in order to follow changes in the GTP-binding protein content of these membranes as they mature. Comparison by isoelectric focusing/SDS 2D gel blots (Figure 1) shows that specific GTP-binding proteins are enriched in each fraction, while several proteins are common to all fractions.

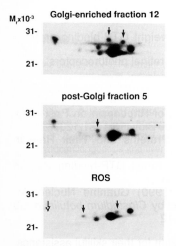

$\alpha$ $^{32}$P GTP bound

$M_r$x10$^{-3}$

Golgi-enriched fraction 12

31-

21-

post-Golgi fraction 5

31-

21-

ROS

31-

21-

pI 4                    6

Figure 1. Golgi-enriched membranes contain several GTP-binding proteins absent from the post-Golgi and ROS membranes (closed arrows).

One of the GTP-binding proteins, also present in the Golgi, is enriched on post-Golgi vesicle membranes (closed arrow).

ROS membranes have a different set of the GTP-binding proteins from the post-Golgi vesicle membranes. Differences include both the presence of specific proteins (closed arrows) and the absence of a protein found on post-Golgi membranes (open arrow).

NATO ASI Series, Vol. H 74
Molecular Mechanisms of Membrane Traffic
Edited by D. J. Morré, K. E. Howell, and J. J. M. Bergeron
© Springer-Verlag Berlin Heidelberg 1993

# Author Index

Adamik, R, 175
Altschuler, Y, 71
Alvarez-Hernandez, X, 319
Aroeti, B, 313
Auger, D, 289
Avrahami, D, 199

Balch, WE, 35
Ball, CL, 229
Barnes, S, 69
Baudhuin, P, 327
Beevers, L, 331, 333
Benos, DJ, 293
Bergeron, JJM, 237
Bernstein, M, 179
Bonatti, S, 53
Bonzelius, F, 203
Borchelt, DR, 199
Bounelis, P, 289
Bourne, J, 35
Bowser, R, 387
Brennwald, P, 387
Brewer, CB, 297
Brightman, AO, 65
Brodsky, FM, 301
Brown, WJ, 367
Bucci, C, 377
Burger, KNJ, 181
Burgess, CC, 323, 325
Burón, M, 329
Butler, JM, 331
Buxbaum, JD, 201

Canalejo, A, 329
Caporaso, GL, 201
Cardone, MH, 203
Cassagne, C, 187
Chapman, RL, 359
Chen, GL, 95
Clague, MJ, 215
Collawn, J, 205
Colombo, MI, 395
Colosimo, ME, 107
Courtoy, PJ, 327
Crawford, JM, 69
Cunningham, SA, 293
Cupers, Ph, 327
Curran, P, 173
Cyr, DM, 91
Czernik, AJ, 201

Davidson, HW, 35
Dean, GE, 375
Deretic, D, 407
Dice, JF, 335
Douglas, MG, 91
Duden, R, 117

Emr, SD, 363
Enrich, C, 309, 311
Erickson, AH, 359
Esquela, A, 127
Evans, WH, 309, 311

Farquhar, MG, 17
Finlay, BB, 321
Fishman, PH, 173
Frizzell, RA, 293

Galili, G, 71
Gandy, SE, 201
Garcia-del Portillo, F, 321
Garrett, MD, 387
Geilen, CC, 65, 197
Glass, J, 319
Godbold, GD, 359
Godleski, JJ, 69
Gonatas, NK, 171
Gordon, PB, 339
Gorvel, J-P, 215
Gould, SJ, 99
Grab, DJ, 315
Greengard, P, 201
Griffing, LR, 111
Griffiths, G, 117
Grim, MG, 127
Gruenberg, J, 215

Hartl, F-U, 81
Hastings, DL, 69
Haun, R, 175
Hennig, D, 145
Herman, GA, 203
Herman, PK, 363
Herskovits, JS, 325
Hiebsch, RR, 45
Hobman, TC, 17
Hoflack, B 377
Holen, I 339
Hollinshead, M, 245
Hong, W, 57

## Subject Index

Printing: Druckhaus Beltz, Hemsbach
Binding: Buchbinderei Schäffer, Grünstadt

# NATO ASI Series H

# NATO ASI Series H

# NATO ASI Series H